T0250446

Parallel Processing in Computational Mechanics

NEW GENERATION COMPUTING

Series Editor
Hojjat Adeli
Department of Civil Engineering
The Ohio State University
Columbus, Ohio

Parallel Processing in Computational Mechanics

edited by

Hojjat Adeli

The Ohio State University
Columbus, Ohio

Marcel Dekker, Inc.   **New York • Basel • Hong Kong**

Library of Congress Cataloging-in-Publication Data

Parallel processing in computational mechanics / edited by Hojjat
  Adeli.
     p.   cm. -- (New generation computing; 2)
  Includes bibliographical references and index.
  ISBN 0-8247-8557-6
  1. Parallel processing (Electronic computers) 2. Mechanics,
Analytic--Data processing. I. Adeli, Hojjat. II. Series.
QA76.58.P37778   1991
005'.35--dc20   91-28349
                                             CIP

This book is printed on acid-free paper.

Copyright © 1992 by MARCEL DEKKER, INC. All Rights Reserved

Neither this book nor any part may be reproduced or transmitted in any form or
by any means, electronic or mechanical, including photocopying, microfilming,
and recording, or by any information storage and retrieval system, without per-
mission in writing from the publisher.

MARCEL DEKKER, INC.
270 Madison Avenue, New York, New York 10016

Current printing (last digit):
10 9 8 7 6 5 4 3 2 1

PRINTED IN THE UNITED STATES OF AMERICA

Preface

This is the second volume in the series New Generation Computing. Contributed by leading researchers, the volumes in the series are aimed at the most recent advances in computer science and new and emerging computer technologies. This volume as well as the first volume of the series, *Supercomputing in Engineering Analysis*, concentrate on high-performance computing using the new generation of computers with vector and parallel processing capabilities. These machines have provided the means to tackle problems not readily tractable on traditional computers. The volumes are introductory in nature and intended primarily for an engineering audience. They are also valuable for computer scientists and software developers.

The first chapter is a brief introduction to parallel processing and parallel machines. An overview of parallel programming languages and techniques is presented in Chapter 2. Various synchronization abstractions, single-assignment languages, various parallel languages, and parallel programming in conventional programming languages are discussed in this chapter.

Chapter 3 discusses the basic concepts for the development of parallel algorithms and parallel solution of linear systems. This topic is central to the solution of many engineering problems that require the solution of a large number of linear equations. Chapter 4 presents asynchronous parallel algorithms for solution of large sets of linear systems. Chapter 5 deals with the parallel execution of ordinary sequential programs.

Many engineering problems are simulated as discrete systems (for example, simulation of air traffic or simulation of the communication system for a message-passing multiprocessor such as hypercube-based Intel iPSC). Chapter 6 presents parallel simulation of discrete systems.

The remaining chapters present applications of parallel processing in several engineering areas and disciplines. Chapter 7 presents finite element analysis on concurrent machines. Chapter 8 discusses applications of parallel processing in structural engineering, with an emphasis on the solution of the eigenvalue problem, which is also found in several other engineering disciplines. Chapter 9 continues with parallel computations in solid mechanics.

Chapter 10 deals with the solution of fluid mechanics problems on parallel computers. The last chapter presents parallel methods for the solution of partial differential equations with applications in groundwater hydrology. Parallel processing is bound to play a significant role in the areas presented in Chapters 7–11. The information presented in these chapters, however, can also be used for development of parallel algorithms and software packages in other engineering fields.

The topics of parallel processing and supercomputing are akin to each other. As such, this volume complements the first volume in the series, *Supercomputing in Engineering Analysis*.

Hojjat Adeli

Contents

Contributors

Hsi-Ya Chang   received his Ph.D. in Civil Engineering from Duke University in 1986. He is currently an assistant professor at the University of Miami. He has authored several papers in the area of parallel processing.

L.S. Chien   received his Ph.D. from Purdue University in 1989. After spending a year as a postdoctoral fellow at George Washington University's Joint Institute of Advanced Flight Science at NASA Langley Research Center, he moved to the Air Force Institute of Technology, where he is a research scientist.

David John Evans   is Professor of Computing at the University of Technology, Loughborough, United Kingdom. He has been associated with parallel processing for many years. He was a member of the early Manchester University design team on ATLAS. More recently he has led a team researching into both architectures and algorithms for parallel processing and has successfully constructed two parallel MIMD systems. He is the European Editor of *Journal of Parallel Computing* and Vice-Chairman of the International Parallel Computing Society. His current interests are in the design of VLSI processor arrays.

Charbel Farhat   received a Ph.D. in computational mechanics from the University of California, Berkeley, in 1986. He is currently an associate professor in the Department of Aerospace Engineering Sciences at the University of Colorado, Boulder. He has authored 45 papers and published extensively in the area of parallel processing.

Richard M. Fujimoto   received his Ph.D. in Computer Science and Electrical Engineering from the University of California, Berkeley, in 1983. He is currently an associate professor of Information and Computer Science at the Georgia Institute of Technology. He has authored about 40 research and technical publications. He was the General Chairman and Program Chairman of the 1990 and 1989 Distributed Simulation Conference, respectively.

Yoichi Muraoka   received the B.S. degree in electronics and engineering communication from Waseda University, Tokyo, Japan, in 1965, and the Ph.D. degree in computer science from the University of Illinois, Urbana-Champaign in 1971. From 1971 to 1985 he worked at NTT Electrical Communication Laboratories. Since 1985, he has been a professor in the Department of Electronics and Communication Engineering, Waseda University. His current fields of interest include parallel processing, electronic books, and neural networks.

Alexander Peters   received his Ph.D. in Civil Engineering from the Aachen University of Technology in 1988. During 1988–1989 he worked in the field of parallel algorithms for solving partial differential equations at Princeton University. He is currently a member of the research staff at the IBM Scientific Center, Institute for Supercomputing and Applied Mathematics in Heidelberg, Germany. His current research interest is development of efficient numerical algorithms for vector/parallel processing.

Garry Rodrigue   received a Ph.D. in mathematics from the University of Southern California in 1971. He is currently a professor in the Department of Applied Science at the University of California, Davis, and a Research Mathematician at the Lawrence Livermore National Laboratory. He is an Associate Editor of *Journal of Parallel Computing*, *Journal of Parallel and Distributed Computing*, *Journal of Supercomputing Applications*, and *Journal of Undergraduate Mathematics and Its Applications*. He has edited seven books in the areas of scientific computing and parallel processing. Professor Rodrigue has been the Co-chairman of twelve conferences since 1974.

Udo Schendel   received a doctorate degree from the Technical University of Berlin in 1969. He has been a professor of Numerical Mathematics and Computer Science at the Free University of Berlin, Germany, since 1972. Prior to that, he was Head of the Computer Center of the Max-Planck Society in Berlin. He is the Editor-in-Chief of *Journal of Parallel Computing*, editor of several conference proceedings, and President of the Parallel Computing Society. Professor Schendel is the author of two books, including *Introduction to Numerical Methods for Parallel Computers*, published in 1984.

C. T. Sun   received his Ph.D. from Northwestern University in 1967. He joined Purdue University in 1968 where he is currently a professor in the School of Aeronautics and Astronautics. He is currently Associate Editor of *Journal of the Astronautical Sciences*, and is on the editorial board of *Journal of Composite Materials*, *Composites Science and Technology* and *Journal of Composites Technology & Research*. Professor Sun has published extensively in the areas of composites, fracture mechanics, dynamics of large space structures, and computational mechanics.

Senol Utku   received the Sc.D. degree from the Masssachusetts Institute of Technology in 1960. He is currently a professor of civil engineering and computer science at Duke University. His general research area is in the numerical analysis and the computer simulation of mechanical systems. His current interests include concurrent processing and robot dynamics.

Prasad R. Vishnubhotla   received a Ph.D. in Computer Science from the Tata Institute of Fundamental Research, Bombay, India, in 1984. Since then he has been an assistant professor in the Department of Computer and Information Science at The Ohio State University. He has been doing research in the area of parallel processing since 1974. He is the co-author of the text *A Multiprocessor Operating System*, published in 1984. He won an outstanding paper award for a contribution on program verification presented at the Fourth International Conference on Distributed Computing Systems held in San Francisco in 1984.

Hayato Yamana   received the B.S. and M.S. degrees in electronics and engineering communication from Waseda University, Tokyo, Japan, in 1987 and 1989, respectively. Since 1989, he has been a research assistant in the Center for Informatics, Waseda University. He is currently interested in parallel processing.

Nadia Y. Yousif   is currently Head of Department of Computer Science, College of Science, Basrah University, Basrah, Iraq. After her initial university. education in Iraq, Dr. Yousif came to England for postgraduate studies at Essex and Loughborough Universities. She completed her Ph.D. degree on the design and analysis of parallel algorithms under the supervision of Professor D. J. Evans. Her interests and publications include the design of both numerical and non-numerical parallel algorithms.

About the Editor

Hojjat Adeli received his Ph.D. from Stanford University in 1976. He taught at Northwestern University and the University of Utah before joining The Ohio State University in 1983. He is the Editor-in-Chief of the international journal *Microcomputers in Civil Engineering*, which he founded in 1986. A contributor to 25 research journals, he has authored more than 200 research and technical publications in various fields of computer science and engineering. He is the author of *Interactive Microcomputer-Aided Structural Steel Design* and coauthor of *Expert Systems for Structural Design—A New Generation*. He is the editor of *Expert Systems in Construction and Structural Engineering, Knowledge Engineering: Volume One—Fundamentals* and *Volume Two—Applications*, and *Parallel Processing in Computational Mechanics* (Marcel Dekker, Inc.), and coeditor of the two-volume *Mechanics Computing in 1990's and Beyond*. Professor Adeli has received numerous academic, research, and leadership awards, honors, and recognition. His recent awards include The Ohio State University College of Engineering 1990 Research Award in Recognition of Outstanding Research Accomplishments and The Ohio State University 1990 Lichtenstein Memorial Award for Faculty Excellence. Most recently, in December, 1990, he received the American Biographical Institute's **Man of the Year Award**.

1
Parallel Processing and Parallel Machines

Hojjat Adeli and Prasad R. Vishnubhotla *The Ohio State University, Columbus, Ohio*

1 INTRODUCTION

Sequential mathematical models and methods have been around for over three centuries. This has strongly influenced the theory of algorithms and FORTRAN programming during the last 30 years. However, mathematicians and computer scientists have been doing research in parallel processing for many years. Some of these results were reported in the literature in the late 1960s and early 1970s (Pease, 1967; Chazan and Miranker, 1970; Miranker, 1971; Lorin, 1972; Enslow, 1974). Going back farther in time to the early days of computers, von Neumann (1956) in fact predicted the opportunity for highly parallel brainlike processing. The technology for implementing this idea simply did not exist at that time.

Recent advances in computer hardware and software, however, have made multiprocessing in general and parallel processing in particular a viable and attractive technology. Parallel processing provides an opportunity to improve computing efficiency by orders of magnitude. Today's large mainframes can perform 5–10 million instructions per second (MIPS), and supercomputers perform 50–100 MIPS. By combining 100 microprocessors each with performance of 1 MIPS it is possible to gain supercomputer performance for some applications at a fraction of the cost of a supercomputer. Parallel processing appears a promising approach to large-scale engineering computations. In this chapter we present a brief review of parallel processing and parallel machines.

1

2   PARALLEL PROCESSING AND INTELLIGENT SYSTEMS

A number of cognition scientists and artificial intelligence researchers have argued that human intelligence is due to the interaction of a large number of simple processing units. This idea has not been very persuasive in the past, but there is a new resurgence of interest in modeling human thought and intelligence with parallel distributed processing (PDP) (Hinton and Anderson, 1981; Jorgensen and Matheus, 1986; McClelland et al., 1986; Rumelhart et al., 1986a,b). This research is also known as connectionism, a notion used to describe the importance of interactions among neurons in modeling intelligent systems. The PDP models of human thought processes are also appealing from a physiologic point of view.

A study of the mechanics of mind reveals that the human brain is made of a large number of highly interconnected elements that send excitatory and inhibitory signals to each other (Blakemore, 1977). More specifically it is estimated that the human brain is made of $10-100$ billion brain cells or neurons. Each neuron is capable of switching approximately a thousand times each second. In comparison, a large computer consists of about 1 m^2 of silicon containing about a billion transistors. Each transistor is capable of switching thousands of times per second. At present enough knowledge is not available about the human brain, but the apparent similarities between the brain and modern computers have encouraged some researchers to choose the brain as their model of a "thinking machine" (Hillis, 1986).

Today's powerful supercomputers can be a million times faster in processing a task than the neuron can react to a sensory signal. The most powerful computers do not exhibit the intelligence of a small child, however. The ability of brain to perform certain functions is greatly superior to that of computers, and this is attributed to the brain's extensive interconnection of neurons.

Rumelhart et al. (1986a) present a general framework for parallel distributed processing. The major elements of a general PDP model are a set of processing units, a state of activation for representing the system at any time, a pattern of connectivity among units, a propagation rule for propagating a pattern of activities through the network of connectivities, an activation rule, and a learning rule through which patterns of connectivity are modified by experience. The fundamental difference between a PDP model and other models of cognitive processes is that in a PDP model knowledge is not stored as a static representation of patterns, but rather the connection strength between units is stored, and the connectivities are used to create the patterns. In contrast to AI (artificial intelligence) rule-based systems, there is no rule here; a PDP model simply learns from examples.

Philosopher Hubert Dreyfus (PC AI, 1987) argues that artificial intelligence research has not been able to solve some of the most fundamental problems of intelligence, such as visual image recognition and learning. He suggests that using parallel processing to create neural nets is a more promising approach to the problem of machine learning. He cautiously predicts that the real action in the future will be in parallel machines capable of simulating neural networks.

3 CLASSIFICATION OF COMPUTER ARCHITECTURES

Multiprocessors are classified as tightly coupled systems and loosely coupled systems (Chambers et al., 1984). In a tightly coupled system processors cooperate closely on the solution to a problem. A loosely coupled system consists of a number of independent and not necessarily identical processors that communicate with each other via a communication network.

A generally accepted taxonomy for multiprocessor computer architecture has yet to be developed. Flynn (1966) classifies computers into four categories in terms of the sequence of instructions performed by the machine and the sequence of data manipulated by the instruction stream.

1 In single instruction–single data stream (SISD) architecture instructions are executed sequentially but may be overlapped in their execution stages (pipelining). Von Neumann uniprocessor architecture falls in this category. Instructions are fetched from the memory in serial fashion and executed in a single processor.

2 In single instruction–multiple data stream (SIMD) architecture multiple processing elements are all supervised by the same control unit. All processors receive the same instructions broadcast from the control unit but operate on different data sets from distinct data streams. These are usually array processors (for example, Cray-1). An SIMD machine exploits the following type of parallelism:

DO 1 I = 1,1000

1   X(I) = A(I) + B(I) + C(I) + D(I)

Note that in this FORTRAN code a common operation is applied to a large set of data.

3 With multiple instruction–single data stream (MISD), multiple processors, each receiving distinct instructions, operate over the same data stream. The output of one processor becomes the input of the next processor in the form of a pipe. This structure is considered impractical, and no real machine has ever been built in this category.

4   In multiple instruction–multiple data stream (MIMD) each processor has its own control unit and the processors execute independently. The processors interact with each other either through shared memory or by using message passing to execute an application. An MIMD machine exploits the following type of parallelism in which different operations or sets of operations are performed on separate or common sets of data:

$$X = A + B - C$$
$$Y = AB + C$$

For true parallel processing an MIMD architecture is desirable.

It should be noted that Flynn's taxonomy is deficient in at least two respects (Davidson, 1990). First, it suffers from oversimplification. Although pipelined computers (such as Cray-1) are usually considered SIMD, Flynn's taxonomy does not address pipelining (overlapping parts of operations in time) satisfactorily. Second, Flynn's taxonomy does not address the question of memory (local or global) and its accessibility by the processor.

Another classification of multiprocessor computers is in terms of the number of processors used in the system. In a coarse-grained architecture a small number of powerful processors are linked together. An example within this category is the four-processor Cray X-MP48 (with performance of 1000 million floating-point operations per second, or MFLOPS).

At the other end of the spectrum are the fine-grained machines or massively parallel machines in which several thousands of relatively weak processors are combined. Examples of machines within this category are the massively parallel processor (Batcher, 1980) and the Connection Machine (Hillis, 1985). In medium-grained machines dozens or hundreds of general-purpose microprocessors, such as the Intel 80386 or Motorola 68020, are linked together.

SIMD, MISD, and MIMD machines are tightly coupled systems. Maximum utilization of parallel processing is achieved on MIMD machines and loosely coupled systems.

4   INTERCONNECTION NETWORKS IN MULTIPROCESSORS

The processors and memories in a multiprocessor can be connected in several ways (Satyanarayanan, 1980; Siegel, 1985). Based on the interconnection used multiprocessors may be classified as shared-memory multiprocessors and distributed multiprocessors.

4.1   Shared-Memory Multiprocessors

In a shared-memory system the network connects each processor to each memory bank but neither the processors nor the memories are connected directly to each other. The shared-memory is globally addressable by all the processors. The possible interconnections for these machines are (Bhuyan, 1987) single shared bus, multiple-bus network, crossbar switch, and multistage switch. These interconnection types are described briefly here. Irrespective of the type of the interconnection used, the network delays result in high memory access times that affect the performance. The shared-memory systems, however, are much easier to program compared with distributed systems, discussed in a subsequent section.

Single Shared Bus. A single shared bus (Fig. 1) is the simplest in design and hardware cost but it is also the most inefficient; a processor accesses shared memory by competing with other processors for the bus and by holding the bus during the memory transaction. Thus at any time only one processor can access the shared memory. Early designs (for example the DEC KI-10 dual processor) used a master-slave relation between the processors for scheduling reasons. The more recent designs, such as Sequent and Encore (discussed in a subsequent section), are symmetric systems in which no master-slave relationship exists between the processors.

Machines in this category are usually equipped with a special kind of hidden memory called a *cache* (Smith, 1982). A cache can be viewed as a collection of fast registers in which a register stores both the address and the value of a memory location. Typically a separate cache is associated with each processor. As the processor makes memory references the frequently referenced memory locations (address and value) are copied into the cache. As a result subsequent read requests to these locations are satisfied by the cache itself and do not require main memory references. This not only reduces the access time for an individual memory location but also improves the overall performance by reducing the contention on the bus. When a location copied into one or more caches is updated, however, there is a need to update all caches to maintain consistency.

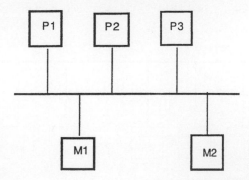

Figure 1   Single shared bus (P1, P2, and P3 are processors and M1 and M2 are memory banks).

The overheads due to these protocols are still a limiting factor in the performance of bus-based multiprocessors.

Multiple-Bus Network.   A possible enhancement to the single shared bus is to use a multiple-bus network (Fig. 2) connecting processors and memories so that multiple processors accessing different memory banks can proceed simultaneously. This technology is very new and is currently under investigation (Mudge et al., 1987).

Crossbar Switch.   A crossbar switch (Fig. 3) allows arbitrary interconnection between processors and memories without contention, and consequently it is the most desirable interconnection. Some experimental shared-memory systems were designed in the mid-1970s using crossbars such as C.mmp (Wulf et al., 1981) and CCN (Joseph et al., 1984), but the prohibitive costs of these switches make them unattractive for commercial use. Some researchers believe that optical communication technology will renew interest in crossbar networks (Sawchuk et al., 1987).

Multistage Switch.   Multistage interconnection networks strike a balance between cost and performance. The most commonly used multistage connection is the Omega network (Figure 4) (Lawrie, 1975). To connect n inputs to n outputs, it uses $\log_k n$ stages of switching with n/k $k \times k$ switches at each stage (note that each switching node has k inputs and k outputs). Actual designs use $k = 2$ (shown in Fig. 4) or $k = 4$. Several experimental systems have been built using multistage switches, such as TRAC (Sejnowski et al., 1980), PASM (Siegel, 1985), Ultracomputer (Gottlieb et al., 1983), Cedar (Gajski et al., 1983), and IBM RP3 (Pfister et al., 1985). BBN Butterfly (with $k = 4$) is the first commercial shared-memory multiprocessor that uses a multistage network.

Machines based on multistage switches suffer less contention in memory access compared to bus-based machines. As a result they can accommodate a larger number of processors without performance degradation. However, delays due to multiple switches result in a higher memory access time. In view of this, some designs (Butterfly and IBM RP3) keep each memory module *local* to a

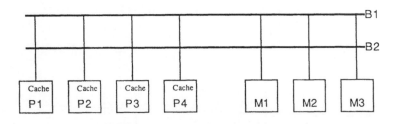

Figure 2   Multiple bus network (B1 and B2 are busses, P1, P2, P3, and P4 are processors, and M1, M2, and M3 are memory banks).

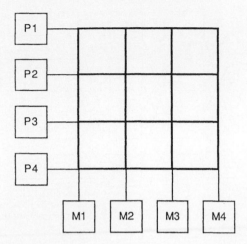

Figure 3 Crossbar switch (P1, P2, P3, and P4 are processors and M1, M2, M3, and M4 are memory banks).

particular processor but at the same time allow global memory accesses from all other processors. Thus the local memory can be used as a programmable cache. Issues related to achieving consistency in such machines are discussed in Dubois et al. (1988) and Hill and Larus (1990).

4.2 Distributed Multiprocessors

In a distributed multiprocessor each processor has a private or local memory but there is no global shared memory in the system. The processors are connected using an interconnection network, and they communicate with each other only by passing messages over the network. A variety of interconnection schemes may be used depending on the application's requirements. A few of the possible interconnection schemes are ring (Fig. 5a), mesh (Fig. 5b), hypercube (Fig. 5c and d), tree (Fig. 5e), X-tree (Fig. 5f) (Despain and Patterson, 1978), torus (Fig. 5g) (Wittie, 1981), and cube-connected cycles (Fig. 5h) (Preparata and Vuillemiu, 1981).

Two parameters characterize an interconnection. One is the diameter of the network, which is the maximum distance required for a message to travel between any two arbitrary nodes. The other is the degree of a node, which is the number of other processors to which the node is connected. It measures the cost of hardware interfaces at each node. Thus a fully interconnected network has every pair of nodes directly connected to each other and therefore has a diameter of 1, but it is the most expensive interconnection in terms of hardware cost. A ring is the least expensive but has a diameter of $n/2$ for an n-node network.

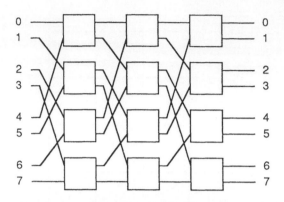

Figure 4   An 8×8 Omega network with $k = 2$ (numbers of input and output are 8).

Hypercube architecture (Heath, 1986) appears to strike a balance between these two extremes. An n-dimensional hypercube (or binary n cube) has 2^n identical nodes, a diameter of n, and a nodal degree of n for all nodes. Figures 5c and 5d show a three-dimensional and a four-dimensional hypercube. Hypercube architecture has drawn a lot of attention because various other interesting structures, like ring, mesh, and tree, can be easily embedded in a hypercube. The first experimental hypercube, called Cosmic Cube (Seitz, 1985), was designed at Caltech in collaboration with the Jet Propulsion Laboratory. The Intel Personal SuperComputer (iPSC), described in a subsequent section, was the first commercially available hypercube (Rattner, 1985). Other commercial hypercubes are the Ametek System/14, the NCube/10, and the FPS T Series. Critics of hypercube architecture contend that even the logarithmic increase in node connectivity may be unacceptable as the number of nodes increases because the interface hardware cannot be placed on a standardized processor chip. This appears to be the primary disadvantage of hypercube architecture.

5   MULTIPROCESSING CLASSIFICATION

Multiprocessing can be classified into three categories: (1) throughput-oriented multiprocessing, (2) availability-oriented multiprocessing, and (3) response-oriented or parallel processing (Patton, 1985). In throughput multiprocessing the goal is to maximize the number of independent jobs done in parallel in a general-purpose computing environment. This is achieved by taking advantage of the inherent central processing unit (CPU) and I/O (input/output) balance to produce a uniformly loaded system with scheduled response. The system is required to be fail-soft (that is, to merely record its last operational state).

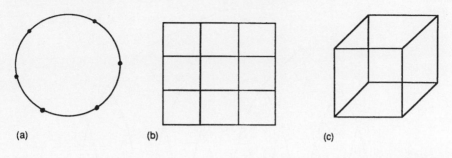

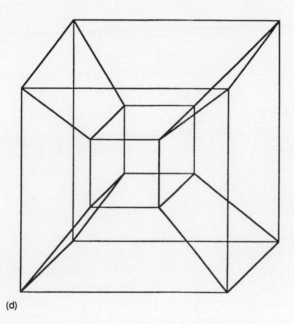

Figure 5 Distributed multiprocessors: (a) ring, (b) two-dimensional mesh, (c) three-dimensional hypercube, (d) four-dimensional hypercube, (e) tree, (f) X-tree, (g) torus, and (h) three-dimensional cube-connected cycles.

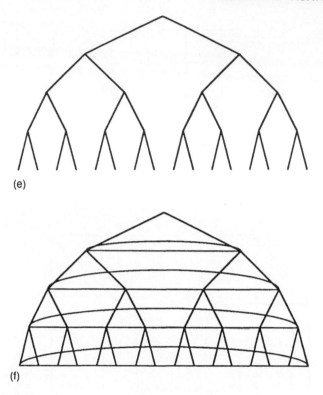

(e)

(f)

Figures 5 e,f

In availability-oriented multiprocessing the goal is to maximize the number of interdependent tasks done in parallel in an interactive computing environment. The system is required to be fail-safe (that is, to preserve its integrity in the event of failure) or never-fail. These systems are I/O intensive.

In response-oriented multiprocessing or parallel processing the goal is to maximize the number of cooperating processes done in parallel or to minimize the response time. This means keeping as many processors as possible busy for as long as possible. Applications for these systems can normally be partitioned into subtasks to be processed concurrently. Their performance is failure sensitive.

The ideal performance of a multiprocessor system should be linear. That is, the computational efficiency of a system made of N processors should be N times the efficiency of a single processor (Fig. 6). In the early 1970s Minsky and Papert (1971) pessimistically argued that the performance of a highly parallel system is $\log_2 N$ at best. Using the rule that the speed of a multiprocessor is equal

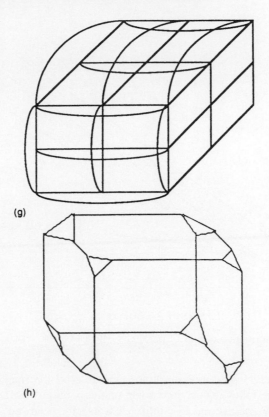

(g)

(h)

Figures 5 g,h

to the speed of its slowest processor, Amdahl (1967) proposed the performance curve $N/\log N$. Extracting the parallelism from the FORTRAN DO loops, Kuck et al. (1984) report a performance curve of $0.3N$, which exceeds Amdahl's law. Through adroit manual extraction of parallelism it is possible to gain an efficiency of over 90% (Uhr, 1984).

Fifth-generation parallel processing may be considered supercomputing, but not in the same sense as the application of the Cyber 205, Cray X-MP, and their subsequent versions. In the former, high performance is achieved through the use of high-volume–low-cost components. In the latter, high performance is achieved through the use of low-volume–high-cost components (Patton, 1985).

6　COMMERCIAL PARALLEL PROCESSORS

Some 30 vendors have introduced "parallel machines" during the last 3 years. Some of these machines may not perform "true" parallel processing, but the

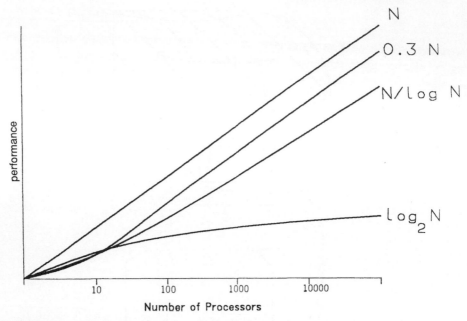

Figure 6   Performance curves of parallel systems.

number of parallel processor machines available on the market is rapidly increasing. In the following sections some of the popular MIMD machines (both shared-memory and distributed multiprocessors) are reviewed.

6.1   Sequent S Series

In spring 1986 Sequent Computer Systems introduced two UNIX-based S (Symmetry) series machines (S27 and S81) in their line of practical parallel computers (Chandler, 1987). These S series machines are single-bus shared-memory systems that use the Intel 16 MHz 80386 CPUs. Each CPU includes a 80387 floating-point coprocessor, a 64 KB two-way set-associative cache, and a 64 bit bus interface logic operating at 80 MB/s. For a long time the consensus has been that shared-memory multiprocessor computers are impractical because the addition of each processor improves the overall performance marginally, not linearly, because of the increased competition for the shared resources, such as the memory and the system bus.

To overcome this difficulty Sequent claims achieving close to linear performance through the special design of the local processor cache, memory, system bus, synchronization mechanisms, and other components. Sequent's S machines are not based on master-slave organization. They are based on a tightly coupled symmetric system that distributes the processing load dynamically to an arrangement of 2–30 CPUs. All CPUs can execute application programs as the users perform I/O or other service functions. The multiprocessing operating system is Dynix 3.0 (dynamic load-balancing UNIX) that can accommodate FORTRAN, PASCAL, C, and Ada.

The S27 configuration includes 2–10 of the 80386 CPUs, 8–80 MB of main memory, 150 MB to 4.3 GB of disk storage, and up to 96 RS-232 serial connections. The S81 configuration includes 2–30 of the 80386 CPUs, 8–240 MB of main memory, 264 MB to 17.2 GB of disk storage, and up to 256 serial connections. The S machines are not intended to compete with Cray-like machines for vector-intensive applications.

6.2 Encore Multimax

In 1985 Encore Corporation introduced its Multimax single-bus shared-memory multiprocessor (Encore, 1985). The current version of Multimax, known as Multimax 320, uses up to 20 National Semiconductor 32332, 32-bit processors with LSI (large-scale integrated) memory management units and floating-point coprocessors. Each processor is capable of executing 2 MIPS. Thus the total processing power of a high-end Multimax is 15 MIPS. It can accommodate up to 32 MB of shared memory. The system is configured in terms of dual-processor cards, each containing two processors, and a 32 KB cache shared between the two processors. Multimax uses a fast bus, called Nanobus, which can carry 96 bits of new information every 80 ns, even if previous requests are still in progress. The result is a true data-transfer bandwidth of 100 MB/s. Encore is in the same price range as Sequent.

Encore Multimax supports two operating systems, UMAX 4.2 and UMAX V, which are based on UNIX 4.2 and UNIX V, respectively. Programs running on different processors can obtain simultaneous access to operating system services from a single shared copy of UMAX. By using a technique called multithreading, which implements multiple simultaneous streams of control, UMAX can process many requests in parallel.

6.3 FLEX/32

Because of the problem of traffic congestion, bus-based machines can use only a limited number of processors (usually less than 20). To overcome this difficulty Flexible Computer Corporation has developed FLEX/32 on a design concept referred to as a multicomputer (Flexible Computer Corporation, 1985). In this architecture, in addition to a global shared memory each processing node has its own local memory and I/O facility and runs its own operating system. At present the largest FLEX/32 in the market uses 40 processors or nodes, but Flexible Computer Corporation claims that this machine can theoretically use up to

20,000 processors. Another feature of FLEX/32 is that different types of processors can be mixed in the machine. At present two processor modules are available, one based on the Motorola 68020 and the other based on Semiconductor 32032 microprocessors.

6.4   Intel Personal SuperComputer

The Intel Personal SuperComputer or iPSC consists of 16, 32, 64, or 128 high-performance microcomputers (Intel Corporation, 1986). Each microcomputer (an Intel 80286 CPU; an 80386 version is also available) and its associated local memory (512 KB of dynamic random-access memory, DRAM), together with a numeric processing unit (Intel 80287), is called a node. The interconnection network is a hypercube.

The nodes are connected with each other through high-speed bidirectional communication channels forming a self-contained "cube" in a freestanding enclosure. A "cube manager" connected to the cube includes the XENIX operating system that supports program development tools (e.g., C and FORTRAN compilers) and commands to manage the cube and to monitor its status. There is no global shared memory. Each node has its local memory and is therefore completely independent. Nodes communicate with their neighbors through queued message passing. A built-in timer measures the processing (computation plus communication) time of each processor.

iPSC was developed primarily for researchers involved in the development of parallel processing applications. It can be used as a stand-alone concurrent machine or as a server in a distributed processing environment through an Ethernet interface. Intel has recently introduced iPSC-VX in which a vector processor is attached to each CPU. This machine is claimed to have a peak performance of 424 MFLOPS. In comparison, the peak performance of the four-processor supercomputer Cray X-MP48 is 1000 MFLOPS.

The efficiency of interprocessor communication is a major factor in the performance of processor networks. The iPSC/1 (with the 80286 processor) provided hardware support only for communication between neighboring processors. This was done simply by using the standard Ethernet chip for each communication port. On this machine communication between remote processors was supported by the operating system using software protocols. As a result remote communication on this machine has been slow. The problem appeared to be more acute with the prospect of using the faster 80386 processor for the iPSC/2. This prompted the designers of iPSC/2 to support remote communication in hardware using the wormhole routing protocol (Dally and Seitz, 1987).

We describe the wormhole routing briefly. The normal routing method, known as the store-and-forward method, does not exploit parallelism in transmitting a single message. To send a message of N bytes from node A to node B, the system (or the program) first identifies a path $A = A_0, A_1, \ldots, A_m = B$ of nodes and then sends the complete N byte message from node A_i to node A_{i+1} successively until it reaches node B. In this approach there is a significant loss of efficiency because node A_i must wait until it receives all the N bytes of the message from node A_{i-1} before it can start transmitting the message to node A_{i+1}. Moreover, a significant amount of buffer space is needed in each node to store the complete message. In the wormhole routing strategy node A_i does not have to wait until it receives the whole message from node A_{i-1} before it can start transmitting the message to node A_{i+1}. As node A_i receives bytes from node A_{i-1} it can ship them to node A_{i+1}, thereby achieving a significant amount of parallelism in message transmission.

6.5 Transputers

The parallel machines already described have fixed architectures. To buy a parallel machine the customer specifies the desired configuration (such as the number of nodes and the amount of memory in each node) and the manufacturer hardwires the network interconnections before delivery. Simple alterations, such as replacing a malfunctioning board, can be done by the customer, but major changes, such as adding more processors beyond the current capacity of the machine, may require enhancements to the network that need rewiring and perhaps the presence of a field engineer. These machines cannot be reconfigured in a way that is best suited for a given application. They are meant to be used by mapping the application onto the fixed architecture.

INMOS transputers (Pountain, 1984; Walker, 1985) have been designed to facilitate the reconfiguration of parallel machines to meet the interconnection requirements of a given application. A transputer is a single chip containing a 32 bit RISC-like microprocessor, 2 KB of on-chip memory, a memory interface to access up to 4 GB of off-chip memory, and four bidirectional communication links that provide point-to-point communication between the transputers. The four communication links coming out of a transputer look like telephone jacks, and different transputers can be connected simply by plugging the links together.

Although transputers can be connected in arbitrary ways, one type of interconnection is naturally suggested by the transputer architecture and has certain advantage over the hypercube. INMOS supplies a board with four transputers and 256 KB of memory local to each transputer. On this board the four transputers are hardwire connected in a square. If the diaonals are also connected (manually), the resulting board looks like a supertransputer because it has four communication links that can be used to connect it to other transputer boards (Fig. 7a). Four such boards may be connected in a similar fashion to yield a 16-node supertransputer (Fig. 7b). If we use the term T *net* (T for transputer) for this type of interconnection, then a single transputer is a zero-dimensional T net,

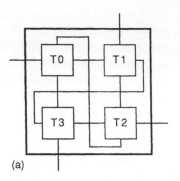

(a)

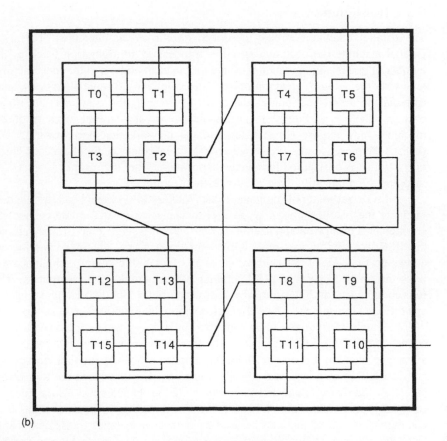

(b)

Figure 7   Transputers: (a) a 4-node T net; (b) a 16-node T net.

the four-transputer board is a one-dimensional T net, and the 16-node transputer system is a two-dimensional T net. It should be noted that an n-dimensional T net has 4^n nodes and its diameter d_n is given by $d_n = 2d_{n-1} + 1$. The diameter of this network grows faster than that of a hypercube, but the advantage of T net is that it has a constant connectivity (which is currently 4) irrespective of its dimension.

Current versions of transputers do not provide hardware support for remote communication. Earlier, remote communication was not supported even in software (INMOS, 1984; Logical Systems, 1989). Each user application was required to encode a remote communication protocol as required for that application. Subsequent operating systems for transputers, for example Helios (Perihelion, 1988) and Trollius (Burns et al., 1990), do support remote communication in software. However, for many applications the remote communication performance may not be satisfactory because of software protocol overheads.

The importance of hardware routing is being felt in the transputer community as well. INMOS has been working on a new generation of transputer chip, called the H1, that supports remote communication in hardware. At present the hardware characteristics of H1 have not yet been published.

REFERENCES

Adeli, H., and Vishnubhotla, P. (1987). Parallel processing. Microcomputers in Civil Engineering, Vol. 2, No. 3, pp. 257–269.

Amdahl, G. M. (1967). Validity of the single processor approach to achieving large scale computing capabilities. Proceedings of the AFIPS, Vol. 30, Thompson, Washington, D.C., pp. 483–485.

Batcher, K. E. (1980). Design of a massively parallel processor. IEEE Transactions on Computers, Vol. C-29, No. 9, September 1980, pp. 836–840.

Bhuyan, L. N. (1987). Interconnection networks for parallel and distributed processing. IEEE Computer, June, pp. 9–12.

Blakemore, C. (1977). *Mechanics of the Mind*. Cambridge University Press, Cambridge, England.

Burns, G., Radiya, V., Daoud, R., and Machiraju, R. (1990). All about Trollius. OCCAM User Group Newsletter, Vol. 13, pp. 54–65.

Chambers, F. B., Duce, D. A., and Jones, G. P., Eds. (1984). *Distributed Computing*. Academic Press, New York.

Chandler, D. (1987). The business of parallel processing is business. UNIX Review, 5(6):17–27.

Chazan, D., and Miranker, W. L. (1970). A nongradient and parallel algorithm for unconstrained minimization. SIAM Journal of Control, 8(2):207–217.

Dally, W. J., and Seitz, C. L. (1987). Deadlock-free message routing in multiprocessor interconnection networks. IEEE Transactions on Computers, Vol. 36, No. 5, pp. 547–553.

Davidson, D. B. (1990). A parallel processing tutorial. IEEE Antennas and Propagation Society Magazine, April, pp. 6–19.

Despain, A. M., and Patterson, D. A. (1978). X-TREE: A tree structured multiprocessor computer architecture. Proceedings of the 5th Symposium on Computer Architecture, Stanford, California, April 1978, pp. 144–151.

Dubois, M., Scheurich, C., and Briggs, F. A. (1988). Synchronization, coherence, and event ordering in multiprocessors. IEEE Computer, February, pp. 9–21.

Encore (1985). *Multimax Technical Summary.* Encore Computer Corporation, Marlboro, Massachusetts.

Enslow, P. H., Jr., Ed. (1974). *Multiprocessors and Parallel Processing.* Wiley Interscience, New York.

Flexible Computer Corporation (1985). FLEX/32 Multicomputer—System Overview, Document 030-000-001, Dallas, Texas.

Flynn, M. J. (1966). Very high-speed computing systems. Proc. IEEE, Vol. 54, pp. 1901–1909.

Gajski, D., Kuck, D., Lawrie, D., and Sameh, A. (1983). CEDAR—A large scale multiprocessor. Proceedings of the International Conference on Parallel Processing, West Lafayette, Indiana, IEEE Computer Society Press, August 23–26, pp. 524–529.

Gottlieb, Grishman, R., Kruskal, C. P., McAuliffe, K. P., Rudolph, L., and Snir, M. (1983). The NYU ultracomputer—Designing an MIMD shared memory parallel computer. IEEE Transactions on Computers, Vol. C-32, No. 2, pp. 175–189.

Heath, M. T. (1986). *Hypercube Multiprocessors.* SIAM, Philadelphia.

Hill, M. D., and Larus, J. R. (1990). Cache considerations for multiprocessor programmers. Communications of the ACM, Vol. 33, No. 8, pp. 97–102.

Hillis, W. D. (1985). *The Connection Machine.* MIT Press, Cambridge, Massachusetts.

Hillis, W. D. (1986). Parallel computers for AI databases. In: Brodie, M. L., and Mylopoulos, J., Eds., *On Knowledge Base Management Systems.* Springer-Verlag, New York.

Hinton, G. E., and Anderson, J. A., Eds. (1981). *Parallel Models of Associative Memory.* Erlbaum, Hillsdale, New Jersey.

INMOS (1984). *OCCAM Programming Manual,* Prentice-Hall International, Englewood Cliffs, New Jersey.

Intel Corporation (1986). iPSC System Overview Manual.

Jorgensen, C., and Matheus, C. (1986). Catching knowledge in neural nets. AI Expert, 1(4):30–38.

Joseph, M., Prasad, V. R., and Natarajan, N. (1984). *A Multiprocessor Operating System.* Prentice-Hall International, Englewood Cliffs, New Jersey.

Kuck, D. J., Sameh, A. H., Cytron, R., Veidenbaum, A. V., Polychronopoulos, C. D., Lee, G., McDaniel, T., Leasure, B. B., Bechman, C., Davies, J. R. B., and Kruskal, C. P. (1984). The effects of program restructuring, algorithm change, and architecture choice on program performance. Proceedings of the International Conference on Parallel Processing, IEEE Computer Society Press, pp. 129–138.

Lawrie, D. (1975). Access and alignment of data in an array processor. IEEE Transactions on Computers, Vol. C-24, December 1975, pp. 1145–1155.

Logical Systems (1989). *Transputer Tool Set—Version 88.4.*

Lorin, H. (1972). *Parallelism in Hardware and Software.* Prentice-Hall, Englewood Cliffs, New Jersey.

McClelland, J. L., Rumelhart, D. E., and the PDP Research Group, Eds. (1986). *Parallel Distributed Processing—Explorations in the Microstructure of Cognition. Volume 1, Psychological and Biological Models.* MIT Press, Cambridge, Massachusetts.

Minsky, M., and Papert, S. (1971). On some associative, parallel and analog computations. In: Jacks, E. J., Ed., *Associative Information Techniques.* Elsevier, New York.

Miranker, W.L. (1971). A survey of parallelism in numerical analysis, SIAM Review, 13(4):524–547.

Mudge, T. N., Hayes, J. P., and Winsor, D. C. (1987). Multiple bus architectures. IEEE Computer, June 1987, pp. 42–48.

Patton, P. C. (1985). Multiprocessors: Architecture and application. Computer, IEEE, June 1985, pp. 29–40.

PC AI (1987). Interview with philosopher Hubert Dreyfus. Summer 1987, pp. 54–57.

Pease, M.C. (1967). Matrix inversion using parallel processing, *Journal of the Association for Computing Machinery*, 14(4):757–764.

Perihelion (1988). *The Helios Operating System.*

Pfister, G. F., Brantley, W. C., George, D. A., Harvey, S. L., Kleinfelder, W. J., McAuliffe, K. P., Melton, E. A., Norton, V. A., and Weiss, J. (1985). The IBM research parallel processor prototype. Proceedings of the International Conference on Parallel Processing, August 1985, pp. 764–771.

Pountain, D. (1984). Microprocessor design. Byte, August 1984, pp. 361–365.

Preparata, F. P., and Vuillemiu, J. (1981). The cube connected cycles: A versatile network for parallel computations. Communications of ACM, 24(5):300–309.

Rattner, J. D. (1985). *Concurrent Processing: A New Direction in Scientific Computing.* AFIPS Press, Montvale, New Jersey, Vol. 54.

Rumelhart, D. E., Hinton, G. E., and McClelland, J. L. (1986a), A general framework for parallel distributed processing. Chapter 2 in Rumelhart, D. E., McClelland, J. L., and the PDP Research Group, Eds. *Parallel Distributed Processing—Explorations in the Microstructure of Cognition.* Volume 1. Foundations. MIT Press, Cambridge, Massachusetts pp. 46–76.

Rumelhart, D. E., McClelland, J. L., and the PDP Research Group, Eds. (1986b). Parallel Distributed Processing—Explorations in the Microstructure of Cognition. Volume 1. Foundations. MIT Press, Cambridge, Massachusetts.

Stayanarayanan, M. (1980). *Multiprocessors—A Comparative Study.* Prentice-Hall, Englewood Cliffs, New Jersey.

Sawchuk, A. A., Jenkins, B. K., Raghavendra, C. S., and Varma, A. (1987). Optical crossbar networks. IEEE Computer, June 1987, pp. 50–60.

Seitz, C. L. (1985). The cosmic cube. Communications of ACM, 28(1):22–33.

Sejnowski, M. C., et al. (1980). *An Overview of the Texas Reconfigurable Array Computer.* AFIPS Press, Montvale, New Jersey, pp. 631–641.

Siegel, H. J. (1985). *Interconnection Networks for Large-Scale Parallel Processing: Theory and Case Studies.* Lexington Books, Lexington, Massachusetts.

Smith, A. J. (1982). Cache memories. ACM Computing Surveys, Vol. 14, No. 3, pp. 473–530.

Uhr, L. (1984). *Algorithm-Structured Computer Arrays and Networks*. Academic Press, Orlando, Florida.

von Neumann, J. (1956). Probabilistic logics and the synthesis of reliable organisms from unreliable components. *Automata Studies*, Princeton University Press, Princeton, New Jersey.

Walker, P. (1985). The transputer—A building block for parallel processing. Byte, May 1985, pp. 219–235.

Wittie, L. D. (1981). Communication structures for large multimicrocomputer systems. IEEE Transactions, Computers, 30(4):64–273.

Wulf, W. A., Levin, R., and Harbison, S. P. (1981). *C.mmp/Hydra: An Experimental Computer System*. McGraw-Hill, New York.

2
Parallel Programming Languages and Techniques

Prasad R. Vishnubhotla and Hojjat Adeli   *The Ohio State University, Columbus, Ohio*

1   INTRODUCTION

Parallel programming can be generally defined as the art of programming a collection of computers to execute a single application efficiently. The application may be a numerical or a symbolic computation, in which case efficiency means achieving a faster execution speed. On the other hand, the application may be a real-time system or an operating system, in which case efficiency means being able to meet the real-time requirements imposed by the devices and the users.

There are various approaches to programming an application to achieve speedup in execution. One approach is to program the application in a sequential language like FORTRAN and to use a restructuring compiler like Paraphrase (Kuck et al., 1986) to generate parallel code for a given machine. Another approach is to program in a functional language (Hudak, 1989) or a data flow language (Arvind and Nikhil, 1990) and to use a compiler and hardware architecture that can exploit the fine-grained parallelism in the application. A third approach is to program in a parallel programming language in which the programmer introduces parallelism explicitly. Although the first two approaches relieve the programmer from some of the difficulties of explicit parallel programming, they do not extend easily for programming in real-time environments. Moreover, the applicability of these methods for parallel processing on distributed architectures is still being studied. Thus programming languages with explicit parallelism

21

seem to be more generally applicable, and in this chapter we concentrate our attention on this class.

The following features are generally desirable in a parallel programming language:

Ability to express execution
Interprocess communication
Process synchronization

2   PARALLEL PROCESSES

In sequential programming the whole computation is specified as a single stream of execution flow. In parallel programming the programmer is required to decompose the computation into several processes that may be executed in parallel on different processors by the operating system. Such decomposition may be easy for some applications but difficult for others. The programmer is required to have a thorough understanding of the application to obtain a useful decomposition. Furthermore, the programmer should not make any assumptions about how fast the processes will execute because such details depend on the actual speeds of the processors, which may be dissimilar, and on the scheduling algorithms used by the operating system.

3   COMMUNICATION AND
SYNCHRONIZATION

When a computation is decomposed into parallel processes, there is a need for the various processes to (1) communicate with each other so that they can exchange information, and (2) synchronize with each other so that processes can wait if the information that they need is not yet produced by the other processes. The desirable features in the design of an interprocess communication are efficiency, transparency, and reliability (van der Wal and van Dijk, 1990).

A common form of communication is to use shared memory: multiple processes are given access to shared variables they can both read and write. Communication is established by one or more processes writing into shared variables, which are then read by the other processes. In this case synchronization is required to prevent one process from overwriting the information generated by another process before that information is read by all other processes that need it. Such synchronization can be implemented using *locks* or the so-called semaphores (Dijkstra, 1968a; Brawer, 1989), in which the locking and unlocking is done indivisibly: if multiple processes try to lock at the same time then only one of them succeeds in locking.

Another form of communication and synchronization is *message passing*, in which processes communicate by sending messages to each other. Message passing is usually considered more general than shared memory because shared memory cannot be easily implemented on distributed systems, whereas message passing can be implemented on shared-memory machines as well as distributed systems. Another interesting feature of message passing is that communication and synchronization are combined into a single mechanism. Typically the sending of a message is *asynchronous:* the sender simply sends the message and continues but does not wait until the message is received. It is the responsibility of the operating system to buffer the messages and to deliver them to the receivers when they are ready. In most systems that support message passing, the receiving of a message is *synchronous:* a receiver waits until the desired message is available. All synchronization can be programmed using the send-and-receive operations appropriately. If sending is also synchronous, that is, if the sender continues only after the message is received by the receiver, this reduces the burden on the operating system to maintain buffers. Such synchronous send-and-receive operations are supported in CSP (communicating sequential processes) (Hoare, 1978) and its variation OCCAM (Jones, 1987).

4 SYNCHRONIZATION ABSTRACTIONS

In any programming language it is desirable to provide a small number of primitive mechanisms that are easy to learn and are flexible enough to program all interesting applications. However, history has shown us that the simplicity in primitive mechanisms can be deceptive. For example, the IF statement and the GOTO statement are adequate to implement all control flow in sequential programs. They are simple to understand as primitives. As pointed out by Dijkstra (1968b), however, programs that are constructed by composing these primitives arbitrarily result in what is commonly referred to as the spaghetti effect and therefore be difficult to understand. This led to the awareness of structured programming and resulted in the design of languages that support higher level control abstractions, like those in PASCAL (Wirth, 1971). A more puritanical approach would be to completely disallow any form of GOTO statement, as proposed in Dijkstra's guarded command language (Dijkstra, 1976).

The communication and synchronization primitives discussed in Section 3 are certainly primitive. Much research was done during the mid-1970s to design higher level synchronization abstractions for the shared-memory model. The most popular among these was the concept of a monitor proposed by Hoare (1974). The purpose of a monitor is to collect a few shared variables and the procedures that operate on these variables into a module. The users of the module can make calls to its procedures but cannot directly refer to the internal variables of the module. This provides information hiding. A monitor enforces *mutual exclusion* on the execution of its procedures: when a procedure is being executed other calls to any of the monitor's procedures are delayed until the currently executing procedure quits the monitor. This provides a synchronization abstraction because the code to enforce mutual exclusion is generated by the compiler and the programmer of the monitor need not use semaphores explicitly for this purpose. Several experimental concurrent programming languages have been designed to support monitors, for example, Concurrent PASCAL (Brinch Hansen, 1975), Modula (Wirth, 1982), PASCALPlus (Welsh and McKeag, 1980), and CCNPASCAL (Joseph et al., 1984).

A monitor is a passive object whose procedures are executed by the processes making the calls. This makes monitors unsuitable for programming resources like remote file servers in a distributed system. Languages based on message passing, such as CSP, can in principle be used to program distributed resources. However, the point-to-point nature of message passing makes it cumbersome to program a resource that must accept calls from several users. This prompted researchers to consider the *remote procedure call* as a useful programming paradigm. A remote procedure is executed by a process running on the node on which the procedure is located rather than by the process that makes the call. The calling process is simply delayed until the call is completed. A remote procedure call can be implemented by first passing the parameters as a message to the target node and then receiving the results from the target node as another message.

The remote procedure call is supported in several languages, for example, DP (distributed processes) (Brinch Hansen, 1978), Ada (United States Department of Defense, 1983), SR (Andrews et al., 1988), Argus (Liskov, 1988), Emerald (Black et al., 1987), and ALPS (Vishnubhotla, 1988). Interested readers are referred to the survey articles on concurrent and distributed programming languages by Andrews and Schneider (1983) and Bal et al. (1989).

5   SHARED TUPLE SPACES

In the conventional use of shared memory processes are allowed to modify the existing value of a memory location. Such updates are normally required in programming, but it is the update facility that needs synchronization. The need for synchronization creates bottlenecks and limits the amount of parallelism that can be exploited. An alternative view of memory is to treat it as a collection of data packets, or *tuples,* in which the basic operations defined on the tuple space are insert, remove, and read. This approach has been used in Linda (Carriero and Gelernter, 1989). However, it cannot benefit from the random-access nature of conventional memory and requires pattern matching to locate the tuples. The *insert* operation adds a new tuple to the tuple space but does not modify an existing tuple. A pattern to be used later to locate the tuple is stored in the tuple. The *remove* operation deletes a tuple that matches a specified pattern and returns that tuple. Since it is often desirable to read a tuple without deleting it, a *read* operation is also provided. Shared data spaces based on pattern matching are also supported in other systems, for example Associons (Rem, 1981), OPS5 (Brownston et al., 1985), and Swarm (Roman and Cunningham, 1989).

6 SINGLE-ASSIGNMENT LANGUAGES

Although the tuple space approach does not allow multiple updates directly, it does allow indirect programming of updates by first removing a tuple and then reinserting the updated value as another tuple. This gives the power of updating but it also creates interference between readers and writers. If interference is the main concern then this can be completely avoided by providing random-access memory in which each location can be assigned at most once. Such variables are called single-assignment variables and are commonly used in functional and data flow languages. A recent notation for composing concurrent processes supports single-assignment variables as well as ordinary variables (Mani Chandy and Taylor, 1989). Because of the single-assignment restriction, multiple variables (or an array structure) may be required to communicate several values to other processes, resulting in storage overheads. However, the programs would be much easier to understand and debug than when ordinary variables are used. The ALPS programming language (Vishnubhotla, 1990) combines the ideas of tuple spaces with the notion of single-assignment variables.

7 OBJECT-ORIENTED PARADIGM

Object-oriented programming languages, such as C++ or Object PASCAL, have become popular for sequential programming in recent years. An object-oriented program consists of a hierarchy of cooperating objects. The characteristics of object-oriented languages, such as inheritance, modularity, polymorphism, and encapsulation, support the development of reusable and highly modular software packages. In a similar fashion distributed operating systems and problem solvers can be modeled as distributed objects.

An example of a parallel object-oriented language is ConcurrentSmalltalk (Yokote and Tokoro, 1986, 1987a,b). Parallelism in ConcurrentSmalltalk is achieved by assigning a parallel process to an object. To achieve further parallelism it also supports asynchronous message passing (Bal et al., 1989). For a review of research on concurrent object-oriented programming models see Agha (1990).

The following problems must be further investigated before concurrent object-oriented programming languages are widely used (Yau et al., 1990):

Synchronization and parallelism support
Integration of inheritance and parallelism
Tools for supporting object-oriented decomposition

8 PARALLEL SYMBOLIC LANGUAGES

Parallel symbolic processing languages, such as Concurrent Lisp, require a different architecture than procedural numerical processing languages as a result of irregular memory access and a high processing power requirement (Chowkwanyun and Hwang, 1989).

Recently Top Level Common Lisp was introduced to the market by Top Level, Inc., as an extended implementation of Common Lisp for parallel processing based on research performed at the University of Massachusetts in Amherst. Its developers boast of a number of attractive features, including implicit synchronization, multiple-grain size capability, concurrent compilation and loading of many files, and an external debugger that can run on uniprocessor Symbolics Lisp machines. Currently it runs on the Encore Multimax and Sequent Symmetry shared-memory machines.

9 PARALLEL LOGIC LANGUAGES

Parallel logic programming languages retain many characteristics of the logic programming paradigm, such as representation of knowledge and data in logical terms, a built-in inferential mechanism, and metaprogramming. In addition they provide dynamic concurrent processes and communication synchronization mechanisms. The root of practically all the well-known parallel logic programming languages can be traced to PROLOG. Interprocess communication is through instantiation of shared logical variables, which is usually achieved by unification.

Three different methods have been used as synchronization mechanisms in parallel logic languages: input matching (or input unification), read-only unification, and determinacy conditions (Shapiro, 1989). The basic idea behind all of them is to delay the reduction of a goal atom until the atom's arguments are sufficiently instantiated.

10 PARALLEL PROGRAMMING
IN CONVENTIONAL
PROGRAMMING LANGUAGES

Although a significant amount of research is being done in parallel programming language design, a universally accepted language for parallel processing has yet to emerge. At present parallel programming on most parallel machines is done in

FORTRAN or C using operating system primitives for synchronization and communication (Adeli and Vishnubhotla, 1987; Adeli and Kamal, 1989). The operating system on Encore Multimax and the DYNIX operating system on Sequent both have parallel programming support packages. Parallel programming on the hypercube machine is done by organizing the application as a collection of cooperating C (or FORTRAN) programs, each of which runs on a separate processor (Heath, 1986). The different C (or FORTRAN) programs communicate by using operating system primitives for message passing.

The native language for transputers is OCCAM (Jones, 1987), which is based on the model of communicating sequential processes. The language is supported by the transputer architecture; the transputer is at its best when programmed in OCCAM. However, C, FORTRAN, and PASCAL are also supported on transputers. Parallel programming on a transputer network in any of these high-level languages, say C, is done in a way quite similar to that on a hypercube, that is, by organizing the application program as a collection of C programs each running on a separate transputer. These C programs can communicate using a channel interface supported by the transputer's C compiler.

An example of a parallel processing support package is the Encore Parallel Threads (EPT) package available on the Encore Multimax shared-memory machines (Encore, 1988; Adeli and Kamal, 1989). A thread is a unit of execution that is independent of other similar units (threads), yet it can execute concurrently with them. The concept of threads was first developed by Doeppner (1987). The notion of a thread is quite different from and independent of that of a processor. For example, one can have many threads running on one processor or concurrently on several processors. This results in a high level of abstraction to the programmer. It separates him or her from such details as how many processors are available. Concern must be limited to developing a certain number of threads. Encore Parallel Threads provides a set of constructs necessary for implementing the threads on an Encore Multimax. It can be used with the C programming language under the UMAX operating system. EPT provides groups of constructs that support the creation of threads, synchronization of threads through monitors or semaphores, creation of thread control blocks, raising exceptions, handling interrupts, and performing shared I/O. Adeli and Kamal (1990a,b, 1991a,b) developed parallel algorithms for partitioning, analysis, and optimization of large structures and implemented them in C on an Encore Multimax using the EPT.

Another system that supports the notion of threads is the Mach operating systems (Accetta et al., 1986). It is a UNIX-like operating system designed for shared-memory multiprocessors. Processes in UNIX are known to be heavy in the sense that a significant amount of information is associated with processes that must be saved and restored each time a processor is switched between processes. These save-and-restore operations, together known as *context switch*, are time consuming and result in performance degradation. Thus UNIX processes do not seem to be appropriate for parallel processing, particularly if the amount of computation done in each process is small.

This problem is addressed in Mach by supporting two kinds of processes: *tasks,* which are heavyweight processes, and *threads,* which are lightweight processes. Tasks correspond to UNIX processes; that is, interaction between tasks involves an expensive context switch. On the other hand, threads run within a task and share the data space of the task. The only private data associated with a thread is its stack. Thus switching a processor between the threads of the same task involves very little overhead. This can be exploited in parallel processing applications by organizing heavily interfacing processes as threads within the same task.

Mach is available on a variety of parallel machines, including Encore Multimax, Sequent Balance, and BBN Butterfly. Mach can be implemented on most machines that have hardware memory management units (MMUs). The MMU is required because Mach supports virtual memory (Rashid et al., 1988).

11 PARALLEL PROCESSING ON UNIPROCESSOR MACHINES

For the purpose of research and software development it is possible to perform parallel processing on a uniprocessor machine using software that simulates a parallel machine. Thus far simulators have been developed for hypercube machines. An example is the Intel Personal SuperComputer (iPSC) hypercube simulator (Intel, 1986). Using the iPSC simulator one can develop iPSC programs in FORTRAN or C in a controlled environment before executing the final software on an actual hypercube. The simulator provides an error-debugging facility. Thus use of the simulator speeds program development. The iPSC simulator package runs on the DEC VAX 11/780 and Sun machines with the UNIX 4.2 BSD operating system and on the IBM PC AT with the XENIX operating system.

The programming language Modula-2 (Wirth, 1982) also provides limited concurrent programming features. The inclusion of concurrent programming constructs in Modula-2 was for simulating problems naturally expressed in parallel processes rather than improving efficiency (Wilson and Clark, 1988). Communication in Modula-2 is through shared memory. Monitors are used to guarantee mutual exclusion.

12 THE FUTURE OF PARALLEL PROGRAMMING LANGUAGES

Most commercial parallel machines support parallel processing either by providing parallelizing FORTRAN compilers or by supporting primitive process communication and synchronization mechanisms of the kind described in Section 3. These may be adequate to implement small applications on specific machines, but use of such low-level architecture-dependent mechanisms creates a software crisis for parallel systems and causes portability problems when we are faced with large-scale parallel applications that must run on varying architectures. It is then imperative that we focus our attention on high-level architecture-independent parallel programming languages before we develop too many parallel applications using low-level tools.

Each of the approaches described in Sections 4 through 7 supports high-level architecture-independent programming. However, each takes an extreme view of how the programs should be organized. Although each approach is very attractive for a certain class (and granularity) of applications, attempting to fit all applications in a single mold often results in awkward and inefficient programs. A useful approach would be to integrate these different methods into a single language. When this is done there is ample possibility to trim and refine the individual mechanisms to make them more elegant and more efficient.

There is also another school of research, a system-oriented approach in which hardware designers and software engineers interact with the objective of finding an integrated solution. An example is the interprocessor communication in the Eindhoven multiprocessor system (EMPS) in which the problem of bus contention in a tightly coupled multiprocessor system is solved by a distributed interrupt mechanism using hardware communication registers (van der Wal and van Dijk, 1990).

REFERENCES

Accetta, M., Baron, R., Bolosky, W., Golub, D., Rashid, R., Tevanian, A., Jr., and Young, M. (1986). Mach: A new kernel foundation for UNIX development. Proceedings of the Summer Usenix, June.

Adeli, H., and Kamal, O. (1989). Parallel structural analysis using threads. Microcomputers in Civil Engineering, Vol. 4, No. 2, pp. 133–147.

Adeli, H., and Kamal, O. (1990a). Parallel stratagems for optimization of large structural systems. Proceedings of the IEEE International Conference on Systems Engineering, Pittsburgh, Pennsylvania, August 9–11, pp. 21–24.

Adeli, H., and Kamal, O. (1990b). Optimization of large structures on multiprocessor machines. Proceedings of Second IEEE Workshop on Future Trends of Distributed Computing Systems, Cairo, Egypt, September 30–October 2, pp. 290–295.

Adeli, H., and Kamal, O. (1991a). Concurrent optimization of large structures. Part I. Algorithms. ASCE Journal of Aerospace Engineering, Vol. 4.

Adeli, H., and Kamal, O. (1991b). Concurrent optimization of large structures. Part II. Applications. ASCE Journal of Aerospace Engineering, Vol. 4.

Adeli, H., and Vishnubhotla, P. (1987). Parallel processing. Microcomputers in Civil Engineering, Vol. 2, No. 3, pp. 257–269.

Agha, G. (1990). Concurrent object-oriented programming. Communications of the ACM, Vol. 33, No. 9, pp. 125–141.

Andrews, G. R., and Schneider, F. B. (1983). Concepts and notations for concurrent programming. ACM Computing Surveys, Vol. 21, No. 3, pp. 261–322.

Andrews, G. R., Olsson, R. A., Coffin, M., Elshoff, I., Nilsen, K., Purdin, T., and Townsend, G. (1988). An overview of the SR language and implementation. ACM Transactions on Programming Languages and Systems, Vol. 10, No. 1, pp. 51–86.

Arvind and Nikhil, R. S. (1990). Executing a program on the MIT tagged-token dataflow architecture.'' IEEE Transactions on Computers, Vol. 39, No. 3, pp. 300–318.

Bal, H. E., Steiner, J. G., and Tanenbaum, A. S. (1989). Programming languages for distributed computing systems. ACM Computing Surveys, Vol. 21, No. 3, pp. 261–322.

Black, A., Hutchinson, N., Jul, E., Levy, H., and Carter, L. (1987). Distribution and abstract types in emerald.'' IEEE Transactions on Software Engineering, Vol. SE-13, No. 1, pp. 65–76.

Brawer, S. (1989). *Introduction to Parallel Programming*. Academic Press, New York.

Brinch Hansen, P. (1975). The programming language concurrent Pascal. IEEE Transactions on Software Engineering, Vol. SE-1, February Issue, pp. 199–206.

Brinch Hansen, P. (1978). Distributed processes: A concurrent programming concept. Communications of the ACM, Vol. 21, No. 11, pp. 934–941.

Brownston, L., Farrell, R., Kant, E., and Martin, N. (1985). *Programming Expert Systems in OPS5: An Introduction to Rule-Based Programming*. Addison-Wesley, Reading, Massachusetts.

Carriero, N., and Galernter, D. (1989). How to write parallel programs: A guide to the perplexed. ACM Computing Surveys, Vol. 21, No. 3, pp. 323–357.

Chowkwanyun, R., and Hwang, K. (1989). Multicomputer load balancing for concurrent lisp execution. In: Hwang, K., and DeGroot, D., Eds. *Parallel Processing for Supercomputers and Artificial Intelligence*. McGraw-Hill, New York, pp. 325–366.

Dijkstra, E. W. (1968a). Cooperating sequential processes. In: Genuys, F., Ed. *Programming Languages*. Academic Press, New York, pp. 43–112.

Dijkstra, E. W. (1968b). GOTO statement considered harmful. Communications of the ACM, Vol. 11, No. 3, pp. 147–148.

Dijkstra, E. W. (1976). *A Discipline of Programming*. Prentice-Hall, Englewood Cliffs, New Jersey.

Doeppner, T. (1987). Threads: A system for the support of concurrent programming. Department of Computer Science, Brown University, Providence, Rhode Island, Report CS-87-11.

Encore Computer Corporation (1988). *Encore Parallel Threads Manual*. Marlboro, Massachusetts.

Heath, M. T. (1986). *Hypercube Multiprocessors*. SIAM, Philadelphia.

Hoare, C. A. R. (1974). Monitors: An operating system structuring concept. Communications of the ACM, Vol. 17, No. 10, pp. 549–557.

Hoare, C. A. R. (1978). Communicating sequential processes. Communications of ACM, Vol. 21, No. 8, pp. 666–677.

Hudak, P. (1989). Conception, evolution, and application of functional programming languages. ACM Computing Surveys, Vol. 21, No. 3, pp. 359–411.

Intel Corporation (1986). *iPSC Hypercube Simulator*.

Jones, G. (1987). *Programming in Occam*. Prentice-Hall, Englewood-Cliffs, New Jersey.

Joseph, M., Prasad, V. R., and Natarajan, N. (1984). *A Multiprocessor Operating System*. Prentice-Hall International, London.

Kuck, D. J., Davidson, E. S., Lawrie, D. H., and Sameh, A. H. (1986). Parallel supercomputing today and the Cedar approach. Science, February, pp. 967–974.

Liskov, B. (1988). Distributed programming in Argus. Communications of the ACM, Vol. 31, No. 3, pp. 300–312.

Mani Chandy, K., and Taylor, S. (1989). The composition of concurrent programs. Proceedings of the Supercomputing '89, Reno, Nevada, November 13–17, pp. 557–561.

Rashid, R., Tevanian, A., Jr., Young, M., Golub, D., Baron, R., Black, D., Bolosky, W. J., and Chew, J. (1988). Machine-independent virtual memory management for paged uniprocessor and multiprocessor architectures. IEEE Transactions on Computers, Vol. 37, No. 8, pp. 896–908.

Rem, M. (1981). Associons: A program notation with tuples instead of variables. ACM Transactions on Programming Languages and Systems, Vol. 3, No. 3, pp. 251–262.

Roman, G.-C., and Cunningham, H. C. (1989). A shared dataspace model of concurrency—Language and programming Implications. Proceedings of the 9th International Conference on Distributed Computing System, Newport Beach, June 5–9, pp. 270–279.

Shapiro, E. (1989). The family of concurrent logic programming languages. ACM Computing Surveys, Vol. 21, No. 3, pp. 413–4510.

United States Department of Defense (1983). *The Ada Programming Language Reference Manual*. U.S. Government Printing Office, ANSI/MIL-STD-1815A, Washington, D.C.

van der Wal, A. J., and van Dijk, G. J. W. (1990). Efficient interprocessor communication in a tightly-coupled homogeneous multiprocessor system. Proceedings of Second IEEE Workshop on Future Trends of Distributed Computing Systems, Cairo, Egypt, September 30–October 2, pp. 362–368.

Vishnubhotla, P. (1988). Synchronization and scheduling in ALPS objects. Proceedings of the 8th International Conference on Distributed Computing Systems, San Jose, California, June 13–17, pp. 256–264.

Vishnubhotla, P. (1990). Fine-grain parallelism in the ALPS programming language. Proceedings of Supercomputing '90, Rockfeller Center, New York, November 12–16.

Welsh, J., and McKeag, R. M. (1980). *Structured System Programming*. Prentice-Hall International, London.

Wilson, L. B., and Clark, R. G. (1988). *Comparative Programming Languages*. Addison-Wesley, Reading, Massachusetts.

Wirth, N. (1971). The programming language Pascal. Acta Informatica, Vol. 1, pp. 35–63.

Wirth, N. (1982). *Programming in Modula-2*. Springer-Verlag, New York.

Yau, S. S., Jia, X., and Bae, D.-H. (1990). Trends in software design for distributed computing systems. Proceedings of Second IEEE Workshop on Future Trends of

Distributed Computing Systems, Cairo, Egypt, September 30–October 2, pp. 154–160.

Yokote, Y., and Tokoro, M. (1986). The design and implementation of Concurrent-Smalltalk. ACM SIG-PLAN Not., Vol. 21, No. 11, pp. 331–340.

Yokote, Y., and Tokoro, M. (1987a). Concurrent programming in ConcurrentSmalltalk. In: Yonezawa, A. and Tokoro, M., Eds. *Object-Oriented Concurrent Programming*. MIT Press, Cambridge, Massachusetts, pp. 129–158.

Yokote, Y., and Tokoro, M. (1987b). Experience and evolution of ConcurrentSmalltalk. ACM SIG-PLAN Not., Vol. 22, No. 12, pp. 406–415.

3
Basic Concepts for the Development of Parallel Algorithms and Parallel Solution of Linear Systems

Udo Schendel   *Free University of Berlin, Berlin, Germany*

1   INTRODUCTION

The effective treatment of large systems makes it desirable to develop new, fast, and efficient methods. In this context parallel algorithms are of increasing importance. In general numerical algorithms can be classified in different ways, for example as algebraic or analytic algorithms, as finite or infinite algorithms, or as direct or iterative algorithms. In the last few years a new classification has become more important: that is, whether an algorithm is serial or parallel. The difference between such algorithms has become highly significant because of the development of parallel and pipeline computers. These machines permit the parallel execution of arithmetic operations. They are able to handle a great deal of information, and there are often hardware possibilities on different levels.

The basic idea of such a parallel computer is that programs using n processors should be n times faster than programs using only one processor. Experience and theory show that the real speedup is less, however.

There is need for a methodology of parallel algorithms aimed at the optimal use of parallel computers. Another more theoretical question is how to solve problems using maximal parallelism. The development of high-speed computers makes it necessary to reconfigure well-known methods for solving large and complex systems and to develop new efficient algorithms. The structure of these algorithms and their software are inherently dependent on the architecture of the computer system used, and vice versa.

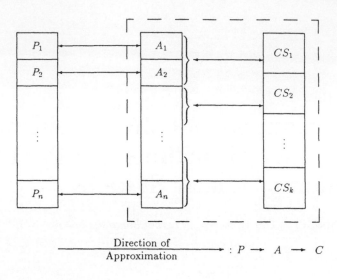

$$: P \rightarrow A \rightarrow C$$

Direction of
Approximation

P : problem
P_i : subproblems
A_i : subalgorithms
CS_k : computer systems

$A := [A_1, A_2, \ldots, A_n]$

$C := [CS_1, \ldots, CS_k]$

Figure 1 Adaptive methods.

Figure 1 gives an idea of the importance of adapting the methods and the architecture of the computer system to the problems under consideration.

Among currently significant applications are curve fitting, weather forecasting, spin model, simplex optimization, physical field evaluation, solution of systems of linear equations, structural analysis, and image processing. Other applications are as follows:

Computational aerodynamics: Wind tunnel experiments have a number of fundamental limitations. These include the model size, wind velocity, density, temperature, wall interference, and other factors. Numerical flow simulations have none of these limitations. The replacement of wind tunnels by supercomputers has been limited only by the processing speed and memory capacity of the computer used.

Artificial intelligence: User-friendly computers must be able to interact with humans at a higher level using speech, pictures, and natural languages (see fifth-generation computers).

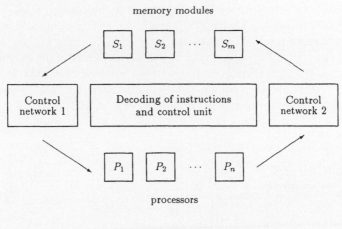

S: stores; P: processors

Figure 2   General configuration with different levels of parallelism.

1.1   Characteristics of Parallel Computers

A well-known classification of computer systems is that of Flynn (1966):

SISD machine: single instruction–single data stream (the conventional von Neumann machine)
SIMD machine: single instruction–multiple data stream
MISD machine: multiple instruction–single data stream
MIMD machine: multiple instruction–multiple data stream

Figure 2 shows a general model of a parallel computer.
Parallelism is possible

Within the control unit
Among the processors
Among the stores
In the data central network

The development of new computer architectures with different levels of parallelisms requires a detailed classification essential for the comparison of computers (Kuck, 1978). Schwartz (1980) made a distinction between paracomputers and ultracomputers based on different methods of memory access.

Examples of modern multiprocessor computers are the Cyber 203/205 (SIMD machine), Cray-1S, Cray X-MP, Cray Y-MP, Cray-2B, HEP-Denelcor

(MIMD machine), Hitachi S9/IAP (integrated array processor); SIMD machine, Alliant (SIMD/MIMD), ETA-10 (SIMD machine), and Sequent (SIMD/MIMD).

The different levels of parallelism in parallel computation are shown in Figure 3, including the measure of the degree of parallelism of the algorithm (A), the language (L), the object code (O), and the machine code (M). In terms of MFLOPS (1 MFLOP = 1 million floating-point operations per second) super-computers are often defined as performing more than 100 MFLOPS.

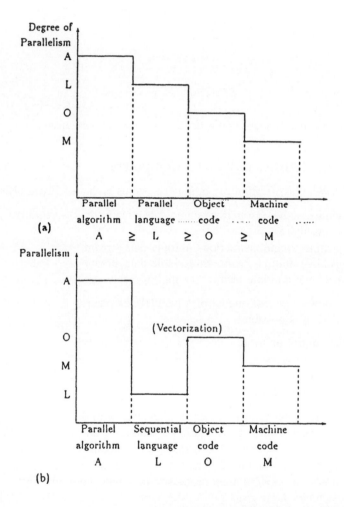

Figure 3   Levels of parallelism. (a) The ideal case of using parallel algorithm language. (b) The case of using vectorizing compiler and sequential language.

Today many different approaches for a classification of computer architecture, especially with respect to parallel processing, are discussed. A classification may be based on quantitative aspects (the degree of parallelism or the granularity), the structure of the control flow (SIMD, MIMD, data driven, or demand driven), the (hardware) technology [very large scale integrated (VLSI), very high speed integrated circuit (VHSIC), air cooled, or liquid cooled], or the topology of the processing elements and the memory units (Fig. 4). The basic classification categories follow:

SIMD and MIMD. The SIMD operation mode means that parallel functional units execute the same instruction sequence on different data. The two best

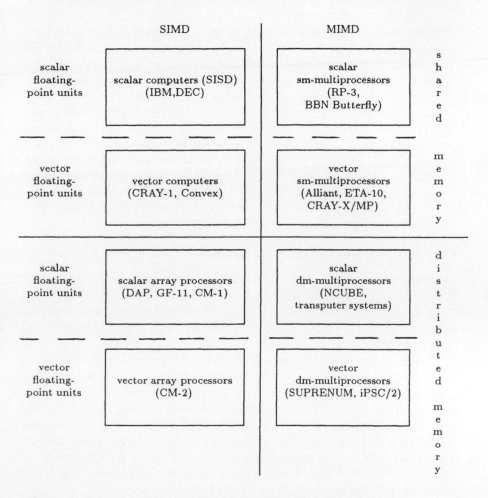

Figure 4 Classification of parallel and supercomputer architectures.

known realizations of the SIMD principle are pipelined floating-point units and array processors. The MIMD principle is the favorite operation mode for multiprocessors based on entire and independent processors. Each processor may execute a different instruction stream within the same application.

Shared and Distributed Memory. One of the central problems to be solved in the design of multiprocessor systems is the memory access. Basically there are two possibilities for organizing this:

1. Shared memory (SM) guarantees fair access to a total memory for each processor.
2. Distributed memory (DM) means that each processor has direct access only to its own private memory.

    Often both memory organization types are combined in hierarchic memory systems. The ETA-10, for instance, has rather fast and small distributed memories and a comparatively slow but large shared background memory.
3. Scalar and Vector Floating-Point Units. Scalar floating-point units to be restricted to a floating-point performance of less than 10 MFLOPS. At present the most cost effective way to achieve higher floating-point rates is vector processing. MIMD multiprocessors that target to the top of supercomputer performance must employ vector processors as basic floating-point units. Therefore the most powerful architectures today are mixed MIMD and SIMD multiprocessor systems. The efficient use of these architectures requires parallelism on two levels: the coarse-grained parallelism related to the global MIMD structure and the fine-grained parallelism that ensures efficient vector processing locally.

1.2   Architectures of Supercomputers

Scalar Computers.   The "traditional" von Neumann computer architecture (SISD) is the basis for mainframes, minicomputers, and microcomputers. Using the current hardware technology the floating-point performance of this architecture seems to be limited to 10 MFLOPS, which is far from supercomputer performance.

Vector Computers.   Historically the first machines to be called supercomputers were vector computers. Their hardware architecture is based on very fast arithmetic pipelines that support the rapid execution of vector instructions operating on all components of the vector operands simultaneously. Vectors in this sense consist of components that can be processed independently. Hence vector processing is a special form of parallel processing based on fine-grained parallelism. The need for vectorization resulted in new vector algorithms and in special compiler tools (vectorizers) for the automatic vectorization of existing codes. Examples of vector machines are the Cray-1, the Cyber 205, the Fujitsu VP, the NEC-SX, the Hitachi S-810, and the IBM 3090-VF.

Scalar SM-Multiprocessors.   Another way to increase computing performance is to combine several single processors with a multiprocessor system and to replace sequential processing by parallel processing. The optimal degree of parallelism (fine or coarse granularity) depends on the number and the power of the single processors as well as on the memory organization. The shared-memory concept restricts the number of central processing units (CPUs) to eight today (e.g., the Alliant). If the memory is accessed via a network a larger number of CPUs can be connected at the cost of longer access times. An example is the Cedar project (clusters of Alliant systems). Further examples in this class are the Sequent, Flexible, Encore, and Concurrent Computers machines.

Vector SM-Multiprocessors.   The step from a single processor to a multiprocessor system is of course also possible and obvious for vector computers. Similarly to scalar multiprocessors, performance is increased by composing several vector CPUs to multiprocessor systems with the same memory access problems. The shared-memory concept limits the number of vector processor at present to eight. The parallelism on these systems is often used to increase the throughput of the system (running different jobs on different CPUs) but not the execution speed of an individual job. MIMD parallel as well as an SIMD-like processing is also possible (e.g., on the Cray X-MP using macrotasking or microtasking constructs). Representatives of this class are the Cray X-MP, the Cray-2, and the ETA-10.

Scalar Array Processors.   Parallel computers started with array processors that perform one instruction simultaneously on an array of operands (in SIMD mode). Recently these systems have been upgraded to massively parallel multiprocessors. Each processor is relatively small and weak, but the enormous degree of parallelism results in supercomputer performance. Typically these systems are used for a restricted class of special applications. We mention here the historical Illiac IV, the Goodyear MPP, the ICL DAP, and the Connection Machine I.

Vector Array Processors.   The combination of (SIMD) array and vector processing has been realized in the Connection Machine 2, which at present is the system with the highest floating-point performance rate for certain applications on very regular data structures.

Scalar DM-Multiprocessors.   Today multiprocessor systems with a large (and principally unlimited) number of processors require that the memory units be physically associated with the processors (distributed memory). The basic unit of such a system, consisting of the CPU, the arithmetic coprocessor, the memory, and the communication unit, is called a *node*. The first prototypes of this class were based on hypercube topologies and were built at the Californian Institute of Technology. Recently Intel came out with its second generation, the iPSC-2. Multiprocessor systems with transputer nodes have also entered the market (Meiko and Parsytec).

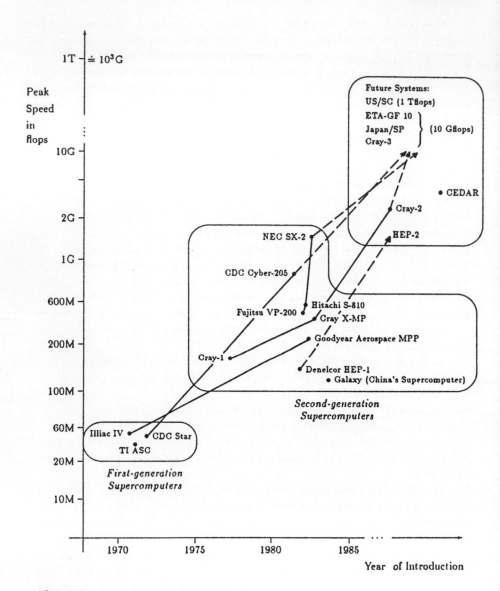

Figure 5   The space of supercomputers.

Vector DM-Multiprocessors. These systems combine the advantages of the vector and the parallel processing concepts. The basic idea is to combine powerful vector nodes with their advantageous cost-performance ratio to a multiprocessor system. Because of the size and the cost of a single node their number is, although on principle unlimited, today practically limited to several hundreds. The computational speed of the nodes, of course, imposes strong requirements on the speed of the communication. If the communication problem is solved satisfactorily, these machines are the most powerful supercomputers existing today. Systems currently entering the market are Suprenum, the Intel iPSC-VX, and Ametek 2010.

1.3 MIMD Computer Systems

The development of supercomputers is shown in the Figure 5 as a graph relating the rise in peak speeds. MIMD computers are classified at the highest level of parallelism (see Fig. 6); they are defined as control-flow computers capable of

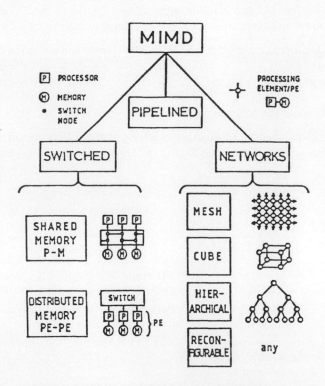

Figure 6   A structural taxonomy of MIMD computer systems.

processing more than one stream of instructions. The different instruction streams may be processed by separate instruction processing units, as in a multimicroprocessor design, or they may time-share a single instruction processing unit.

Because of the great importance in the application area we examine here a little bit more in detail the physical units that make up the MIMD computer and how they are interconnected. Figure 6 illustrates the structural taxonomy of MIMD computer systems.

2 BASIC CONCEPTS FOR THE DEVELOPMENT OF PARALLEL ALGORITHMS

In a parallel algorithm, because more than one task module can be executed at a time concurrency control is needed to ensure the correctness of the concurrent execution. Concurrency control enforces desired interactions among task modules so that the overall execution of the parallel algorithm is correct. With respect to the architecture one can classify the corresponding algorithms (Table 1).

The following connection exists between algorithms and hardware (Table 2):

Systolic algorithms are designed for direct hardware implementation (Mead and Conway, 1980).

MIMD algorithms are designed to be executed on general-purpose multiprocessors.

SIMD algorithms are between the two other types.

In a one-dimensional linear array machine data can flow simultaneously in both directions depending on the algorithm (Fig. 7). In a two-dimensional array

Table 1 Characterization of Algorithms

Type	Concurrent control	Remarks
SIMD	Central control unit SIMD	SIMD machines correspond to synchronous algorithms that require central control units
MIMD	Asynchronous, shared-memory MIMD	MIMD machines correspond to asynchronous algorithms with relatively large granularities
Systolic	Distributed control achieved by simple local control	LSI and VLSI machines[a] for special algorithms

[a]LSI, large scale integrated; VLSI, very large scale integrated.

Table 2 Examples of Applications in Parallel Algorithm Space

Application	Examples
SIMD algorithm	Numerical relaxation for image processing, partial differential equations, Gaussian elimination with pivoting
MIMD algorithm	Concurrent data base algorithms (concurrent accesses to binary search trees), chaotic relaxation, dynamic scheduling algorithms, algorithms with large module granularities
Systolic algorithm using	
One-dimensional linear arrays	Discrete Fourier transform (DFT), solution of triangular linear systems, recurrence evaluation
Two-dimensional square arrays	Dynamic programming, image processing, numerical relaxation, graph algorithms

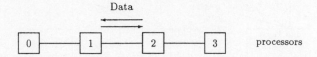

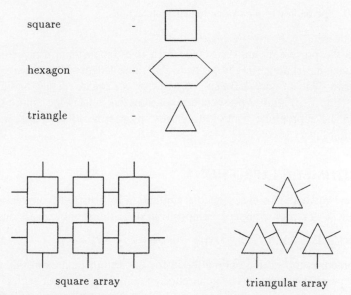

Figure 7 One-dimensional array.

Figure 8 Two-dimensional arrays.

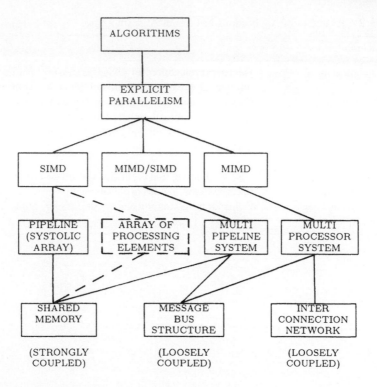

Figure 9   Taxonomy of parallel computer architectures.

machine the processors are distributed in a two-dimensional array so that the connections are symmetric and all have the same length. They must cover the whole array. Three figures have such properties: the square, the hexagon, and the triangle (Fig. 8). Figure 9 presents a taxonomy that relates the nature of parallelism to the appropriate control structures, processor architectures, and communication links.

3   ARITHMETIC EXPRESSIONS

Numerical methods are based on the evaluation of arithmetic expressions. An algorithm $\mathbf{A}$ is represented as a composition of arithmetical expressions $\mathbf{E}_k$

$$\mathbf{A} = \mathbf{E}_1 \circ \mathbf{E}_2 \circ \ldots \circ \mathbf{E}_n \qquad \text{with} \qquad \circ \in \{+,-,\cdot,/\} \tag{1}$$

The problem is to find the optimal code for the arithmetical expression $\mathbf{E}_k$ in a serial or parallel computer. The generation of a minimal code for such a problem is called an *np*-complete problem or a node-parse–complete problem. It is possible to find a minimal evaluation of arithmetic expressions with common subexpressions without using associativity and commutativity.

Example 1

Consider the expressions

$$\mathbf{E}_1 := a_3(a_4 + a_5)$$

$$\mathbf{E}_2 := a_1 a_2 + \{[(a_3(a_4 + a_5) + a_3(a_4 + a_5)] -$$
$$[a_3(a_4 + a_5) + (a_4 + a_5)]\}$$

$$\mathbf{E}_3 := [a_3(a_4 + a_5) + a_3(a_4 + a_5)] - \frac{a_3(a_4 + a_5) + (a_4 + a_5)}{a_1 a_2}$$

Then there are the following subexpressions:

$$\mathbf{T}_1 := (+, a_4, a_5) \qquad \mathbf{E}_1 := (\cdot, a_3, \mathbf{T}_1)$$

$$\mathbf{T}_2 := (-, \mathbf{T}_4, \mathbf{T}_5) \qquad \mathbf{T}_3 := (\cdot, a_1, a_2)$$

The additional subexpressions $\mathbf{T}_4$ and $\mathbf{T}_5$ are illustrated by Figure 10.

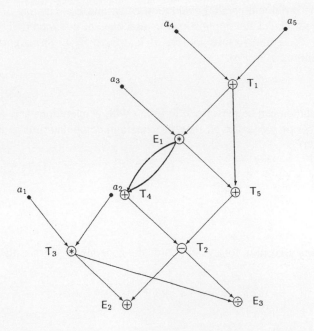

Figure 10 Example of minimal evaluation.

Generally the cost of solving np-complete problems is exponential. On the other hand, it can be shown that the so-called feedback-node problems of the np class can be solved at polynomial cost when this is treated as an enumeration problem (Borodin and Munro, 1975). Other authors try to solve this problem of optimal node generation by using directed, acyclic graphs, so-called dags.

Algorithms for the parallel evaluation of arithmetic expressions are based on the construction of equivalent arithmetic expressions.

Definition

Two arithmetic expression E and $\tilde{E}$ are said to be equivalent if it is possible to pass from E to $\tilde{E}$ by a finite number of applications of commutative, distributive, or associative laws.

Example 2

The following two arithmetic expressions E_1 and $\tilde{E}_1$ are equivalent ($E_1 \sim \tilde{E}_1$):

$$E_1 := [(a_1a_2 + a_3)a_4 + a_5]a_6 + a_7$$
$$\tilde{E}_1 := (a_1a_2a_4a_6 + a_3a_4a_6) + (a_5a_6 + a_7)$$

E_1 is called a generalized Horner algorithm. In Figure 11 the dags $\mathcal{G}_1$ and $\tilde{\mathcal{G}}_1$ describe the serial and parallel evaluation of E_1 and $\tilde{E}_1$. The dag of $\mathcal{G}_1$ of E_1 is a binary tree. Under the applications of associative, commutative, and distributive laws $\mathcal{G}_1$ is transformed into $\tilde{\mathcal{G}}_1$ of $\tilde{E}_1$ ($E_1 \sim \tilde{E}_1$). The algorithm for $\mathcal{G}_1$ should minimize $\tilde{t} = \tilde{t}(p)$.

Lemma

If a computation can be performed in time t with w operations and a sufficiently large number of processors $\bar{p}$, then the maximal computation time according to Brent (1974) is

$$\tilde{t} := \frac{t + (w - t)}{\bar{p}}$$

Theorem

Let A be any algorithm for evaluation of the generalized Horner algorithm (Fig. 11):

$$H := \{[\ldots (a_1a_2 + a_3)a_4 + a_5]a_6 + \ldots \} a_{2n} + a_{2n+1} \tag{2}$$

then

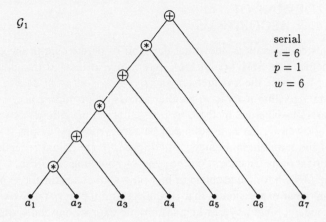

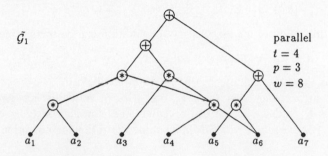

$t :=$ number of serial or parallel time steps
$p :=$ number of processors
$w :=$ number of operations performed by the algorithm

Figure 11 Generalized Horner algorithm.

$$w \geq 3n - \frac{t}{2}$$

This is equivalent to $t < 2n \Rightarrow w > 2n$ (Hyafil and Kung, 1974).

For a hypothetical MIMD machine it is assumed that

$op \in \mathcal{M} := \{+, -, \cdot, /\}$ can be performed by each processor.
Different processors perform different operations $op \in \mathcal{M}$.
$op \notin \mathcal{M}$ requires no time.
$op \in \mathcal{M}$ requires one time unit.

4 DEVELOPMENT OF PARALLEL ALGORITHMS

The previous considerations show that the kind of parallelism of the computer, namely whether it is SIMD or MIMD, essentially influences the nature of the algorithm. Moreover, the arrangement and dynamic sorting of data in memory, to which the algorithm must have parallel access, is of importance.

The basis of numerical methods is the evaluation of arithmetic expressions. The path followed by the evaluation can be represented by graphs (trees). Application of the laws of real numbers often leads to tree height reduction whose possible utilization is dependent on the type of computer considered.

In general we have the problem of mapping a given arithmetic expression $\mathbf{E}$ into an equivalent expression $\tilde{\mathbf{E}}$ that can be performed in parallel on SIMD or MIMD computers. One way in which arithmetic expressions can be parallelized is illustrated by the simple expression

$$\mathbf{E} := a_4 + [a_3 + (a_2 + a_1)] \qquad a_i \in \mathbb{R} \qquad i = 1, \ldots, 4 \tag{3}$$

which by utilization of the associativity of addition can be transformed into

$$\tilde{\mathbf{E}}_1 := (a_4 + a_3) + (a_2 + a_1) \qquad a_i \in \mathbb{R} \qquad i = 1, \ldots, 4 \tag{4}$$

This means that two additions can be made in parallel.

Given an arbitrary expression $\mathbf{E}_0$ one tries to split this into two smaller expressions $\mathbf{E}_1$ and $\mathbf{E}_2$, which can be calculated simultaneously, each by one processor. For execution on a SIMD machine the following conditions must be valid:

A function f exists with $\mathbf{E}_0 = f(\mathbf{E}_1, \mathbf{E}_2)$.
$\mathbf{E}_1$ and $\mathbf{E}_2$ are computed independently of each other and are of the same complexity.
$\mathbf{E}_1$ and $\mathbf{E}_2$ require the same series of computation.

Further splitting of $\mathbf{E}_1$ and $\mathbf{E}_2$ according to these conditions leads to $\mathbf{E}_0$ by recursive doubling (Kogge, 1974). To describe this method consider a set

$$\mathbb{S} = \{a_1, a_2, \ldots, a_N \mid N = 2^k \qquad k \in \mathbb{N}\} \subset \mathbb{R}$$

and an associative operation op $\in \mathcal{M} := \{+, *, \max, \ldots\}$ in S. Now the expression

$$\mathbf{E} = a_1 \text{ op } a_2 \text{ op } \ldots \text{ op } a_N$$

can be calculated. Associativity gives $\tilde{\mathbf{E}} := (a_1 \text{ op } a_2) \text{ op } (a_3 \text{ op } a_4) \text{ op } \ldots \text{ op } (a_{N-1} \text{ op } a_N)$, which can be performed in parallel.

Example 3

Consider the expression

$$\mathbf{E} = a_1 + a_2 + a_3 + a_4$$

By using the associative property of addition we obtain

$$\tilde{\mathbf{E}} = (a_1 + a_2) + (a_3 + a_4)$$

Figure 12 compares both these expressions. Generally, recursive doubling with $N = 2^n$ elements requires $\log_2 N$ parallel steps. Serial implementation requires $N - 1$ steps.

Example 4

Suppose that it is required to compute

$$\mathbf{E} = a_1 + a_2 \times a_3 + a_4$$

The parse tree $\mathcal{G}$ of $\mathbf{E}$ (Fig. 13) is not a unique tree, and no tree height reduction can be obtained by applying the associative law. By using the commutative property of addition $\mathbf{E}$ is transformed into

$$\tilde{E} = (a_1 + a_4) + a_2 a_3$$

and the parse trees shown in Figure 13 are obtained.

Definition

The speedup of a parallel algorithm is $S_p = T_1 / T_p$, where T_p is the execution time using p processors. The efficiency is defined by $E_p = S_p / p$.
Tree height reduction on a MIMD machine is possible. The following questions arise, among others:

How many tree height reductions can be achieved for a given arithmetic expression?

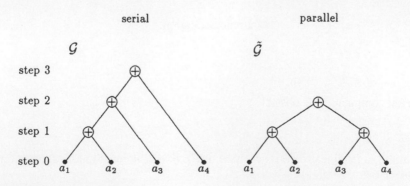

Figure 12   SIMD structure.

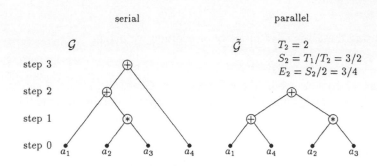

serial   parallel

S: speed-up; T: number of time units; E: efficiency

Figure 13   MIMD structure.

Can general algorithms be developed for tree height reduction?
How many processors are needed for optimality?

5   RECURRENCE RELATIONS

Recurrence relations are appropriate for problems whose solutions are expressed
in the form of a sequence $x_1, x_2, \ldots, x_n$, $x_i \in \mathbb{R}$, where each $x_i, i = 1, 2, \ldots, n$
may depend on x_j, $j < i$. A general example of recurrence relations is a time-
dependent linear system in which the state x_t of the system at time t is a linear
function of the state at time $t-1$:

$$x_1 := c_1 \qquad\qquad\text{initial condition}$$
$$x_2 := a_2 x_1 + c_2$$
$$\vdots$$
$$x_t := a_t x_{t-1} + c_t \qquad\text{recurrence equation} \qquad\qquad (5)$$
$$\vdots$$
$$x_n := a_n x_{n-1} + c_n$$

Serial computation is possible.

Definition

An mth order linear recurrence system $R\langle n, m \rangle$ for n equations is defined for m
$\leq n-1$ by

$$R < n,m> : x_k := \begin{cases} 0 & \text{for } k \le 0 \\ c_k + \displaystyle\sum_{j=k-m}^{k-1} a_{kj}x_j & \text{for } 1 \le k \le n \end{cases} \tag{6}$$

If $m = n-1$ this system is called an ordinary linear system of recurrence equations and is denoted $R\langle n \rangle$.

Example 5

The inner product of two vectors $x = (x_1, x_2, \ldots, x_n)$ and $y = (y_1, y_2, \ldots, y_n)$ can be written as a linear recurrence equation of first order,

$$z := z + x_k y_k \qquad 1 \le k \le n$$

with the initial condition $z := 0$.

Example 6

The Fibonacci sequence

$$f_k := f_{k-1} + f_{k-2} \qquad 3 \le k \le n$$
$$f_1 := f_2 := 1$$

may be considered a second-order recurrence equation.

Using matrix-vector notation Eq. (6) can be written as

$$x = c + Ax$$

where

$$x = (x_1, \ldots, x_n) \qquad \text{and} \qquad c = (c_1, \ldots, c_n)$$

and

$$A = [a_{ik}] \qquad i,k = 1,2, \ldots, n$$

is a strictly lower triangular matrix with $a_{ik} = 0$ for $i \le k$ or $i - k \ge m$. A is a band matrix for $m < n - 1$.

Example 7

For $n = 4$ we have the following ordinary recurrence system $R < 4 >$:

$$\begin{aligned} x_1 &= c_1 \\ x_2 &= c_2 + a_{21}x_1 \\ x_3 &= c_3 + a_{31}x_1 + a_{32}x_2 \\ x_4 &= c_4 + a_{41}x_1 + a_{42}x_2 + a_{43}x_3 \end{aligned} \tag{7}$$

Definition

A general mth order recurrence system $R<n,m>$ is defined by

$$R<n, m>: x_k := H[\bar{a}_k; x_{k-1}, x_{k-2}, \ldots , x_{k-m}] \qquad 1 \le k \le n$$

with m initial conditions $x_{-m+1}, x_{-m+2}, \ldots , x_0$. H is called the recursion function, and $\bar{a}_k$ is a vector of parameters that are independent of the x_i. This definition is suited to SIMD machines.

Example 8

A simple example of a first-order recurrence system is given by Eq. (5):

$$x_1 := c_1 \qquad \text{initial condition}$$
$$x_k := a_k x_{k-1} + c_k \qquad \text{for } 2 \le k \le n$$
$$= H[\bar{a}_k; x_{k-1}]$$

where the parameter vector $\bar{a}_k$ is of length 2 and of the form $\bar{a}_k = [a_k, c_k]$. H is defined by addition and multiplication.

Example 9

Special linear recurrent relations of first order are of the following kind:

$$x_1 = b_1$$
$$x_2 = -a_2 x_1 + b_2$$
$$\vdots$$
$$x_k = -a_k x_{k-1} + b_k$$
$$\vdots$$
$$x_n = -a_n x_{n-1} + b_n$$

finally in matrix-vector notation $Lx = b$ with

$$\left.\begin{aligned} l_{ii} &= 1 & i = 1, \ldots, n \\ l_{i,i-1} &= a_i & i = 2, \ldots, n \\ l_{ik} &= 0 & else \end{aligned}\right\} L = \begin{bmatrix} 1 & & & & & 0 \\ a_2 & 1 & & & & \\ & a_3 & \ddots & & & \\ & & & 1 & & \\ & & & & \ddots & \\ 0 & & & & a_n & 1 \end{bmatrix}$$

$$b = [b_1, \ldots, b_n]^T$$

The solution x is given by

$$x = L^{-1}b = L_n^{-1} L_{n-1}^{-1} \ldots L_3^{-1} L_2^{-1} b$$

with

$$L_i^{-1} = \begin{bmatrix} I_{i-2} & & & 0 \\ & 1 & & \\ & -a_i 1 & & \\ 0 & & I_{n-i} \end{bmatrix} \longleftarrow row\ i \qquad I_k = Identity\ matrix\ of\ order\ k$$

$$\uparrow$$

$$column\ i\ -\ 1$$

The serial and parallel factorization is shown in Figure 14. Remark: look for the inner and global parallelism. For a parallel solution of these systems the associativity of the recursion function H is required. If the given H function is not associative, there is often a so-called companion function G with associative properties.

Definition

A function G is said to be a companion function to the recursion function H if for all $x \in \mathbb{R}$ and for all parameter vectors $\bar{a}, \bar{b} \in \mathbb{R}^p$.

$$H[\bar{a};H[\bar{b};x]] = H[G(\bar{a},\bar{b});x] \tag{8}$$

holds, where $G: \mathbb{R}^p \times \mathbb{R}^p \rightarrow \mathbb{R}^p$. All companion functions have the following property (Schendel, 1984).

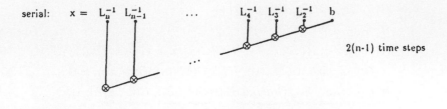

serial: $x = L_n^{-1}\ L_{n-1}^{-1}$ $\cdots$ $L_4^{-1}\ L_3^{-1}\ L_2^{-1}$ b

$2(n-1)$ time steps

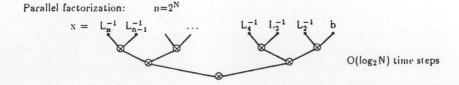

Parallel factorization: $n = 2^N$

$x = L_n^{-1}\ L_{n-1}^{-1}$ $\cdots$ $L_4^{-1}\ L_3^{-1}\ L_2^{-1}$ b

$O(\log_2 N)$ time steps

Figure 14   Recursive doubling.

Theorem

Every companion function G is associative with respect to its recursion function H; that is, for all $x \in \mathbb{R}$ and $\bar{a}, \bar{b}, \bar{c} \in \mathbb{R}^p$ we have

$$H[G(\bar{a}, G(\bar{b},\bar{c}));x] = H[G(G(\bar{a},\bar{b}),\bar{c});x]$$

If such functions G can be found, the parallelization of the recurrence relation can be reduced to the construction of a companion function and the construction of a parallel algorithm is possible by the log sum algorithm.

Example 10

Consider the first-order recurrence

$$x_k := H[\bar{a}_k; x_{k-1}]$$

where $\bar{a}_k \in \mathbb{R}^p$ and x_0 is the initial value.

$$
\begin{aligned}
x_2 &= H[\bar{a}_2; x_1] \\
&= H[\bar{a}_2; H[\bar{a}_1; x_o]] \\
&= H[G(\bar{a}_2,\bar{a}_1); x_o] \qquad \text{where } G \text{ is the companion function} \\[4pt]
x_4 &= H[\bar{a}_4; x_3] \\
&= H[\bar{a}_4; H[\bar{a}_3; x_2]] \\
&= H[G(\bar{a}_4,\bar{a}_3); x_2] \\
&= H[G(\bar{a}_4, \bar{a}_3); H[G(\bar{a}_2, \bar{a}_1); x_o]] \\
&= H[G(G(\bar{a}_4, \bar{a}_3), G(\bar{a}_2, \bar{a}_1)); x_o] \\[4pt]
x_8 &= H[\bar{a}_8; x_7] \\
&= H[\bar{a}_8; H[\bar{a}_7;H[\bar{a}_6; H[\bar{a}_5; x_4]]]] \\
&= H[G(\bar{a}_8, \bar{a}_7); H[G(\bar{a}_6, \bar{a}_5); x_4]] \\
&= H[G(\bar{a}_8, \bar{a}_7); H[G(\bar{a}_6, \bar{a}_5); H[G(G(\bar{a}_4, \bar{a}_3), G(\bar{a}_2 \bar{a}_1); x_o)]]] \\
&= H[G(G(\bar{a}_8, \bar{a}_7), G(\bar{a}_6, \bar{a}_5)); H[G(G(\bar{a}_4, \bar{a}_3), G(\bar{a}_2, \bar{a}_1); x_o)]] \\
&= H[G(G(G(\bar{a}_8, \bar{a}_7), G(\bar{a}_6, \bar{a}_5)), G(G(\bar{a}_4, \bar{a}_3), G(\bar{a}_2, \bar{a}_1))); x_o]
\end{aligned}
$$

In Figure 15 the graph shows the parallel evaluations of the inner G function values; finally one must apply the given H function. The log sum algorithm is applicable to numerical integration and the numerical solution of ordinary differential equations; here it is possible to compute the value y_N without knowing explicitly the previous values.

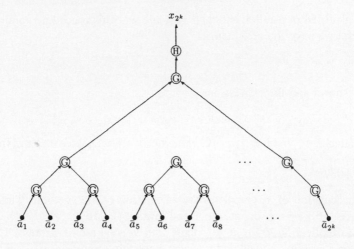

Graph for x_{2^k}

Figure 15   Parse tree for first-order recurrence.

6   PARALLEL SOLUTION OF LINEAR SYSTEMS ON SIMD MACHINES

The column-sweep algorithm serves for solving an $R<n>$ system of recurrence relations by a fast and efficient method such that with $N = O(n)$ processors the system can be computed in $O(n)$ time unit steps. Referring back to the set of equations of Example (7) we present the following algorithm.

Step 1

x_1 is known; then the expressions

$$c_i^{(1)} := a_{i1}x_1 + c_i, \qquad i = 1, \ldots, n$$

can be calculated in parallel. Now x_2 is known and there remains an $R<n-1>$ system.

Step 2

x_2 is known; the expressions

$$c_i^{(2)} := a_{i2}x_2 + c_i^{(1)} \qquad \text{with } c_i^{(1)} := a_{i1}x_1 + c_i$$

are calculated in parallel, and then x_3 is known. This procedure continues and in general we have the next step.

Step k

x_k is known and the expressions

$$c_i^{(k)} := a_{ik}x_k + c_i^{(k-1)} \text{ with } c_i^{(k-1)} := a_{i,k-1}x_{k-1} + c_i^{(k-2)}$$

are calculated, yielding x_{k+1}. Continuing, we finally arrive at the final step.

Step n − *1*

The calculation of x_n completes the calculation of the vector x. This algorithm requires $n-1$ processors at step 1 and fewer thereafter.

In application $R<n, m>$ systems with $m \ll n$ are of interest. The solution of such a system by the column-sweep algorithm requires at most m processors. Thus if a large number of processors is available the procedure is disadvantageous (Schendel, 1984).

One of the fastest procedures for the solution of an $R<n, m>$ system having small m is the so-called recurrent product form algorithm. The algorithm is developed from the product form representation of the solution of the $R<n>$ system:

$$x = Ax + c \tag{9}$$

where A is a strictly lower triangular matrix. The solution of Eq. (9) is

$$x = (I-A)^{-1}c = L^{-1}c \tag{10}$$

As shown in Kogge (1974) L^{-1} can be expressed in the product form

$$L^{-1} = M_n \cdot M_{n-1} \cdot \ldots \cdot M_1 \tag{11}$$

with

$$M_i = \begin{bmatrix} 1 & & & & & & & \\ & \ddots & & & & 0 & & \\ & & 1 & & & & & \\ & & & \dfrac{1}{a_{ii}} & & & & \\ & & & \dfrac{-a_{i+1,i}}{a_{ii}} & 1 & & & \\ & 0 & & \vdots & & \ddots & & \\ & & & \dfrac{-a_{ni}}{a_{ii}} & & & 1 & \end{bmatrix}$$

and in this case $a_{ii} = 1$ throughout. The solution vector

$$x = M_{n-1} \cdot M_{n-2} \cdot \ldots \cdot M_1 \cdot c$$

can be calculated in parallel mode by use of the recursive doubling procedure in $O(\log_2 n)$ time steps (Schendel, 1984).

Example 11

Consider the $R<4, 2>$ system with the recurrence relations

$$x_1 = c_1$$
$$x_2 = c_2 + a_{21}x_1$$
$$x_3 = c_3 + a_{31}x_1 + a_{32}x_2$$
$$x_4 = c_4 + a_{42}x_2 + a_{43}x_3$$

or

$$x = c + Ax$$

where

$$A := \begin{bmatrix} 0 & 0 & 0 & 0 \\ a_{21} & 0 & 0 & 0 \\ a_{31} & a_{32} & 0 & 0 \\ 0 & a_{42} & a_{43} & 0 \end{bmatrix}$$

Thus

$$L := \begin{bmatrix} 1 & 0 & 0 & 0 \\ -a_{21} & 1 & 0 & 0 \\ -a_{31} & -a_{32} & 1 & 0 \\ 0 & -a_{42} & -a_{43} & 1 \end{bmatrix}$$

Hence

$$x = L^{-1}c = M_3 M_2 M_1 c$$

$$= \begin{bmatrix} 1 & 0 & 0 & 0 \\ 0 & 1 & 0 & 0 \\ 0 & 0 & 1 & 0 \\ 0 & 0 & a_{43} & 1 \end{bmatrix} \begin{bmatrix} 1 & 0 & 0 & 0 \\ 0 & 1 & 0 & 0 \\ 0 & 0 & a_{32} & 0 \\ 0 & 0 & a_{42} & 1 \end{bmatrix} \begin{bmatrix} 1 & 0 & 0 & 0 \\ a_{21} & 1 & 0 & 0 \\ a_{31} & 0 & 1 & 0 \\ 0 & 0 & 0 & 1 \end{bmatrix} \begin{bmatrix} c_1 \\ c_2 \\ c_3 \\ c_4 \end{bmatrix}$$

$$= \begin{bmatrix} 1 & 0 & 0 & 0 \\ 0 & 1 & 0 & 0 \\ 0 & a_{32} & 1 & 0 \\ 0 & (a_{43}a_{32} + a_{42}) & a_{43} & 1 \end{bmatrix} \begin{bmatrix} c_1 \\ a_{21}c_1 + c_2 \\ a_{31}c_1 + c_3 \\ c_4 \end{bmatrix}$$

giving

$$x = \begin{bmatrix} c_1 \\ a_{21}c_1 + c_2 \\ a_{32}(a_{21}c_1 + c_2) + a_{31}c_1 + c_3 \\ (a_{43}a_{32} + a_{42})(a_{21}c_1 + c_2) + a_{43}(a_{31}c_1 + c_3) + c_4 \end{bmatrix}$$

The parallel evaluation with $m = 2$ and $n = 4$ requires at most five time steps.

Another method for a linear system to be solved in parallel mode is called parallelization by permutation. Consider the tridiagonal linear system

$$Ax = d \tag{12}$$

with

$$d := [d_1, \ldots, d_n]^T \in \mathbb{R}^n$$

and

$$A := \begin{bmatrix} a_1 & b_1 & & & \\ c_2 & a_2 & b_2 & & 0 \\ & & \ddots & & \\ & & \ddots & \ddots & b_{n-1} \\ 0 & & & c_n & a_n \end{bmatrix}$$

At first sight the SOR method (successive overrelaxation) (Schendel, 1989),

$$a_1 x_1^{(k+1)} := (1 - \omega)a_1 x_1^{(k)} - \omega b_1 x_2^{(k)} + \omega d_1$$
$$a_i x_i^{(k+1)} := (1 - \omega)a_i x_i^{(k)} - \omega c_i x_{i-1}^{(k+1)} - \omega b_i x_{i+1}^{(k)} + \omega d_i$$
$$a_n x_n^{(k+1)} := (1 - \omega)a_n x_n^{(k)} - \omega c_n x_{n-1}^{(k+1)} + \omega d_n$$

for $i = 2, \ldots, n-1$ does not seem to allow parallelization. By using the permutation (see also Golub, 1983),

$$PAP^T y := \tilde{A}y = z = \tilde{d} \tag{13}$$

with

$$y = [x_1, x_3, \ldots, x_{2^k-1}, x_2, \ldots, x_{2^k}]^T$$
$$\bar{d} = [d_1, d_3, \ldots d_{2^k-1}, d_2, \ldots, d_{2^k}]^T$$

and

$$\bar{A} := \begin{bmatrix} a_1 & & & & b_1 & & \\ & a_3 & & & c_3 b_3 & & \\ & & \ddots & & & \ddots & \ddots \\ & & & a_{2^k-1} & & & c_{2^k-1} \, b_{2^k-1} \\ \hline c_2 \, b_2 & & & & a_2 & & \\ & c_4 \, b_4 & & & & a_4 & \\ & & \ddots & \ddots & & & \ddots \\ & & & b_{2^k-2} & & & \\ & & & c_{2^k} & & & a_{2^k} \end{bmatrix}$$

The SOR method applied to $\bar{A}y = \bar{d} = z$ gives

$$a_i y_i^{(k+1)} = (1-\omega)a_i y_i^{(k)}$$

$$a_{2i-1} y_i^{(k+1)} = (1-\omega)a_{2i-1} y_i^{(k)} - \omega c_{2i-1} y_{n/2+i-1}^{(k)} - \omega b_{2i-1} y_{(n/2)+i}^{(k)} + \omega z_i$$

$$\text{for } i = 2, \ldots, \frac{n}{2} \tag{14}$$

$$a_{2i} y_{(n/2)+i}^{(k+1)} \, 1 = (1-\omega)a_{2i} y_{(n/2)+i}^{(k)} - \omega c_{2i} y_i^{(k+1)} - \omega b_{2i} y_{i+1}^{(k+1)} + \omega z_{n/2+i}$$

$$a_n y_n^{(k+1)} = (1-\omega)a_n y_n^{(k)} - \omega c_n y_{n/2}^{(k+1)} + \omega z_n$$

$$\text{for } i = 1, 2, \ldots, \frac{n}{2} - 1 \tag{15}$$

If $y_1^{(k)}, \ldots, y_n^{(k)}$ are known, then $y_1^{(k+1)}, \ldots, y_{n/2}^{(k+1)}$ are computed in parallel by (14) and $y_{(n/2)+1}^{(k+1)}, \ldots, y_n^{(k+1)}$ are computed in parallel by Eq. (15).

Parallelization by permutation is the basic concept behind the odd-even reduction method (Hockney and Jesshope, 1981).

7 PARALLEL SOLUTION OF LINEAR SYSTEMS ON MIMD MACHINES

To illustrate how a parallel solution method can be designed for an MIMD machine (for example, HEP), consider solving a set of symmetric algebraic equations

$$a_{11} + a_{12}x_2 + \ldots + a_{1n}x_n = b_1$$
$$a_{21} + a_{22}x_2 + \ldots + a_{2n}x_n = b_2$$
$$\vdots \qquad \vdots \qquad \vdots \qquad \vdots$$
$$a_{n1} + a_{n2}x_2 + \ldots + a_{nn}x_n = b_n$$

which is equivalent to the system

$$Ax = b \qquad\qquad\qquad (16)$$

with

$$A := [a_{ik}] \qquad i,k = 1,2, \ldots ,n$$
$$b := [b_1, \ldots ,b_n]^T$$
$$x := [x_1, \ldots ,x_n]^T$$

For simplicity it is assumed that row or column interchange for numerical stability is not needed. Also considered is only a factorization of the matrix $A = [a_{ik}]$. A sequential program that produces the lower triangular factor of the matrix A might be as follows:

```
for     k := 1 to n − 1 do
for     j := k + 1 to n do

begin
          a(j,k)
c :=    ───────
          a(k,k)                           } T_j^k
for i := j to n do
a(i,j) := a(i,j) − a(i,k)*c
end;
```

The computational task between begin and end is denoted T_j^k for given values of k and j. Figure 16 illustrates the tasks T_j^k to be executed and their temporal precedence constraints.

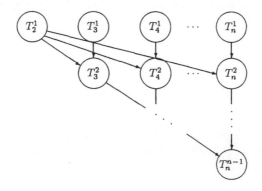

Figure 16 Executing tasks.

It follows in particular that the tasks T_3^3 cannot be executed before T_3^1 is completed since both tasks change the values of the third column of A. Also T_3^2 requires completion of T_2^1 whose elements are used to evaluate components of T_3^2. However, the tasks $T_j^2, j = 3,4, \ldots ,n$, can be computed in parallel once the tasks $T_j^1, j = 2,3, \ldots ,n$, have been finished. It can be shown that even if the number of available processors is high (approximately $n/2$) an efficient parallel algorithm can be constructed. Such an algorithm would keep busy 66% of the parallel processors during the execution period required to decompose the matrix A. If the number of processors p is considerably smaller than n the computational efficiency would quickly increase (above 90% for the HEP). This means that the speedup provided by the HEP working in parallel mode is roughly $S_p = 8r$, where r is the number of process execution modules (PEMS).

8 DOMAIN DECOMPOSITION METHODS

In the application area the efficient numerical solution of PDEs is a very important subject. Besides multigrid methods (Dornscheidt and Schendel, 1988), domain decomposition methods are very suitable for developing parallel algorithms for PDEs.

The class of techniques that has received much attention recently is the class of domain decomposition techniques in which the physical domain is divided into separate subdomains, each handled by a different processor.

Let us consider the numerical solution of the partial differential equation

$$\frac{\partial u}{\partial t} = F(x, u, t, D_x u, D_x^2 u, \ldots)$$

where $x \in \Omega \in \mathbb{R}^m$, $0 \le t \le \tau$, and $u(x, t) \in \mathbb{R}^m$ satisfy given boundary and initial conditions. Then at t*,

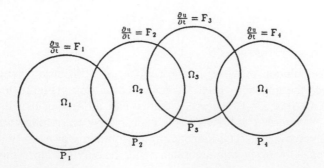

Figure 17 Domain Ω.

$$y\Omega \;=\; \bigcup_{j=1}^{k(t^*)} \Omega_j(t^*)$$

and on each Ω_j,

$$\frac{\partial u}{\partial t} \;=\; F_j(x,\,u) \qquad j \;=\; 1,2,\,\ldots,\,k(t^*)$$

is defined. Each processor P_j solves one of these partial differential equations over a prespecified time interval (see Fig. 17).

Discretization of the PDE by finite differences or finite elements gives the linear system

$$L^h \bar{u} \;=\; \begin{bmatrix} A_1 & B_1 & & & \\ B_1^T & A_2 & B_2 & & \\ & & \ddots & \ddots & \\ & & & & B_{n-1} \\ & & & B_{n-1}^T & A_n \end{bmatrix} \begin{bmatrix} \bar{u}(1) \\ \bar{u}(2) \\ \vdots \\ \vdots \\ \bar{u}(n) \end{bmatrix} = \begin{bmatrix} \bar{b}(1) \\ \bar{b}(2) \\ \vdots \\ \vdots \\ \bar{b}(n) \end{bmatrix} = \bar{b}$$

where $\bar{u}(i)$ is the numerical approximation of $\bar{u}$ on the points $(x_i,\,y_j)$, for $j = 1,2,$ $\ldots,n$, and A_i and B_i are $(n,\,n)$-tridiagonal matrices with $A_i = A_i^T$. A solution can be obtained by an iterative process such as the SOR method or the conjugate gradient method.

Briefly we repeat the conjugate gradient method (CG method) to be used in solving $Ax = b$, where A is a symmetric positive definite matrix (Golub and van Loan, 1983). Choose x_0; set $p_0 := r_0 = b - Ax_0$ (residual). Compute for $k = 0,1,\,\ldots$:

$$\alpha_k \;:=\; -\frac{r_k^T p_k}{p_k^T A p_k} \qquad \text{(i)}$$

$$x_{k+1} \;:=\; x_k - \alpha_k p_k \qquad \text{(ii)}$$

$$r_{k+1} \;:=\; r_k + \alpha_k A p_k \qquad \text{(iii)}$$

$$\beta_k \;:=\; -\frac{p_k^T A r_{k+1}}{p_k^T A p_k} \qquad \text{(iv)}$$

$$p_{k+1} \;:=\; r_{k+1} + \beta_k p_k \qquad \text{(v)}$$

Let $\hat{x}$ be the solution of $Ax = b$ and A an $(n,\,n)$ symmetric positive definite matrix. Then the direction vectors p_k generated by Eqs. (i) through (v) satisfy $p_k^T A p_j = 0$, $0 \le j < k$, $k = 1,\,\ldots,n-1$ (that is, they are conjugate with respect to A), and $p_k \neq 0$ unless $x^k = \hat{x}$. Hence $x^m = \hat{x}$ for some $m \le n$. It follows that the CG method converges, in exact arithmetic, to the solution $\hat{x}$ of $Ax = b$ in no more than n steps. Obviously this algorithm relies heavily on inner products and matrix-vector multiplication Ap; therefore the algorithm is suitable for parallel and vector implementation. You can improve this algorithm by preconditioning A (Golub and Mayers, 1983).

The following Poisson Equation is considered a special example for solving elliptic PDEs by decomposition methods.

$$\Delta u = u_{xx} + u_{yy} = f \quad \text{in } \Omega := \{(0, 1) \times (0, 1)\} \subset \mathbb{R}^2$$

$u = g$ on the boundary $\partial\Omega$ with $u \in C^2(\Omega), f \in C^1(\Omega)$, and $g \in C^1(\partial\Omega)$.

Discretization on the grid G over Ω,

$$G := \{(x_i, y_i) \mid x_i = ih, y_j = jh; i = 1, \ldots, m; j = 1, \ldots, n\}$$

gives the linear system $Lu = b$ as an approximation of the PDE:

$$L = \begin{bmatrix} A_1 & B_1 & & & & \\ B_1^T & A_2 & B_2 & & 0 & \\ & B_2^T & A_3 & B_3 & & \\ & & \ddots & \ddots & \ddots & \\ 0 & & & B_{m-2}^T & A_{m-1} & B_{m-1} \\ & & & & B_{m-1}^T & A_m \end{bmatrix} \in \mathbb{R}^{nm,nm}$$

with

$$A_i = \begin{bmatrix} 4 & -1 & & & & \\ -1 & 4 & -1 & & 0 & \\ & -1 & 4 & -1 & & \\ & & \ddots & \ddots & \ddots & \\ 0 & & & -1 & 4 & -1 \\ & & & & -1 & 4 \end{bmatrix} \in \mathbb{R}^{n,n}$$

and $B_i = -I \in \mathbb{R}^{nn}$, $u, b \in \mathbb{R}^{nm}$.

Special Method

Numerical Schwarz algorithms consider a decomposition of Ω into overlapping subdomains Ω_i and Ω_j of Ω

$$\Omega = \bigcup_{i-1}^{p} \quad \Omega_i \in \mathbb{R}^2 \quad \text{and} \quad \Omega_i \cap \Omega_j \neq \phi$$

For $p = 2$ this follows for Ω_1 and Ω_2 (Fig. 18). After discretization of the Poisson equation by finite differences we have the following system for the domain Ω:

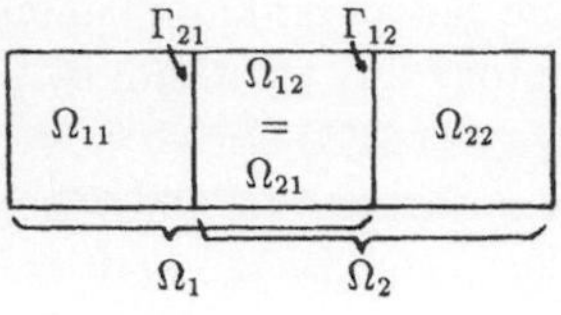

$$\Omega_1 = \Omega_{11} \cup \Gamma_{21} \cup \Omega_{12}$$
$$\Omega_2 = \Omega_{21} \cup \Gamma_{12} \cup \Omega_{22} \text{ and}$$
$$\Omega = \Omega_1 \cup \Omega_2$$

Figure 18   Domain decomposition.

$$Lu = \begin{bmatrix} K_{11} & K_{12} & 0 & 0 & 0 \\ K_{12}^T & K_{22} & K_{23} & 0 & 0 \\ 0 & K_{23}^T & K_{33} & K_{34} & 0 \\ 0 & 0 & K_{34}^T & K_{44} & K_{45} \\ 0 & 0 & 0 & K_{45}^T & K_{55} \end{bmatrix} \begin{bmatrix} u_1 \\ u_2 \\ u_3 \\ u_4 \\ u_5 \end{bmatrix} = \begin{bmatrix} b_1 \\ b_2 \\ b_3 \\ b_4 \\ b_5 \end{bmatrix} = b$$

The matrix L and the vectors u and b are partitioned according to the decomposition of Ω. The submatrices K_{ij} describe the coupling of the unknown u_i and u_j in the different subdomains. We then have the following systems to be solved in the subdomains Ω_1 and Ω_2 (Bjorstad and Widlund, 1984):

$$L_1 v_1 = \begin{bmatrix} K_{11} & K_{12} & 0 \\ K_{12}^T & K_{22} & K_{23} \\ 0 & K_{23}^T & K_{33} \end{bmatrix} \begin{bmatrix} v_{11} \\ v_{12} \\ v_{13} \end{bmatrix} = \begin{bmatrix} e_{11} \\ e_{12} \\ e_{13} \end{bmatrix} = e_1 \qquad \text{with } v_{1i} = u_i$$

$$i = 1,2,3$$

and

$$L_2 v_2 = \begin{bmatrix} K_{33} & K_{34} & 0 \\ K_{34}^T & K_{44} & K_{45} \\ 0 & K_{45}^T & K_{55} \end{bmatrix} \begin{bmatrix} v_{21} \\ v_{22} \\ v_{23} \end{bmatrix} = \begin{bmatrix} e_{21} \\ e_{22} \\ e_{23} \end{bmatrix} = e_2 \qquad \text{with } v_{2i} = u_{2+i}$$

$$i = 1,2,3$$

The right-side vectors e_1 and e_2 are computed by

$$e_1 = \begin{bmatrix} e_{11} \\ e_{12} \\ e_{13} \end{bmatrix} = \begin{bmatrix} b_1 \\ b_2 \\ b_3 - K_{34}u_4 \end{bmatrix} = \begin{bmatrix} b_1 \\ b_2 \\ b_3 \end{bmatrix} - \begin{bmatrix} 0 & 0 & 0 \\ 0 & 0 & 0 \\ 0 & K_{34} & 0 \end{bmatrix} \begin{bmatrix} v_{21} \\ v_{22} \\ v_{23} \end{bmatrix}$$

$$=: f_1 - G_2 v_1$$

and

$$e_2 = \begin{bmatrix} e_{21} \\ e_{22} \\ e_{23} \end{bmatrix} = \begin{bmatrix} b_3 - K_{23}^T u_2 \\ b_4 \\ b_5 \end{bmatrix} = \begin{bmatrix} b_3 \\ b_4 \\ b_5 \end{bmatrix} - \begin{bmatrix} 0 & K_{23}^T & 0 \\ 0 & 0 & 0 \\ 0 & 0 & 0 \end{bmatrix} \begin{bmatrix} v_{11} \\ v_{12} \\ v_{13} \end{bmatrix}$$

$$=: f_2 - E_1 v_1$$

Then we have the system

$$Fv := \begin{bmatrix} L_1 & G_2 \\ E_1 & L_2 \end{bmatrix} \begin{bmatrix} v_1 \\ v_2 \end{bmatrix} = \begin{bmatrix} f_1 \\ f_2 \end{bmatrix} =: f$$

which can be solved by the Block-Jacobi method:

$$\begin{bmatrix} L_1 & 0 \\ 0 & L_2 \end{bmatrix} \begin{bmatrix} v_1^{(K+1)} \\ v_2^{(K+1)} \end{bmatrix} = \begin{bmatrix} f_1 \\ f_2 \end{bmatrix} - \begin{bmatrix} 0 & G_2 \\ E_1 & 0 \end{bmatrix} \begin{bmatrix} v_1^{(K)} \\ v_2^{(K)} \end{bmatrix}$$

This iterative algorithm converges (Bjorstad and Widlund, 1984) against the solutions v_1^* and v_2^* and it holds for the solution of u^* of $Lu = b$:

$$v_1^* = [v_{11}^*, v_{12}^*, v_{13}^*]^T = [u_1^*, u_2^*, u_3^*]^T$$

and

$$v_2^* = [v_{21}^*, v_{22}^*, v_{23}^*]^T = [u_3^*, u_4^*, u_5^*]^T$$

On a MIMD system with two processors each processor solves a linear system of the kind

$$L_i v_i^{(K)} = e_i^K \quad i = 1,2, \quad \text{with } e_1^K = f_1 - G_2 v_2^{(K-1)}$$

$$\text{and} \quad e_2^K = f_2 - E_1 v_1^{(K-1)}$$

Here we have a global and an inner parallelism of the algorithm.

These solutions correspond to the solution of the given PDE on the subdomains Ω_1 and Ω_2, respectively, with the boundary values $v_2^{(K-1)}$ and $v_1^{(K-1)}$, respectively, on Γ_{12} and Γ_{21}, respectively. Convergence results even for the approximation solution on Ω_1 and Ω_2 (Schendel and Schyska, 1987).

An alternative domain decomposition method is the class of capacity matrix methods. In this case the domain Ω is partitioned into no-overlapping subdomains (Fig. 19); that is, for $\Omega \subset \mathbb{R}^2$,

$$p = 2: \quad \boxed{\begin{array}{c|c} \Omega_1 & \Omega_2 \end{array}} \quad \Omega_1 \cap \Omega_2 = \phi, \quad \Omega = \Omega_1 \cup \Omega_2 \cup \Gamma_{12}$$

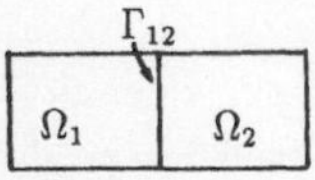

Figure 19 Nonoverlapping subdomains.

$$\Omega = \bigcup_{i=1}^{p} \Omega_i \qquad \Omega_i \subset \mathbb{R}^2 \qquad \text{and} \qquad \Omega_i \cap \Omega_j = \emptyset$$

Example 12

Discretization of the PDE by the grid G on Ω gives the linear system

$$Lu = \begin{bmatrix} K_{11} & 0 & K_{13} \\ 0 & K_{22} & K_{23} \\ K_{13}^T & K_{23}^T & K_{33} \end{bmatrix} \begin{bmatrix} u_1 \\ u_2 \\ u_3 \end{bmatrix} = \begin{bmatrix} b_1 \\ b_2 \\ b_3 \end{bmatrix} = b$$

The matrix L and the vectors u and b are partitioned according to the decomposition of Ω (Schendel and Schyska, 1987). Finally, the following capacity matrix system must be solved:

$$(K_{33} - K_{13}^T K_{11}^{-1} K_{13} - K_{23}^T K_{22}^{-1} K_{23})u_3 = b_3 - K_{13}^T K_{11}^{-1} b_1 - K_{23}^T K_{22}^{-1} b_2$$

This can be computed by block-Gauss elimination. It then follows for u_1 and u_2:

$$u_1 = K_{11}^{-1}(b_1 - K_{13}u_3) \qquad and \qquad u_2 = K_{22}^{-1}(b_2 - K_{23}u_3)$$

That is, u_1 and u_2 are solutions of the PDE on the subdomains Ω_1 and Ω_2 with the given boundary values b_1 and b_2, respectively, on the outer boundaries $\partial\Omega_1$ and $\partial\Omega_2$, respectively, and the boundary values u_3 on Γ_{12}. For the right side e holds:

$$e = b_3 - K_{13}^T y_1 - K_{23}^T y_2 \qquad \text{with } y_1 = K_{11}^{-1} b_1 \qquad and \qquad y_2 = K_{22}^{-1} b_2$$

y_1 and y_2 are the solutions of the PDE on Ω_1 and Ω_2 with homogeneous boundary conditons on Γ_{12}. The capacity matrix system can be solved by the preconditioned CG method. The solutions y_1 and y_2 on the subdomains can be solved on a MIMD system with two processors in which one again has inner and outer parallelism. In this case the preconditioners of Dryja (1984) and Golub and Mayers (1983) have been applied effectively.

REFERENCES

Bjorstad, P. E., and Widlund, O. B. (1984). Solving elliptic problems on regions partitioned into substructures. In: Birkhoff, G., and Schoenstadt, A., Eds. *Elliptic Problem Solves II*. Academic Press, New York.

Borodin, A., and Munro, I. (1975). *The Computational Complexity of Algebraic and Numeric Problems*. American Elsevier, New York.

Brent, R. P. (1974). The parallel evaluation of general arithmetic expressions. J. ACM, 21(2):201–206.

Dornscheidt, G., and Schendel, U. (1988). Lösung einer elliptischen Randwertaufgabe mit dem Mehrgitterverfahren. Serie A Mathematik, A-88-02, FU Berlin.

Dryja, M. (1984). A finite-element capacitance method for elliptic problems partitioned into sub-regions. Numer. Math. 44:153–168.

Feilmeier, M., Joubert, G., and Schendel, U., Eds. (1984). *Parallel Computing 83.* North-Holland, Amsterdam.

Feilmeier, M., Joubert, G., and Schendel, U., Eds. (1986). *Parallel Computing 85.* North-Holland, Amsterdam.

Flynn, M. J. (1966). Very high speed computing systems. Proc. IEEE, 14:1901–1909.

Golm, K., and Schendel, U. (1986). On special parallel algorithms for parabolic differential equations. In: Feilmeier, M., Joubert, G., and Schendel, U., Eds. *Parallel Computing 85.* North-Holland, Amsterdam.

Golub, G. H., and Mayers, D. (1983). The use of preconditioning over irregular regions. Lecture at Sixth Int. Conf. on Computing Methods in Applied Sciences and Engineering, Versailles, France.

Golub, G. H., and van Loan, C. F. (1983). *Matrix Computations.* Johns Hopkins University Press, Baltimore.

Hockney, R. W., and Jesshope, C. R. (1981). Parallel Computers. Adam Hilger Limited.

Hyafil, L., and Kung, H. T. (1974). The complexity of parallel evaluation on linear recurrences. J. ACM, 24(3):513–521.

Kogge, P. M. (1974). Parallel solutions of recurrence problems. IBM J. Res. Develop., 18:138–148.

Kubig, E., and Schendel, U., (1989). VLSI-Implementierungen von numerischen Algorithmen. Preprint A-89-7, Fachbereich Mathematik, Serie A, FU Berlin.

Kuck, D. J. (1978). *Structures of Computers and Computations.* John Wiley & Sons, New York.

Mead, C. A., and Conway, L. A. (1980). *Introduction to VLSI systems.* Addison-Wesley, Reading, Massachusetts.

Metropolis, N., Sharp, D. H., Worlton, W. J., and Ames, K. R., Eds. (1986). *Frontiers of Supercomputing.* University of California Press, Berkeley.

Rathmann, S., and Schendel, U. (1988). Über Präkonditionierer bei Dekompositionsmethoden. Serie A Mathematik, A-88-13, FU Berlin.

Schendel, U. (1984). *Introduction to Numerical Methods for Parallel Computers.* Ellis Horwood Series, John Wiley and Sons, New York.

Schendel, U., (1989). *Sparse Matrices.* Ellis Horwood Series, John Wiley & Sons, New York.

Schendel, U., and Schyska, M. (1987). Dekompositionsmethoden zur Lösung elliptischer partieller Differentialgleichungen. Preprint 269/87, Fachbereich Mathematik Serie A, FU Berlin.

Wouk, A., Ed. (1986). *New Computing Environments: Parallel, Vector and Systolic.* SIAM, Philadelphia.

Young, D. M. (1971). *Iterative Solution of Large Linear Systems.* Academic Press, New York.

4

Asynchronous Parallel Algorithms for Linear Equations

David John Evans   *University of Technology, Loughborough, England*

Nadia Y. Yousif   *Basrah University, Basrah, Iraq*

1   INTRODUCTION

The importance of asynchronous parallel algorithms is becoming more and more evident as the exploitation of multiprocessor systems progresses, for it is becoming extremely difficult to balance the computational work equally among processors in any given algorithm, which then introduces synchronization, which is another potential source of overhead. Thus algorithms with no synchronization, that is, asynchronous, seem to be the ideal solution.

In this chapter some basic and new results are presented for solving large sets of linear systems on multiprocessors. Asynchronous numerical methods are introduced that are suitable for multiprocessor systems. The motivation for this type of parallel algorithms stems from the fact that no parallel algorithm balances the work load for all processors equally. It then seems reasonable to look for algorithms that work on different processors without synchronization, that is, asynchronous algorithms. These algorithms have achieved importance in parallel computing because the coordination of processors is an overhead in the execution of the algorithms, which often destroys any gains from having multiple processors cooperating on algorithm execution. This means a reduction in parallelism and decreases the maximum speedup one expects to achieve in using a multiprocessor (Enslow, 1974; Dunbar, 1978).

A wide variety of problems in numerical mathematics can be solved by solving sets of linear equations, for example the computer solution of problems have continuous analytic solutions, and entails the replacement of the problem by a series of discrete approximations. In particular, the numerical solution of partial differential equations that is required for the solution of many engineering problems is accomplished by replacing the equations by a banded system of linear equations whose solution yields an approximation of the exact solution in the form of a set of values generated by a function that approximates the true analytic solution. To obtain the required accuracy it may be necessary to solve a set of linear equations in which the right side has been altered slightly. Also, it is expedient to compute the solution of the new set of equations from that of the old set by an iterative process, such as the Jacobi, Gauss-Seidel, and successive overrelaxation (SOR) methods. The coefficient matrix for the set of linear equations is factorized, from which different right-hand sides may be generated to form these methods.

The asynchronous form of these iterative methods has been developed by Baudet (1978a) and Kung (1976), and a theory of convergence for these iterative methods is presented. These asynchronous algorithms have been developed on the Loughborough Neptune, a four-processor multiple instruction–multiple data stream (MIMD) system and Baudet's work extended by implementing the successive overrelaxation method and an explicit block SOR iterative method to solve the Dirichlet problem of the Laplace equation in which some numerical results are obtained and compared.

Algorithms with no synchronization at all are called purely asynchronous (PA) (Baudet, 1978b; Kung, 1980). These algorithms have also been implemented and extended to the purely asynchronous overrelaxation method (PAOR).

2   METHODS FOR SOLVING LINEAR SYSTEMS OF EQUATIONS

A finite difference method for approximating a boundary value problem leads to a system of algebraic simultaneous equations. For linear boundary value problems these equations are always linear but their number is generally large, and for this reason their solution is a major problem in itself.

These problems can be expressed in matrix-vector notation by the equation

$$Ax = b \tag{1}$$

where A is an ($n \times n$) matrix of the coefficients, b is a known n vector, and x is an unknown n vector whose value is to be found.

Provided that det(A) is nonzero the unique solution of the equation is expressed simply as

$$x = A^{-1}b$$

where A^{-1} is the inverse of the matrix A.

However, in normal practice it is uncommon to compute the inverse of A since more efficient ways of solving the problem are available. In addition, if the coefficient matrix is sparse it is unlikely (except in very special cases) that the sparsity is preserved in the inverse. However, methods of solution belong essentially to either the class of direct methods or the class of iterative methods.

2.1   Direct Methods

Direct methods solve the system of equations (1) in a known number of arithmetic operations, and errors in the solution arise entirely from rounding errors introduced in the computation. A common direct method (a variant of the well-known Gaussian elimination method) for the solution of Eq. (1) requires the decomposition of A into a pair of factors L and U, where L is a lower triangular matrix and U is an upper triangular matrix of the same order as A. This decomposition can only be carried out provided that A is nonsingular. Equation (1) can then be replaced by

$$LUx = b \tag{2}$$

The solution of Eq. (1) is calculated from Eq. (2) by assuming $Ux = y$ and then solving $Ly = b$ for y by forward substitution and $Ux = y$ for x by back-substitution.

When solving partial differential equations the coefficient matrix A usually appears in a tridiagonal block structure that is preserved in factors L and U.

2.2   Iterative Methods

In any iterative method the solution is obtained in which an initial approximation is used to calculate a second approximation, which in turn is used to calculate a third, and so on. The iterative procedure is said to be convergent when the difference between the exact solution and the successive approximations tends to zero as the number of iterations increases.

In these methods it is useful to scale and arrange the equations in such a way that the matrix A has the splitting

$$A = D - L - U$$

where L and U are lower and upper triangular matrices, respectively, with null diagonals and D is the diagonal matrix. Equation (1) may now be written

$$(D - L - U)x = b \tag{3}$$

which can be written

$$Dx = (L + U)x + b$$

The Jacobi iterative method, or the method of simultaneous corrections, is defined as

$$Dx^{(k+1)} = (L + U)x^{(k)} + b \qquad k \geq 0 \tag{4}$$

giving

$$x^{(k+=)} = D^{-1}(L + U)x^{(k)} + D^{-1}b \tag{5}$$

The matrix $D^{-1}(L + U)$ or $(I - D^{-1}A)$, where I is the identity matrix, is called the point Jacobi iteration matrix. Each point x_i for $i = 1, 2, \ldots, n$ of the vector x is then iterated as

$$x_i^{(k+1)} = \frac{1}{a_{ii}}\left(b_i + \sum_{j=1}^{i-1} a_{ij}x_j^{(k)} + \sum_{j=i+1}^{n} a_{ij}x_j^{(k)}\right) \qquad k \geq 0 \tag{6}$$

Related to the Jacobi method is the simultaneous overrelaxation method (the JOR method). In this method the displacement vector,

$$d_1^{(k)} = x^{(k+1)} - x^{(k)}$$

of the JOR method is taken to be ω times the displacement vector $d_1^{(k)}$ defined by the Jacobi iteration. Hence, by Eq. (4) we have

$$Dd_1^{(k)} = D(x^{(k+1)} - x^{(k)}) = (L + U)x^{(k)} + b - Dx^{(k)}$$

and the JOR iteration, defined by

$$d^{(k)} = \omega d_1^{(k)}$$

can be written as

$$x^{(k+1)} - x^{(k)} = \omega D^{-1}[(L + U)x^{(k)} + b - Dx^{(k)}]$$

Therefore,

$$x^{(k+1)} = [\omega D^{-1}(L + U) + (1 - \omega)]x^{(k)} + \omega D^{-1}b \qquad k \geq 0 \tag{7}$$

where ω is the underrelaxation factor. If $\omega = 1$ then we have the Jacobi method. Now, for $1 \leq i \leq n$, each component of the vector x is represented as

$$x_i^{(k+1)} = \omega \frac{1}{a_{ii}}\left(b_i + \sum_{j=1}^{i-1} a_{ij}x_j^{(k)} + \sum_{j=i+1}^{n} a_{ij}x_j^{(k)}\right) + \tag{8}$$

$$(1 - \omega)a_{ii}x_i^{(k)} \qquad k \geq 0$$

In these two methods the evaluation of one component in an iteration depends on the values of components that are obtained only from the previous iterations. This means that these methods are inherently sequential.

For most coefficient matrices the Gauss-Seidel iterative method converges more rapidly than the Jacobi matrix (Varga, 1962). The method is defined by the equation

$$Dx^{(k+1)} = Lx^{(k+1)} + Ux^{(k)} + b \tag{9}$$

which is written as

$$x^{(k+1)} = (D - L)^{-1}Ux^{(k)} + (D - L)^{-1}b \tag{10}$$

Since $D - L$ is a nonsingular matrix, Eq. (10) shows that the Gauss-Seidel point iteration matrix is $(D - L)^{-1}U$. From Eq. (9) the iteration of each point x_i is given by

$$x_i^{(k+1)} = \frac{1}{a_{ii}} \left(b_i + \sum_{j=1}^{i-1} a_{ij} x_j^{(k+1)} + \sum_{j=i+1}^{n} a_{ij} x_j^{(k)} \right) \tag{11}$$

$$for \; i = 1, 2, \ldots, n$$

This iterative method has the computational advantage that it does not require the simultaneous storage of two approximations $x_i^{(k+1)}$ and $x_i^{(k)}$ as in the point Jacobi iterative method.

Related to the Gauss-Seidel iteration method is the successive overrelaxation method. In this method the displacement or correction vector $d^{(k)} = x^{(k+1)} - x^{(k)}$ of the SOR method is taken to be ω times the displacement vector $d_1^{(k)}$ defined by the Gauss-Seidel iteration. Therefore, from Eq. (9) we have

$$Dd_1^{(k)} = D(x^{(k+1)} - x^{(k)}) = Lx^{(k+1)} + Ux^{(k)} - Dx^{(k)} + b$$

Hence the SOR iteration, defined by

$$d^{(k)} = \omega d_1^{(k)}$$

is then obtained from

$$x^{(k+1)} - x^{(k)} = \omega D^{-1}(Lx^{(k+1)} + Ux^{(k)} - Dx^{(k)} + b)$$

Therefore,

$$x^{(k+1)} = \omega D^{-1}Lx^{(k+1)} + [(1 - \omega)I + \omega D^{-1}U]x^{(k)} + \omega D^{-1}b \tag{12}$$

Clearly, the choice of $\omega = 1$ yields the Gauss-Seidel method. From Eq. (12) the equation of each component x_i can be formulated as

$$x_i^{(k+1)} = \frac{\omega}{a_{ii}} \left(b_i + \sum_{j=1}^{i-1} a_{ij}x_j^{(k+1)} + \sum_{j=i+1}^{n} a_{ij}x_j^{(k)} \right) - (\omega - 1)x_i^{(k)} \tag{13}$$

It was argued that these methods in their original form required a form of synchronization and hence were not considered suitable to implement on parallel

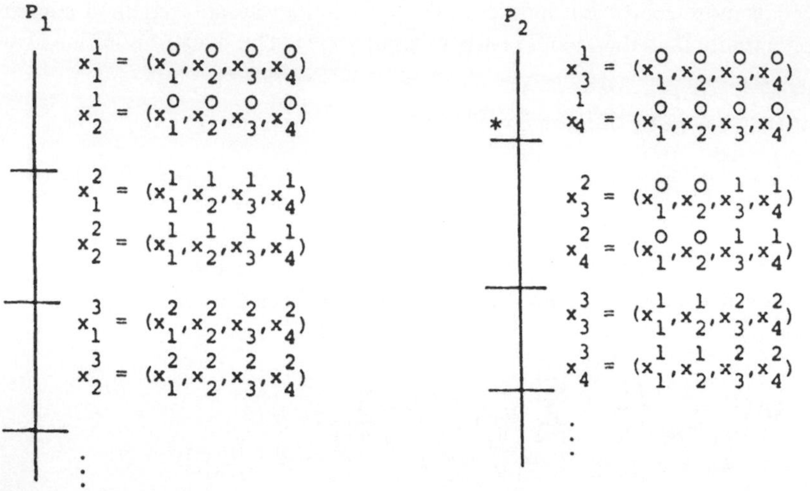

Figure 1 Two processors working asynchronously on four equations. *The bar means the end of execution.

computers. However, many studies of parallel computation have improved this argument by utilizing reordering strategies.

Since we are concerned with a MIMD computer whose processors act asynchronously, iterative methods for solving a system of linear equations must be constructed in an asynchronous form, that is, asynchronous iterative algorithms. By an asynchronous iterative algorithm we mean that in any evaluation the processes always use the values of the components that are currently available at the beginning of the computation. This means that each processor must compute different and independent subsets of the components. The idea of performing iterative methods asynchronously is that each processor in one iteration can perform, for example, one or more components of the vector x using their initial values stored in a shared memory. In the next iteration a processor is required to compute the same components by making use of the values obtained from the previous iteration. Because of the processor hardware, not all the processors are equal in performance time. Therefore each processor can use at any computation time the values of the vector components that are evaluated and released from the previous iteration. On the other hand, when the value of a component is not available the processors can use the current value that was used in the previous iterations, as shown in Figure 1, in which two processors are working asynchronously on a system of four equations.

3   MATHEMATICAL FOUNDATION FOR THE ASYNCHRONOUS ITERATIVE METHODS

This section represents a mathematical form for the asynchronous iterative methods to solve a system of equations.

Suppose that F is a linear operator from $\mathbb{R}^n$ to $\mathbb{R}^n$ and is given by

$$F(x) = Ax + b \tag{14}$$

where x is an unknown n vector, A is an $n \times n$ matrix of coefficients and b is a constant n vector.

Chazan and Miranker (1969) introduced the chaotic relaxation scheme, a class of iterative methods for solving the system of equations (14). They also showed that iterations defined by a chaotic relaxation scheme converge to the solution of Eq. (14) if and only if $\rho(\ |J|\ ) < 1$, where $\rho(\ |J|\ )$ is the spectral radius of the matrix $J = A - I$.

The main motivation for defining chaotic relaxation is that when iterative methods are implemented on a multiprocessor system by using a chaotic relaxation scheme the communication and synchronization between the cooperating processes are significantly reduced. This reduction is obtained by not forcing the processes to follow a predetermined sequence of computations but allowing a process to choose dynamically not only the components to be evaluated but also the previous iterates used in the evaluation. However, this scheme has a restriction that there must exist a fixed positive integer S such that in carrying out the evaluation of the ith iterate a process cannot make use of any value of the components of the jth iterate if $j < i - s$.

When each process evaluates a subset of the components of the solution vector it may wait for the values of the other components that are carried out by the other processes to be released to use them in further iterations, although the amount of work for evaluating these components in each process is the same. In fact, the actual time of computation differs slightly for many reasons, such as communication between processors or the hardware design of each processor. However, this requires a form of synchronization that one wishes to avoid. The penalty factor (Kung, 1976) of synchronizing processes at the end of each iteration is very large, as are the synchronization primitives to make the processors idle; the computational time, which creates an unnecessary overhead, therefore increases accordingly. As a consequence the maximum expected speedup in using a multiprocessor system may be reduced.

To avoid this restriction of chaotic relaxation, Baudet (1978a) introduced a class of asynchronous iterative methods in which chaotic relaxation is considered a special case. The mathematical form of the asynchronous iterative methods is given in the following definition.

Definition 1

Let $F: \mathbb{R}^n \to \mathbb{R}^n$ be a linear operator such that

$$F: = \begin{bmatrix} f_1(x) \\ \vdots \\ f_n(x) \end{bmatrix} \qquad X: = \begin{bmatrix} x_1 \\ \vdots \\ x_n \end{bmatrix}$$

Let $\mathbf{J}: = (J_j)_{j=1}^{\infty}$ be a sequence of nonempty subsets of $(1, \ldots, n)$ and let

$$S: = \begin{bmatrix} s_1(j) \\ \vdots \\ s_n(j) \end{bmatrix}_{j=1}^{\infty}$$

be a sequence of elements in $\mathbb{IN}^n$.

Thus a sequence

$$X(j) = \begin{bmatrix} x_1(j) \\ \vdots \\ x_n(j) \end{bmatrix} \in \mathbb{R}^n \qquad \text{for } 1 \le j \le \infty$$

is called an asynchronous iteration if the sequence $(X(j))_{j=1}^{\infty}$ is determined by a quadruple $(F, X(0), \mathbf{J}, S)$ in the following way:

1. $F, \mathbf{J}, S$ as defined

$$X(0) = \begin{bmatrix} x_1(0) \\ \vdots \\ x_n(0) \end{bmatrix}$$

2. $x_i(j) = \begin{cases} x_i(j-1) & \text{for } i \notin J_j \\ f_i(x_1(s_1(j)), \ldots, x_n(s_n(j))) & \text{for } i \in J_j \end{cases}$ \hfill (15)

3. i occurs infinitely often in the sets $J_j; j = 1, 2, \ldots, 1 \le i \le n$.

4. $\forall_i \in \{1, \ldots, n\}$ $(s_i(j) \le j-1, j = 1, 2, \ldots,$ and $\lim_{j \to \infty} s_i(j) \to \infty$

The sequence $x(j)$ defined by the asynchronous iteration results naturally from the successive approximation if it is performed on a MIMD computer without synchronization of the processors.

By Eq. (15) we mean that to evaluate a component x_i in the jth iteration we might consider its recent value obtained from the $(j-1)$th iteration if it has been released. Otherwise the previous values of the components obtained in the early iterations are considered for further evaluations.

It is obvious that an asynchronous iterative method is subject to the use of the most recent values of the components instead of the values of an early iterate, as stated in the second part of condition 4. However, the first part of condition 4 states that only components of the previous iterates can be used in the evaluation of a new iterate. Finally, condition 3 of Definition 1 states that no component of vector x is ever abandoned during any iterate.

As an example of iterative methods, let us consider the point Jacobi method. Jacobi's method, which is defined on the operator F and an initial vector $x(0)$, is defined by

$$J_j = \{1, \ldots, n\} \quad \text{for } j = 1, 2, \ldots$$
$$S_i(j) = j - 1 \quad \text{for } j = 1, 2, \ldots, \text{ and } i = 1, \ldots, n \quad (16)$$

This means that all the components $x_1, \ldots, x_n$ are evaluated at once by one computational process, with the provision that the components of the new iterate, say j, cannot be evaluated until the values of the components from the iterate $j - 1$ have been obtained.

On the other hand, the asynchronous Jacobi method is defined as

$$J_j = \{1 + (j - 1 \bmod n)\} \quad \text{for } j = 1, 2, \ldots$$

$$S_i(j) = n \left\lfloor \frac{(j-1)}{n} \right\rfloor \quad \text{for } j = 1, 2, \ldots, \quad (17)$$

$$\text{and} \quad i = 1, \ldots, n$$

This means that each component is evaluated by one process and up to n processes can be used to perform the computation. In this method of computation of the components of the new iterate, j does not wait for the values of these components from the previous iterate, $j - 1$. Instead, any recent available value of these components at that time is considered for that computation.

4   THE CONVERGENCE THEOREM OF ASYNCHRONOUS ITERATIVE METHODS

In the convergence theorem of iterative methods in general, the matrix of coefficients of a given problem must be required to satisfy stronger conditions. Before stating the convergence theorem of an asynchronous iteration, which requires a sufficient condition only, we present some characteristics of a nonnegative matrix and the spectral radius ρ.

Theorem 1 (Varga, 1962)

If A is an $n \times n$ matrix then A is convergent if and only if $\rho(A) < 1$. The proof of this theorem is in Varga (1962).

Lemma 1

Let A be a nonnegative square matrix. Then $\rho(A) < 1$ if and only if there exists a positive scalar λ and a positive vector v such that

$$Av \leq \lambda v \quad \text{and} \quad \lambda < 1 \tag{18}$$

By following the concepts of Perron's theorem (e.g., see Varga, 1962) we let $\lambda = \rho(A_t)$, and therefore since $A_t > 0$ there exists a positive eigenvector v corresponding to the eigenvalue λ.

Thus the positive scalar λ and the positive vector v can verify

$$A_v \leq A_t v = \lambda v \quad \text{with} \quad \lambda < 1$$

Thus we have completed the proof of both conditions, and now we can state the convergence theorem and its proof.

Theorem 2: The Convergence Theorem (Baudet, 1978a)

If F is a contracting operator on a closed subset D of IR^n and if $F(D) \subset D$, then any asynchronous iteration $(F, X(0), J, S)$ corresponding to F and starting with a vector $X(0)$ in D converges to the unique fixed point of F in D.

5 PROBLEM FORMULATION

The asynchronous iterative methods introduced in Section 3 and implemented on the Neptune parallel system were used to solve the two-dimensional Laplace equation

$$\frac{\partial^2 \phi}{\partial x^2} + \frac{\partial^2 \phi}{\partial y^2} = 0 \tag{19}$$

applied to a connected planar region R in the x, y plane, where ϕ is a dependent variable with specified boundary conditions. When mesh lines covering the domain are drawn parallel to the x and y axes they intersect in mesh points. Therefore solving a set of linear equations at these points yields an approximation to the partial differential equation.

Now consider a uniform grid of mesh size h on the unit square as shown in Figure 2. Assume that $\phi(x, y)$ is differentiable; then by Taylor's series we obtain approximations to the second-order derivatives in the form

$$\frac{\partial^2 \phi}{\partial x^2} = \frac{\phi(x + h, y) - 2\phi(x, y) + \phi(x - h, y)}{h^2} + 0(h^2) \tag{20}$$

and

$$\frac{\partial^2 \phi}{\partial y^2} = \frac{\phi(x, y + h) - 2\phi(x, y) + \phi(x, y - h)}{h^2} + 0(h^2) \tag{21}$$

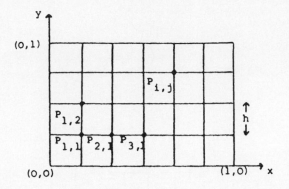

Figure 2   The grid of points in a unit square.

If for the mesh point $(x_i, y_j) = (ih, jh)$ we denote $\phi(x_i, y_j)$ by $\phi_{i,j}$, then Laplace's equation can be replaced at the point (x_i, y_j) by the finite difference equation obtained from adding Eqs. (20) and (21) to give

$$\frac{1}{h^2}(\phi_{i+1,j} + \phi_{i-1,j} + \phi_{i,j+1} + \phi_{i,j-1} - 4\phi_{i,j}) = 0 \tag{22}$$

which is known as the five-point finite difference equation.

From Eq. (22) we obtain a set of simultaneous equations whose solution is a finite difference approximation of the exact solution $\{\phi_{i,j}\}$ at the internal mesh points. In other words, the evaluation of any point $\phi_{i,j}$ is related to the values of its nearest neighbors, and Eq. (22) corresponds to the computational molecule given in Figure 3. The 16×16 matrix illustrated in Figure 4 therefore represents the coefficient matrix derived when a second-order elliptic partial differential equation (i.e., the Laplace equation) is discretized on a network of lines spaced 1/5 apart. Thus using the computational molecule (Fig. 3) we obtain the following 16×16 matrix (Fig. 5). In this matrix each row contains at most five nonzero entries; that is, the matrix is sparse so that the computation in any iterative method concentrates only on these five points in each row.

This completes the brief description of the two-dimensional Laplace equation on which the numerical experiments are based in a later section.

6   DETERMINATION OF THE OPTIMUM RELAXATION FACTOR FOR ITERATIVE METHODS

The JOR and SOR iterative methods described in Section 3 show a reasonable improvement over the Jacobi and the Gauss-Seidel methods, respectively, if the relaxation factor ω is well chosen. However, the definition of the asynchronous

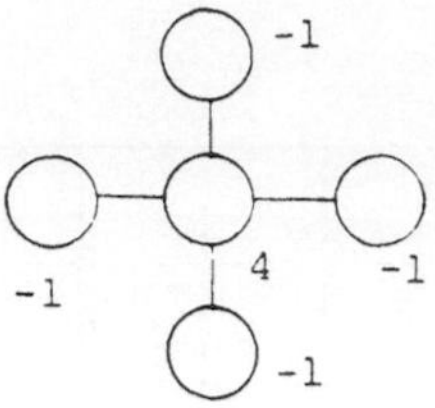

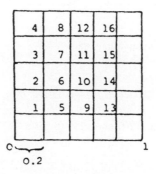

Figure 3

4	8	12	16
3	7	11	15
2	6	10	14
1	5	9	13

0 ⌣ 1
0.2

Figure 4

iterative methods can be extended by introducing a relaxation factor $\omega > 0$ (Baudet, 1978a). Therefore, the operator F defined in Section 4, Theorem 2 can be represented as

$$F_\omega = \omega F + (1 - \omega)E$$

where E is the identity operator of R^n. It follows that

$$| F_\omega(x) - F_\omega(y) | \le \omega | F(x) - F(y) | + (1 - \omega) | x - y |$$

To prove F_ω is a contracting operator with a contracting matrix A, F_ω is a Lipchitzian operator with the Lipchitzian matrix

$$A_\omega = \omega A + | 1 - \omega | I$$

where A_ω is the iteration matrix. Since A is a nonnegative matrix with $\rho(A) < 1$ and we have

$$\rho(A_\omega) = \omega\rho(A) + | 1 - \omega | \tag{23}$$

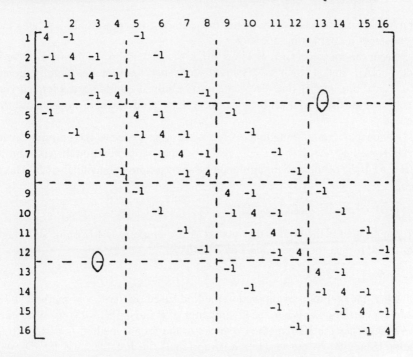

Figure 5

and the range of ω that makes $\rho(A_\omega) < 1$ can now be obtained. Because it is known that the spectral radius of a matrix is the maximum absolute eigenvalue of the matrix, to show $|\rho(A_\omega)| < 1$ we therefore write

$$-1 < \rho(A_\omega) < 1$$

then,

$$-1 < \omega\rho(A) + \omega - 1 < 1$$

By adding 1 to the equality we obtain

$$0 < \omega\rho(A) + \omega < 2$$
$$0 < \omega(\rho(A) + 1) < 2$$

then,

$$0 < \omega < \frac{2}{[1 + \rho(A)]} \tag{24}$$

which gives the range of ω by which the method converges. This indicates that F_ω becomes a contracting operator.

From this range of ω it is known that the maximum value ω can have is $2/(1 + \rho(A))$, so that when $\rho(A)$ is slightly less than 1 ω becomes slightly greater than 1, and this implies that the problem to be solved cannot be accelerated very much; that is, the model problem is not a very good example to illustrate this method.

Given a system of equations $Ax = b$, we want to derive the optimum ω for the JOR method. Let $B = I - D^{-1}A$ denote the Jacobi iteration matrix, and let $m(B)$ and $M(B)$ denote the minimum and the maximum eigenvalues of B. Therefore,

$$\rho(B) = \max(\mid m(B) \mid , \mid M(B) \mid)$$

We remark that if the contracting matrix A has property (A) (Young and Hageman, 1981), then the eigenvalues of B satisfy

$$m(B) = - M(B) \tag{25}$$

However, the optimum extrapolation method based on the Jacobi method (JOR) is always convergent. When the eigenvalues of B satisfy Eq. (25) we have $\omega = 1$. For this case the JOR method reduces to the Jacobi method without extrapolation. Since $B_\omega = \omega B + (1 - \omega)I$ represents the JOR iteration matrix, then in general the optimum ω occurs when the minimum and the maximum eigenvalues of B_ω are equal in their values and opposite in their sign.

Thus from Eq. (23) we have

$$m(B_\omega) = \omega m(B) + 1 - \omega$$

and

$$M(B_\omega) = \omega M(B) + 1 - \omega$$

Hence the optimum ω occurs when

$$m(B_\omega) = -M(B_\omega)$$

that is,

$$\omega m(B) + 1 - \omega = -\omega M(B) - 1 + \omega$$
$$2 = 2\omega - \omega m(B) - \omega M(B)$$

which implies that

$$\omega = \frac{2}{2 - m(B) - M(B)} \tag{26}$$

which is the value of the optimum ω for the JOR method. The diagram in Figure 6 illustrates the values of ω and the optimum ω.

For the SOR method we want to find the optimum value ω_b of ω that minimizes the spectral radius of the SOR iteration matrix and thereby maximizes the rate of convergence of the method. At the present time no formula for ω_b for an arbitrary set of linear equations can be derived. However, it can be calculated for many of the difference equations approximating first- and second-order partial differential equations because their matrices are often of a special type, that is, two-cyclic matrices that possess property (A). Young (1954) proved that when a two-cyclic matrix is translated into what he called a consistently ordered form A, which can be done by a simple reordering of the rows and corresponding columns of the original two-cyclic matrix, then the eigenvalues λ of the point SOR iteration matrix L_ω associated with A are related to the eigenvalues μ of the point Jacobi iteration matrix B associated with A by the equation

$$(\lambda + \omega - 1)^2 = \lambda\omega^2\mu^2 \tag{27}$$

From this equation it can be proved that

$$\omega_b = \frac{2}{1 + \sqrt{1 - \rho^2(B)}} \tag{28}$$

where $\rho(B)$ is the spectral radius of the Jacobi iteration matrix.

To derive Eq. (28) we need to present some theorems and definitions. It is sufficient to know that the block tridiagonal matrices A of the form

$$A = \begin{bmatrix} D_1 & A_1 & & & & \\ B_1 & D_2 & A_2 & & 0 & \\ & B_2 & D_3 & A_3 & & \\ & & \ddots & \ddots & \ddots & \\ & 0 & & & D_{k-1} & A_{k-1} \\ & & & & B_{k-1} & D_k \end{bmatrix} \tag{29}$$

are two-cyclic and consistently ordered, where D_i, $i = 1, 2, \ldots, k$ are square diagonal matrices not necessarily of the same order.

Theorem 6.1

If the block tridiagonal matrix of the form in Eq. (29) is written $A = D - L - U$, where D, L, and U are the diagonal, lower, and upper matrices, respectively, and the matrix $A(\alpha)$, $\alpha \neq 0$, is defined by

$$A(\alpha) = D - \alpha L - \alpha^{-1}U$$

then

$$\det(A(\alpha)) = \det(A) \qquad \square$$

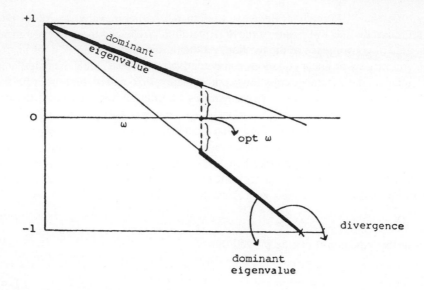

Figure 6   Optimum ω for the JOR method.

This theorem also holds when each D_i is a full matrix because the diagonal blocks of $A(\alpha)$ are independent of α.

Theorem 6.2

The nonzero eigenvalues of the Jacobi iteration matrix corresponding to the matrix A in Eq. (29) occur in pairs $\pm \mu_i$.

Theorem 6.3

Let L_ω and B represent the SOR and the Jacobi point iteration matrices, respectively, corresponding to the block tridiagonal matrix A of Eq. (29). Then if λ is a nonzero eigenvalue of L_ω and μ satisfies the relation

$$(\lambda + \omega - 1)^2 = \lambda\omega^2\mu^2 \qquad \omega \neq 0 \tag{30}$$

then μ is an eigenvalue of B. Conversely, if μ is an eigenvalue of B and satisfies Eq. (30) then λ is an eigenvalue of L_ω.

Theorem 6.4

When the matrix A is block tridiagonal of the form given in Eq. (29) with nonzero diagonal elements and all the eigenvalues of the Jacobi iteration matrix B associated with A are real and are such that $0 < \rho(B) < 1$, then

$$\omega_b = \frac{2}{1 + \sqrt{1 - \rho^2(B)}}$$

and

$$\rho(L_\omega) = \omega_b - 1 \tag{31}$$

Furthermore, the SOR method applied to the equations $Ax = b$ converges for all ω in the range $0 < \omega < 2$ (Young, 1971).

7 RATE OF CONVERGENCE OF THE BASIC ITERATIVE METHODS

If we consider the solution of the system of equations

$$Ax = b$$

where b is a known vector in R^n obtained from the boundary conditions of the Dirichlet problem for the Laplace equation on a square mesh of R^2, we now present the rates of convergence of the basic iterative methods described in Section 4.

Consider, for example, the square region $R: 0 < x < \pi, 0 < y < \pi$, when $h = \pi/n$, where R is divided into n^2 square nets with $N \times N = (n - 1)^2$ interior points to be evaluated by employing the five-point formula of Eq. (22). Let A denote the square matrix formed from the left side of Eq. (22), and the N eigenvectors of A are denoted by x_{pq} (ih, jh) with corresponding positive eigenvalues μ_{pq} for $p, q = 1, 2, \ldots, n - 1$.

From the relation

$$Ax_{pq} = \mu_{pq} x_{pq} \tag{32}$$

we have

$$x_{pq} = \sin pih \sin qjh \tag{33}$$

$$\mu_{pq} = 4h^{-2} \left(\sin^2 \frac{ph}{2} + \sin^2 \frac{qh}{2} \right) \tag{34}$$

Clearly, by inspection, we have

$$\min \mu_{pq} = \mu_{1,1} = 8h^{-2} \sin^2 \frac{h}{2}$$

$$= 8h^{-2} \left\{ \left[\frac{h}{2} - \left(\frac{h}{2} \right)^3 \frac{1}{3!} + \left(\frac{h}{2} \right)^5 \frac{1}{5!} - \cdots \right]^2 \right\}$$

$$= 8h^{-2} \left\{ \frac{h^2}{4} - \left[\left(\frac{h}{2} \right)^3 \frac{1}{3!} \right]^2 + \cdots \right\}$$

$$= 2 - \frac{h^4}{12} + \cdots$$

Thus

$$\mu_{1,1} \to 2 \quad \text{as} \quad h \to 0 \tag{35}$$

Similarly,

$$\max \mu_{pq} = \mu_{n-1,n-1} = 8h^{-2} \left(1 - \sin^2 \frac{h}{2} \right) \approx \frac{8}{h^2} \quad \text{as } h \to 0 \tag{36}$$

The spectral radius of A is therefore $\mu_{n-1,n-1}$.

First we consider the rate of convergence of the Jacobi method, which is controlled by the eigenvalues of the matrix

$$B = D^{-1}(L + U)$$

Since

$$A = D - L - U$$

then

$$B = D^{-1}(D - A)$$
$$= I - D^{-1}A \tag{37}$$

In our problem $D = (4/h^2)I$; therefore the Jacobi iterative matrix becomes

$$B = I - \frac{h^2}{4} A$$

As the eigenvalues of A are given by Eq. (34), the eigenvalues λ_{pq} of B are

$$\lambda_{pq} = 1 - \frac{h^2}{4} \left[4h^{-2} \left(\sin^2 \frac{ph}{2} + \sin^2 \frac{qh}{2} \right) \right]$$

$$= 1 - \sin^2 \frac{ph}{2} - \sin^2 \frac{qh}{2}$$

$$= \cos^2 \frac{ph}{2} - \sin^2 \frac{qh}{2}$$

Since $\cos^2 x = \frac{1}{2} + \frac{1}{2} \cos 2x$ and $\sin^2 x = \frac{1}{2} - \frac{1}{2} \cos 2x$, then

$$\lambda_{pq} = \frac{1}{2}(\cos ph + \cos qh) \qquad \text{for } p,q = 1, 2, \ldots, n^{-1} \qquad (38)$$

Consequently, the spectral radius of B is

$$\lambda(B) = \max | \lambda_{pq} | = \lambda_{n-1,n-1}$$
$$= \frac{1}{2}[\cos (n - 1) h + \cos (n-1) h] = \cos (n-1) h$$

Substituting the value of n in this equation yields

$$\lambda(B) = \cos h$$

$$\therefore \lambda(B) = \cos h - \left(1 - \frac{h^2}{2}\right) \qquad \text{as } h \to 0 \qquad (39)$$

Hence the rate of convergence of the Jacobi method is given by

$$R(B) = -\log \cos h \qquad\qquad\qquad (40)$$
$$= -\log \left(1 - \frac{h^2}{2}\right) = \frac{h^2}{2} + 0(h^4)$$

Thus the rate of convergence of the Jacobi method is $h^2/2$, which is rather slow for small values of h.

Now consider the JOR method with the corresponding matrix

$$B_\omega = \omega B + (1 - \omega)I$$

where ω is the relaxation factor.

Let $\gamma_{p,q}$ denote the eigenvalues of B_ω, where $p,q = 1, 2, \ldots, (n - 1)$. Since the eigenvalues of B are given in Eq. (38), γ_{pq} are

$$\gamma_{pq} = \frac{1}{2}\omega(\cos ph + \cos qh + (1 - \omega)$$

and the spectral radius of B_ω is

$$\gamma(B_\omega) = \max |\gamma_{pq}| = \gamma_{n-1,n-1}$$
$$= 1/2\omega \, [\cos (n - 1) h + \cos (n - 1) h] + (1 - \omega)$$
$$= \omega \cos (n - 1) h + (1 - \omega)$$
$$= \omega \cos \left(\frac{\pi}{h} - 1\right) h + (1 - \omega)$$
$$= \omega \cos h + 1 - \omega$$

$$\therefore \gamma(B_\omega) = \omega \cos h + (1 - \omega) \sim \omega\left(1 - \frac{h^2}{2}\right) + (1 - \omega) \qquad \text{as } h \to 0$$

or
$$\sim \left(1 - \omega\frac{h^2}{2}\right) \qquad \text{as } h \to 0$$

The rate of convergence then becomes

$$R(B_\omega) = -\log(\omega \cos h + 1 - \omega) \sim -\log\left(1 - \omega\frac{h^2}{2}\right) \qquad (41)$$

$$= \omega\frac{h^2}{2} + O(\omega^2 h^4) \qquad \text{as } h \to 0$$

Therefore the rate of convergence of the JOR method is ω times faster than that of the Jacobi method. It follows that unless $\omega > 1$ this method cannot be accelerated and it can be considered the same as the Jacobi method.

Now consider the Gauss-Seidel method, which is controlled by the iteration matrix

$$L = (I - L)^{-1}U$$

For our model problem the maximum eigenvalue of L is given by (Young, 1971)

$$\rho(L) = \cos^2 h \sim 1 - h^2 \qquad \text{as } h \to 0$$

Consequently the rate of convergence is

$$R(L) = -\log(\cos^2 h) = -\log(1 - h)$$
$$\sim h^2 + O(h^4) \qquad (42)$$

which is twice as fast as that for the Jacobi method.

The rate of convergence of the SOR method can now be obtained where the SOR iteration matrix can be obtained after reformulating Eq. (12), which becomes

$$x^{(k+1)} = (I - \omega L)^{-1}[\omega U + (1 - \omega)I]x^{(k)} + (I - \omega L)^{-1}\omega b$$

Thus

$$L_\omega = (I - \omega L)^{-1}[\omega U + (1 - \omega)I]$$

becomes the SOR iteration matrix.

It was proved that the spectral radius of the SOR method is

$$\rho(L_\omega) = \omega_b - 1$$

(see Theorem 6.4), where ω_b is the optimum overrelaxation factor given by (Varga, 1962):

$$\omega_b = \frac{2}{1 + \sqrt{1 - \rho^2 (B)}}$$

where $\rho(B)$ is the spectral radius of the J method. Thus

$$\rho(L_\omega) = \frac{2}{1 + \sqrt{1 - \rho^2 (B)}} - 1 = \frac{1 - \sqrt{1 - \rho^2 (B)}}{1 + \sqrt{1 - \rho^2 (B)}}$$

Since $\rho(B) = \cos h$, then

$$\rho(L_\omega) = \frac{1 - \sin h}{1 + \cos h} = \frac{1 - \left(h - \dfrac{h^3}{3!} + \dfrac{h^5}{5!} - \cdots \right)}{1 + h - \dfrac{h^3}{3!} + \dfrac{h^5}{5!} - \cdots}$$

thus

$$\rho(L_\omega) - \frac{1 - h}{1 + h} \qquad \text{as } h \to 0$$

Therefore,

$$R(L_\omega) = -\log \frac{1 - h}{1 + h} = -\log (l - h) + \log (l + h) \qquad (43)$$

$$= -\left[-h - \frac{(-h)^2}{2} + \frac{(-h)^3}{3} - \cdots \right] + \left(h - \frac{h^2}{2} + \frac{h^3}{3} - \cdots \right)$$

$$= 2h + \frac{2h^2}{3} + \frac{2h^5}{5} + \cdots$$

$$\sim 2h + 0(h^3) \qquad \text{as } h \to 0$$

which is larger than that of the Gauss-Seidel method by the factor $2/h$.

To this end it becomes clear that the SOR method is faster than the Gauss-Seidel method, which is in turn faster than the Jacobi method. However, all the iterative methods have been implemented on the Neptune asynchronous system, and the results obtained clearly identify the differences of the rates of convergence.

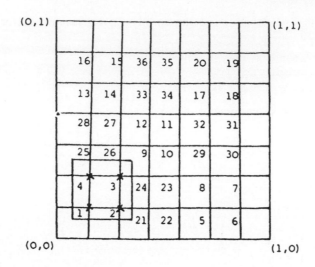

Figure 7

8   EXPLICIT FOUR-POINT BLOCK
ITERATIVE METHOD

The iterative methods studied in Sections 2 through 7 were solved considering each point of the grid mesh individually, which is ideal for use on MIMD computers. A faster rate of convergence may be obtained when a group of points is formed in a way that all the points in one group are evaluated at once in one iteration step rather than solving the individual points. An example of such a method is the line successive overrelaxation (LSOR) method in which the points on one row of the mesh are grouped together and solved implicitly (Varga, 1962). It has been proved that for the solution of the Dirichlet problem with uniform mesh size h in the unit square, the LSOR method is asymptotically faster by a factor of $\sqrt{2}$ over the point successive overrelaxation method (Parter, 1981). The implicit nature of the equations makes it very difficult to apply parallelism efficiently, however. In this section we develop a new block method, which can be implemented explicitly and is therefore more suitable for parallel implementation.

The four-point block iterative method was developed by Evans and Biggins (1982) and applied to solve the model problem of the solution of the Laplace equation in the unit square. In this method the mesh points are ordered in groups of four points and the groups themselves are ordered in red-black ordering, as shown in Figure 7. Suppose that the system of equations to be solved by the

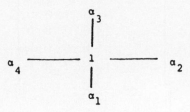

Figure 8

two-dimensional Dirichlet problem, in which the five-point scheme shown in Figure 8 is used, is given by the form

$$Ax = b \tag{44}$$

The left side of the finite difference equation of such a system has the form

$$x_{i,j} + \alpha_1 x_{i-1,j} + \alpha_2 x_{i,j+1} + \alpha_3 x_{i+1,j} + \alpha_4 x_{i,j-1}$$

The coefficient matrix A of Eq. (44), which is illustrated here for an 8×8 mesh, has the block structure

$$A = \begin{bmatrix}
R_0 & & & & & & & R_2 & R_3 & & & & & & & \\
& R_0 & & & & & & R_4 & R_2 & & R_3 & & 0 & & & \\
& & R_0 & & & & & R_1 & & R_4 & R_2 & R_3 & & & & \\
& & & R_0 & & 0 & & & R_1 & & R_4 & & R_3 & & & \\
& & & & R_0 & & & & R_1 & & & R_2 & & R_3 & & \\
& & & & & R_0 & & & & R_1 & R_4 & R_2 & & & R_3 & \\
& 0 & & & & & R_0 & 0 & & & R_1 & & R_1 & R_2 & & \\
& & & & & & R_0 & & & & & R_1 & & & R_4 & \\
R_4 & R_2 & R_3 & & & & & R_0 & & & & & & & & \\
& R_4 & & R_3 & & 0 & & & R_0 & & & & & & & \\
R_1 & & R_2 & & R_3 & & & & & R_0 & & & 0 & & & \\
& R_1 & R_4 & R_2 & & R_3 & & & & & R_0 & & & & & \\
& & R_1 & & R_4 & R_2 & R_3 & & 0 & & & R_0 & & & & \\
& & & R_1 & & R_4 & & R_3 & & & & & R_0 & & & \\
& & & & R_1 & & R_2 & & & & & & & R_0 & & \\
& 0 & & & & R_1 & R_4 & R_2 & & & & & & & R_0 & \\
\end{bmatrix} \tag{45}$$

where

$$R_0 = \begin{bmatrix} 1 & \alpha_2 & 0 & \alpha_3 \\ \alpha_4 & 1 & \alpha_3 & 0 \\ 0 & \alpha_1 & 1 & \alpha_4 \\ \alpha_1 & 0 & \alpha_2 & 1 \end{bmatrix} \qquad R_1 = \begin{bmatrix} 0 & 0 & 0 & \alpha_1 \\ 0 & 0 & \alpha_1 & 0 \\ 0 & 0 & 0 & 0 \\ 0 & 0 & 0 & 0 \end{bmatrix}$$

$$R_2 = \begin{bmatrix} 0 & 0 & 0 & 0 \\ \alpha_2 & 0 & 0 & 0 \\ 0 & 0 & 0 & \alpha_2 \\ 0 & 0 & 0 & 0 \end{bmatrix} \qquad R_3 = \begin{bmatrix} 0 & 0 & 0 & 0 \\ 0 & 0 & 0 & 0 \\ 0 & \alpha_3 & 0 & 0 \\ \alpha_3 & 0 & 0 & 0 \end{bmatrix} \qquad (46)$$

$$R_2 = \begin{bmatrix} 0 & \alpha_4 & 0 & 0 \\ 0 & 0 & 0 & 0 \\ 0 & 0 & 0 & 0 \\ 0 & 0 & \alpha_4 & 0 \end{bmatrix}$$

As can be seen, here the matrix A is block partitioned. Since the blocks are taken in red-black ordering, however, the matrix has block property (A) and is block consistently ordered. Also, if the blocks are taken in natural ordering the matrix is block consistently ordered. For both orderings, therefore, the full theory of block SOR is applicable to this method.

In the point iterative methods discussed in Section 2 each component of the iterative $x^{(k)}$ is determined explicitly. This means it can be calculated by itself using the values of the components, which were evaluated previously, whereas in the block iterative method one can improve the values of the approximate solution simultaneously with a group of points on the mesh. Such methods are called implicit methods since the solution of a whole group of points can be found at a time. In fact, implicit methods have larger convergence rates, at the cost of computation in each iteration in addition to that of the explicit method.

The explicit block SOR method can be considered a point SOR method applied to a transformed matrix A^E. However, it is necessary to calculate the matrix

$$A^E = [\mathrm{diag}\{R_0\}]^{-1} A$$

The matrix $[\mathrm{diag}\{R_0\}^{-1}]$ is simply $\mathrm{diag}\{R_0^{-1}\}$, and R_0^{-1} is given by

$$R_0^{-1} = \frac{1}{d} \begin{bmatrix} 1 - \alpha_1\alpha_3 - \alpha_2\alpha_4 & \alpha(\alpha_2\alpha_4 - \alpha_1\alpha_3 - 1) & 2\alpha_2\alpha_3 & \alpha_3(\alpha_1\alpha_3 - \alpha_2\alpha_4 - 1) \\ \alpha_4(\alpha_2\alpha_4 - \alpha_1\alpha_3 - 1) & 1 - \alpha_1\alpha_3 - \alpha_2\alpha_4 & \alpha_3(\alpha_1\alpha_3 - \alpha_2\alpha_4 - 1) & 2\alpha_3\alpha_4 \\ 2\alpha_1\alpha_4 & \alpha_1(\alpha_1\alpha_3 - \alpha_2\alpha_4 - 1) & 1 - \alpha_1\alpha_3 - \alpha_2\alpha_4 & \alpha_4(\alpha_2\alpha_4 - \alpha_1\alpha_3 - 1) \\ \alpha_1(\alpha_1\alpha_3 - \alpha_2\alpha_4 - 1) & 2\alpha_1\alpha_2 & \alpha_2(\alpha_2\alpha_4 - \alpha_1\alpha_3 - 1) & 1 - \alpha_1\alpha_3 - \alpha_2\alpha_4 \end{bmatrix}$$

where

$$d = (\alpha_1\alpha_3 - \alpha_2\alpha_4)^2 - 2(\alpha_1\alpha_3 + \alpha_2\alpha_4) + 1 \qquad (47)$$

The block structure of A^E is the same as that of the matrix A given in Eq. (45) with the submatrices R_0 replaced by identity matrices I and the submatrices R_i, $i = 1, 2, 3, 4$, replaced by $R_0^{-1}R_i$. As R_i has a column of zeros, so does $R_0^{-1}R_i$,

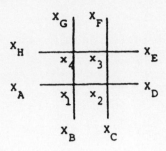

Figure 9

and since an element α_i occurs as the (p, q)th element of R_I, the qth column of $R_0^{-1}R_i$ is the pth column of R_0^{-1} multiplied by α_i. For example, for our model problem the Dirichlet problem

$$\alpha_1 = \alpha_2 = \alpha_3 = \alpha_4 = -\tfrac{1}{4}$$

so that from Eq. (47) we have

$$R_0^{-1} = \frac{1}{6} \begin{bmatrix} 7 & 2 & 1 & 2 \\ 2 & 7 & 2 & 1 \\ 1 & 2 & 7 & 2 \\ 2 & 1 & 2 & 7 \end{bmatrix}$$

and, for example,

$$R_0^{-1}R_1 = -\frac{1}{24} \begin{bmatrix} 0 & 0 & 2 & 7 \\ 0 & 0 & 7 & 2 \\ 0 & 0 & 2 & 1 \\ 0 & 0 & 1 & 2 \end{bmatrix}$$

Now to derive the corresponding equations of the points that form a block of four points, consider the area of a mesh shown in Figure 9. The main difference between the block method and the point method is that in the former method the equations of all the points in the block are formed in a way that the points are not dependent on each other in obtaining the new values. However, the equations required for the calculation of the values of x_1, x_2, x_3, and x_4 of Figure 9 commonly contain certain expressions that involve X_A, X_B, . . . , X_H and these are calculated only once and stored. Thus if we let

$$s_1 = x_A + X_B$$
$$s_2 = X_C + X_D$$

$$s_3 = X_E + X_F$$
$$s_4 = X_G + X_H \tag{48}$$
$$s_5 = s_1 + s_1 + s_3 + s_3$$
$$s_6 = s_2 + s_2 + s_4 + s_4$$

then we have

$$x_1 = \frac{1}{24}(7s_1 + s_6 + s_3)$$

$$x_2 = \frac{1}{24}(7s_2 + s_5 + s_4)$$

$$\tag{49}$$

$$x_3 = \frac{1}{24}(7s_3 + s_6 + s_1)$$

$$x_4 = \frac{1}{24}(7s_4 + s_5 + s_2)$$

When the overrelaxation factor ω is added, these equations take the following form if x_1, for example, is considered:

$$x_1^{(k+1)} = x_1^{(k)} + \omega(x_1^* - x_1^{(k)}) \tag{50}$$

where x_1^* is the value of x_1 in Eq. (49).

Equation (50) requires an average of 6½ additions and 3 multiplications per point per iteration with the assumption that the constant 1/24 is stored. This calculation can be less if Eqs. (48) and (49) are reorganized. Thus the explicit form of this block SOR method takes only about as much work as the implicit form of the four-point method (Biggins, 1980).

Finally we present the formula for the overrelaxation factor ω for the four-point block method. It has been mentioned that the theory for this method is equivalent to the theory of the SLOR method (Parter, 1981). Therefore

$$\omega_{\text{opt}} = \frac{2}{1 + \sqrt{1 - \rho^2(B)}} \tag{51}$$

where $\rho(B)$ is the spectral radius of the Jacobi iteration matrix B that corresponds to the four-point block iterative method. In actual fact the spectral radius $\rho(B)$ cannot be obtained theoretically. Rather, it can be estimated experimentally. Thus Biggins (1980) obtained the value of this quantity experimentally using the power method in which the results turned out to be equivalent to the values obtained for the line Jacobi method. The spectral radius of the line Jacobi matrix is given by Varga (1962) as

$$\rho\,(B) = \frac{\cos\,\pi h}{2\,-\,\cos\,\pi h} \tag{52}$$

Thus ω_{opt} of Eq. (51) can now be easily obtained. The spectral radius and the rate of convergence of this method can then be obtained, which are equivalent to that of the SLOR method. Thus the spectral radius of the SLOR matrix (L'_ω) is given by

$$\lambda' = \omega_{opt} - 1 \tag{53}$$

and the rate of convergence of SLOR is

$$R\,(L'_\omega)\,-\,2\sqrt{2R\,(B)} \tag{54}$$

For our model problem the rate of convergence of the SLOR method is given by Varga (1962) as

$$R\,(L'_\omega) \sim \frac{2\sqrt{2\pi}}{m\,+\,1} \qquad m \to \infty \tag{55}$$

This completes the basic principles of the explicit four-point block iterative method from which some parallel versions are derived and run on the Neptune system.

9 PARALLEL MATRIX ITERATIVE METHODS

The iterative solution of the Dirichlet problem described earlier was first considered for a simulated multiprocessor system by Rosenfeld and Driscoll (1969) using the chaotic relaxation strategy suggested by Chazan and Miranker (1969). Baudet (1978a) and others have also implemented synchronous and asynchronous iterative methods on the C.mmp and Cm*. Here we extend these results and introduce block methods that are more suitable to parallel implementation.

9.1 Experimental Results

The iterative methods described earlier have been implemented experimentally on the Neptune system to solve the two-dimensional Dirichlet problem on a square mesh of points. The approximate solution to this problem is found by solving a large system of equations $Ax = b$, where A is a $n \times n$ sparse matrix and b is the vector usually obtained from the boundary conditions. The spectral radius of the matrix A, $\rho(A)$, should be less than 1 to guarantee that any asynchronous iterative method converges.

Hence we wish to solve an $N \times N$ matrix linear system on a computer of P processors. The natural way to evaluate these points (or components) is to allocate a fixed amount of work to each process. In this case we allocate $S = N/P$ lines (i.e., rows) of the matrix to each process. The number of processes is taken to be less than or equal to the number of processors. The rows are allocated to processes so that the first S rows are assigned to process 1, the second S rows are assigned to process 2, and so on. However, the rows are processed sequentially although the components within each row are treated pairwise. This strategy is termed sequential decomposition.

Each process iterates on its own subset permanently, but it is required to read all its components at the beginning of each iterate and then it releases all the values of the components for the next iterate. However, the reading and the release of the components should be maintained in critical sections, and the data structure holding the obtained values should be accessed by all processes; that is, it should be stored in the shared memory. Since critical sections cannot therefore be accessed by many processes at a time, the processes may complete their work in different times, although they start their work at almost the same time. When a process P is busy releasing the values of its components, other processes cannot use these values when they require them in their computations until the release is completed. This means that the components of process P iterate using the old values of these related components. This undoubtedly increases the number of iterations required to find the solution within a required accuracy.

This stragegy is implemented for all the basic iterative methods discussed in Section 2, which are implemented asynchronously. However, a synchronous version of the Jacobi iterative method is implemented as the basis for our comparisons of the asynchronous iterative methods and to show their effectiveness. The methods are defined as follows.

9.2 The Synchronous Jacobi Method (or the J Method)

The Jacobi iterative method of Eq. (6) is implemented by using the five-point formula given in Eq. (22). This method is convergent since the spectral radius of the Jacobi matrix of Eq. (5) is less than 1 (Varga, 1962).

From Eq. (6) we see that all the components of an iterate are computed simultaneously using the values of the previous iterate; hence parallelism can be introduced easily. Since each process is allocated to a subset of rows to evaluate, when a process completes its computation it must then halt and wait for the other processes to complete their computations and then start the evaluation of the next iterate. The program that implements this method is given in Yousif (1983).

Table 1 Synchronous Jacobi Iterative Method to Solve the Two-Dimensional Laplace Equation Using Sequential Decomposition

Mesh size $(N \times N)$	ϵ	No. of paths	Timing (s)	No. of iterations	Speedup
12×12	10^{-5}	1	238.730	306	1
		2	132.520	306	1.80
		3	100.770	306	2.37
		4	87.380	306	2.73
24×24	10^{-2}	1	498.280	162	1
		2	269.080	162	1.85
		3	199.520	162	2.5
		4	169.340	162	2.94

However, the iteration counter is placed before the parallel paths (processes) are generated and incremented only when all paths are synchronized and the required accuracy has been reached. The experiments are carried out for two mesh sizes: first for a mesh size of 12×12, that is, 144 interior points to evaluate, and with the error difference (i.e., $\epsilon = 10^{-5}$), and second for a mesh size of 24×24, that is, 576 points to evaluate, with $\epsilon = 10^{-2}$. The Neptune timing results, number of iterations, and the speedups are listed in Table 1. However, these results are taken for any number of paths less than or equal to the number of processors, which is four. The boundary values used were 100 on the first column with the other boundaries set to zero. The speedup ratio is taken as the ratio of the timing results when only one path is generated and when k paths are generated. The iteration is implemented so that all the subsets contain an equal number of rows arranged in the following manner. Each row index is separated from the next row index by a distance equal to the number of the generated paths. This means that if four is the number of paths that are generated and the mesh contains 12 rows, then each subset contains 3 rows such that the subset S_i contains the rows indexed as follows:

S_1 contains 1, 5, and 9
S_2 contains 2, 6, and 10
S_3 contains 3, 7, and 11
S_4 contains 4, 8, and 12

We call this strategy nonsequential decomposition. The results of this strategy are listed in Table 2. Both these decomposition strategies are consistently ordered since the matrix for the two-dimensional Laplace equation using the five-point formula has property (A) and this matrix is of two-cyclic form.

Table 2 Synchronous Jacobi Iterative Method Using Nonsequential Decomposition

Mesh size $(N \times N)$	ϵ	No. of paths	Timing (s)	No. of iterations	Speedup
12×12	10^{-5}	1	127.710	168	1
		2	84.960	206	1.50
		3	70.350	232	1.81
		4	65.370	252	1.95

The timing results for sequential decomposition are greater than those for nonsequential decomposition strategies, and this is related to the fact that the components in one row are related to the components in the row neighbors that is, above and below, which are evaluated by other processes. This means, in each iteration, that the processes are sychronized and the values of their components are released and therefore the new values are used by other processes as soon as they are released, whereas in sequential decomposition most of the related components are in one process and are evaluated sequentially within the process. In fact only the boundaries of the subsets in the processes are related, in which case the evaluation of such components can be performed in parallel as in nonsequential decomposition. This is why the number of iterations is large as well.

On the other hand, the speedup ratios in Tables 1 and 2 show that the speedup in sequential decomposition is greater than that in nonsequential decomposition. This is because the components in the sequential strategy are iterated equally whatever the number of paths but in nonsequential decomposition the number of iterations increases when the number of paths increases as a result of subset communication and because the components cannot always obtain the recent values of those related components that are evaluated by other processes. We therefore conclude that sequential strategy can exploit parallelism more than nonsequential strategy in the synchronous iterative method. The full synchronization of the processes in each iteration step is a significant overhead in an asynchronous multiprocessor in general and also in the parallel implementation of the Jacobi iterative method. This overhead can be decreased by introducing the asynchronous Jacobi iterative method (or the AJ method), which is a related Jacobi method, but each process iterates on its subset and never waits for other processes to complete their work. When each process completes the evaluation of the components in its subset, it releases their values to the other processes by updating the corresponding components of the shared data structure (in our case this is an array), and then immediately after it starts a new iteration step on its subset by using in the computation the values of the components as they are known at the beginning of the reevaluation. In the AJ method critical sections are

Table 3   AJ Method Using the Sequential Decomposition Strategy

Mesh size $(N \times N)$	ϵ	No. of paths	Timing (s)	No. of iterations	Speedup
12 × 12	10^{-5}	1	233.040	306	1
		2	125.380	321	1.86
		3	86.860	326	2.68
		4	67.920	333	3.43
24 × 24	10^{-2}	1	470.270	162	1
		2	248.380	164	1.89
		3	174.650	164	2.69
		4	134.830	165	3.49

required to update the components at the end of each iteration step and to copy the values of the components required for the next iteration. In this method the iteration counter is placed within each process (path); therefore each process has its own counter and a convergence test is used to test whether its subset has converged. A global convergence test is required to ensure that all the components of all subsets have converged to the required accuracy. This is implemented by allocating a "flag" on each process to be set when the subset of the process has converged completely; a simple test for all flags of the other processes is then carried out. The process terminates its computation when all the flags are found to be set, that is, when all subsets of other processes have converged. The flags are stored in the shared memory so that they can be read by all processes. The experimental results are listed in Tables 3 and 4, where two different mesh sizes were used with different error distances and also the same boundary conditions as in the J method. The two decomposition strategies of the rows of the mesh as described in the J method were also implemented for the AJ method.

From Tables 3 and 4 we note that the sequential decomposition strategy is better than nonsequential decomposition; the timing results in the former are less than the results in the latter. The speedup ratios are also good and slightly less than a linear speedup in the sequential case, whereas it is not linear when the number of processes increases in the nonsequential case.

The difference in the results is that since the algorithms run asynchronously, nonsequential decomposition allocates nonsequential rows to processes and since the AJ method uses values of the components on the neighboring rows from the old iteration, the greater the number of processes the less the probability of the components to obtain the values of their related components that are evaluated by other processes. This is because processes do not usually complete their computations at the same time, and because of the critical sections within each

Table 4   AJ Method Using the Nonsequential Decomposition Strategy

Mesh size $(N \times N)$	ϵ	No. of paths	Timing (s)	No. of iterations	Speedup
12×12	10^{-5}	1	233.650	306	1
		2	148.250	375	1.58
		3	104.620	378	2.23
		4	79.910	377	2.92
24×24	10^{-2}	1	489.220	162	1
		2	270.700	174	1.81
		3	184.300	174	2.65
		4	140.120	174	3.49

process other processes may wait for the values of the components as they are updated by other processes.

Another observation from Tables 3 and 4 is the number of iterations, which increases slightly in the case of sequential strategy when the number of processes increases from one to four. In the nonsequential strategy a sharp increase in the number of iterations is noted between the two runs when only one process is used and when two to four processes were in use. Because the number of iterations is important in any iterative method, the sequential decomposition strategy is therefore chosen for further discussion. A method faster than the Jacobi iterative method is the Gauss-Seidel method, which is at least twice as fast.

9.3   Asynchronous Gauss-Seidel Method (AGS Method)

This iterative method implements the Gauss-Seidel method shown in Eqs. (10) and (11), which is the same as the AJ method except that the process uses new values for the components in its subset as soon as they are known for further evaluations in the same cycle. The process releases the values of its components at the end of its cycle, however, in which case the other processes can use them when they are required. The sequential decomposition strategy was implemented when the same boundary conditions and convergence test were used. Also, the processes were iterating asynchronously. The experiments were also carried out for the nonsequential decomposition strategy, and the results for both implementations are listed in Tables 5 and 6.

The results of this method also indicate that sequential decomposition is better suited to the Gauss-Seidel method than nonsequential decomposition. This is because of the way the components of the Gauss-Seidel method iterate. To calculate any component of the mesh using the five-point formula requires values

Table 5 AGS Method Using Sequential Decomposition

Mesh size $(N \times N)$	ϵ	No. of paths	Timing (s)	No. of iterations	Speedup
12×12	10^{-5}	1	123.580	162	1
		2	73.840	189	1.67
		3	52.890	199	2.34
		4	42.950	211	2.88
24×24	10^{-2}	1	311.830	107	1
		2	176.630	114	1.77
		3	120.930	116	2.56
		4	97.170	118	3.21

Table 6 AGS Method Using Nonsequential Decomposition

Mesh size $(N \times N)$	ϵ	No. of paths	Timing (s)	No. of iterations	Speedup
12×12	10^{-5}	1	123.320	162	1
		2	118.250	264	1.04
		3	81.990	304	1.50
		4	64.960	306	1.90

from the same iteration step for some neighboring components. When nonsequential decomposition is used the related components therefore go into different processes. Thus when more than one process is cooperating at the same time and when the subset's size is sufficiently large, the components in each row may not obtain the most recent values of the "north" and the "south" components situated in the other process; instead they obtain the old values and iterate on them until the new values are released. This in fact reduces the speedup; it also slows the process and reduces the efficiency of the processor carrying out this process. Thus many more iterations are required to solve this system of equations than the sequential algorithm. Another factor that reduces the efficiency is the number of critical sections that read all the components and release them at the end of the evaluation. However, if the number of processes k is 2 and each is a large subset, then the update of the shared data structure of the corresponding components performed within the critical section takes a longer time than for a smaller size. The other process therefore waits longer to enter the critical section to update its own components, since access to a critical section is made sequential.

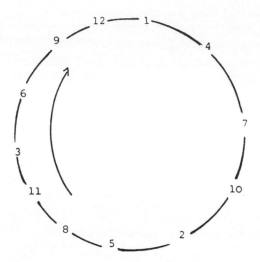

Figure 10

On the other hand, if the mesh has 12 rows then each process allocates 6 rows. If we imagine that the two subsets form a circular list, then in this list the newer dependent rows are 5 rows apart. Thus if two processors are used, one can be four times slower than the other even though there is no interference between them.

When $k = 3$ the subsets contain odd and even rows in the alternative form. Obviously the sizes of the subsets are smaller than that of $k = 2$. Also, any row in any process depends on the north and the south rows, which are allocated to different processes, not the same process. If three processors of different speeds are cooperating, then the probability of the row to obtain the most recent values from the neighboring rows is therefore low. However, if we arrange the subsets in a circular list, as in Figure 10, when three processors are cooperating, then each row is three rows apart from the neighboring rows, and in any cycle two processors can be slower than the third. A speedup cannot therefore be achieved; similarly when $k = 4$ and four processors are cooperating.

In sequential decomposition all neighboring components that are related are computed sequentially in one subset, except for the rows on the boundary of the subset, which have their related rows in different subsets that are carried out by other processes. In this case the component can obtain the new values of most of its related components in the same subset. This, however, explains why the number of iterations is less than that in nonsequential decomposition when the parallel running time is compared with the sequential running time (i.e., when

Table 7 PA Method Using Sequential Decomposition

Mesh size $(N \times N)$	ϵ	No. of paths	Timing (s)	No. of iterations	Speedup
12 × 12	10^{-5}	1	117.870	162	1
		2	59.730	163	1.97
		3	39.690	163	2.97
		4	29.840	163	3.95
24 × 24	10^{-2}	1	310.900	107	1
		2	160.900	108	1.93
		3	104.710	107	2.97
		4	79.200	107	3.93

$k = 1$). Speedup is achieved accordingly, and this is slightly improved when a larger mesh size is used. The reason is that the potentiality of a parallel system is exploited when a large mesh problem is considered.

This asynchronous iterative method can be improved if it is implemented without using critical sections, that is, the synchronization overhead can be decreased. Such a method is called the purely asynchronous method.

9.4 Purely Asynchronous Iterative Method (PA Method)

This method implements the Gauss-Seidel concept [Eq. (11)], but each process obtains the new values of its components by using the most recent values of the components and releases each new value immediately after its evaluation. In other words, the method is implemented as in the AGS method, in which each process allocates a subset of components but when each component is evaluated it is released immediately without waiting for other components of the same subset to be evaluated. This update can be made directly to the corresponding component of the shared array with the use of a critical section. Therefore each process always obtains the most recent value of any component it requires in its computation, which implies that each process iterates as many times as the others.

The sequential decomposition strategy for this algorithm is implemented when all the conditions used in the previous programs are used here as well. The results of this strategy together with those of the nonsequential strategy are listed in Tables 7 and 8, respectively.

It is obvious from these two tables that in this method the two decomposition strategies behave equally, so that the same timing results, the same number of iterations, and the same speedup are obtained. This comes from the nature of the

Table 8 PA Method Using Nonsequential Decomposition

Mesh size $(N \times N)$	ϵ	No. of paths	Timing (s)	No. of iterations	Speedup
12 × 12	10^{-5}	1	118.140	162	1
		2	59.660	163	1.98
		3	39.990	163	2.95
		4	30.050	163	3.93
24 × 24	10^{-2}	1	311.420	107	1
		2	157.270	107	1.98
		3	104.720	107	2.97
		4	78.790	105	3.95

method in updating the components. It therefore does not matter by which process the related components are computed as long as they are released for other processes. Linear speedup is achieved in this method, however, because of the absence of critical sections that generate substantial overheads.

When a component is updated by a process direct access to the shared memory to update the corresponding component is made, which means that in this case there is no need for a local array to hold the values; instead only the shared array is used in a process to read from and to update its components. Obviously one can think of multiple access to the same location in the shared data. For example, one process is busy updating a component in the shared array while the other is trying to read the value of this component to be used in its computation. Since the processes are at different speeds, however, the probability of such an occurrence could be very low. Also, because of the sparsity and the special form of the matrix associated with our system of equations, access to the global array by a given process is mostly confined to access of components within its own subset and only a few means of access to the components in the two adjacent subsets.

On conclusion of all these results, the PA method is considered the best. This is because of the absolute absence of any critical section, which implies less overhead, and because from the space or memory point of view the method uses only one data structure, whereas the AJ method implements critical sections as well as a larger space. This is because it requires from each process not only a complete duplication of all components (as at the beginning of its cycle) but yet another copy of the components in its own subset. Although this is a major concern in practice, the experimental results show useful comparisons between the AJ method and the J method. The AGS method requires only one copy of the components to be local to the processes in addition to the shared data structure that holds the results. However, both the AJ and AGS methods require a critical

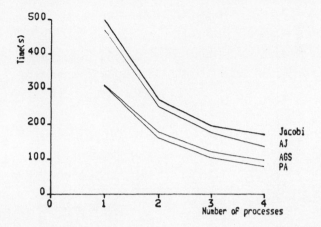

Figure 11   Timing results of the J, AJ, AGS, and PA methods solving the two-dimensional Dirichlet problem.

section to read in all the components to be used in each process at the beginning of the cycle and a critical section to release all components evaluated by a process to be available to other processes at the end of each cycle of the iteration step.

Now, by considering the sequential decomposition strategy, the timing results, speedup ratios, and the number of iterations obtained for the four methods, the synchronous Jacobi, AJ, AGS, and the PA methods are illustrated in Figures 11, 12, and 13, respectively, where the 24 × 24 mesh size is chosen. Figure 12 illustrates that a full linear speedup is achieved in the PA method; the AJ method achieves a better speedup than the AGS, although the AJ method is slower to converge than the AGS. This is due to the behavior of the method on iterating the components. On the other hand, Figure 13 illustrates that the Jacobi and PA methods produce a constant number of iterations for any number of processes, but in the AJ method the number of iterations increases slightly with an increase in the number of processes. The AGS method shows a sharp increase with the increase in the number of processes. However, when the error difference is small the increase in the number of iterations becomes clearer, as the results of the 12 × 12 mesh size indicate (see Table 5). This increase in the number of iterations reduces the speedup and explains why the AJ method can be introduced easily to a parallel system rather than the AGS method, which is an inherently sequential method.

All these asynchronous methods converge successfully and give the correct answer for the given boundary conditions when the algorithms are run sequentially or in parallel.

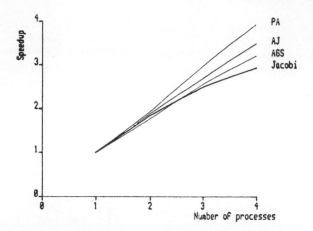

Figure 12   Speedup achieved by the J, AJ, AGS, and PA methods.

9.5   Over- and Underrelaxation Strategies

The asynchronous iterative methods discussed previously can be accelerated if an overrelaxation or an underrelaxation factor is added to the system of equations to accelerate the convergence of these methods.

The JOR and SOR methods of Eqs. (8) and (13), respectively, are introduced in this section together with the purely asynchronous overrelaxation (PAOR) method that solves the two-dimensional Laplace equation. Experimentally the PAOR method is implemented on the Neptune system in which the same Eq. (13), is applied but the values of the mesh components are obtained as described for the purely asynchronous method. The experiments were carried out to solve a mesh of size 12 × 12, that is, 144 components, with the same convergence test and boundary conditions as used in the previous asynchronous iterative methods. The results of this algorithm are listed in Table 9, where different numbers of processes are carried out by four processors. The parameter ω represents the overrelaxation factor that accelerates the method. In these experiments only sequential decomposition is implemented.

A linear speedup is achieved for this method as in the PA method, but the timing results of this method are less than those of the PA method. Figure 14 illustrates the difference in the timing results of these two methods when a 12 × 12 mesh size is implemented. This difference can be calculated from the theory of the rate of convergence of the Gauss-Seidel method and the SOR method, since the PA and PAOR methods implement the Gauss-Seidel equations for computing the mesh components, as mentioned earlier.

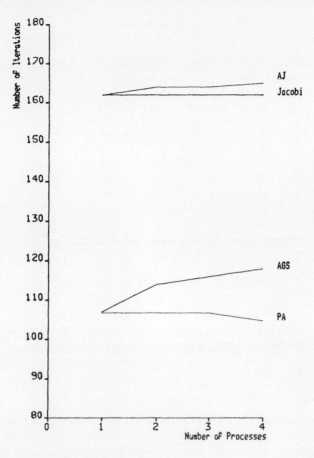

Figure 13 Number of iterations required to solve the system of equations by the Jacobi, AJ, AGS, and PA methods.

The rate of convergence of the Gauss-Seidel method shown in Eq. (42) is given as h^2, where $h = \pi/n$ and $n = 13$ for the 12×12 model problem. If we let R denote the rate of convergence,

$$R \, (PA) \approx h^2 = \left(\frac{\pi}{13}\right)^2 = 0.0584$$

The rate of convergence of the PAOR method is given as the SOR rate of convergence as in Eq. (43); thus

$$R \, (PAOR) = 2h = 0.48332$$

Table 9 PAOR Method Using Sequential Ordering

Mesh size ($N \times N$)	No. of processors	ω	Time (s)	No. of iterations	Speedup
12 × 12	1	1.0	136.800	162	1
	2	1.0	69.930	163	1.96
	3	1.0	47.210	163	2.90
	4	1.0	36.010	163	3.80
12 × 12	1	1.62	26.270	28	1
	2	1.62	13.840	29	1.90
	3	1.62	8.980	28	2.93
	4	1.62	6.650	28	3.95
16 × 16	1	1.0	393.840	263	1
	2	1.0	207.010	265	1.9
	4	1.0	105.090	270	3.75
16 × 16	1	1.70	55.600	37	1
	2	1.70	29.460	38	1.89
	4	1.70	15.130	38	3.70

The ratio of R(PAOR) over the R(PA) then gives the factor of improvement that is obtained. This factor is given by

$$\frac{R \text{ (PAOR)}}{R \text{ (PA)}} = \frac{0.48332}{0.0584} \approx 8.3$$

This shows that there is a factor of improvement of about 8.3 for the PAOR method compared with the PA method. This factor increases as h decreases, that is, as the order of the system of equations to be solved increases. For example, if $N = 32$ then the factor becomes 20.5.

Now, if we compare the experimental results of the PA and PAOR methods listed in Tables 7 and 9, respectively, when a 12 × 12 mesh size is used, we see that the factor of improvement is about 5 when only the timing results for different number of paths are compared. Thus when

$$\text{No. of paths} = 1 \rightarrow \frac{117.870}{26.270} = 4.5$$

$$\text{No. of paths} = 2 \rightarrow \frac{59.730}{13.840} = 4.3$$

$$\text{No. of paths} = 3 \rightarrow \frac{39.690}{8.980} = 4.4$$

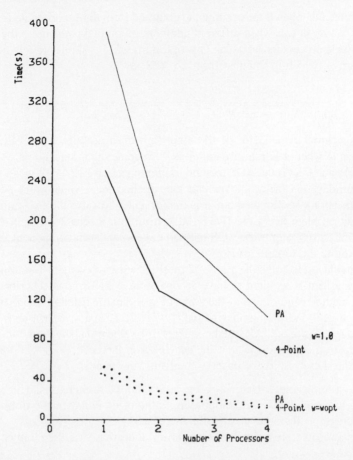

Figure 14   Timing results of the PA and the four-point block methods for 16 × 16 mesh size.

$$\text{No. of paths} = 4 \rightarrow \frac{29.840}{6.550} = 4.6$$

These figures show that the experimental results are not sufficiently adequate to explain the theoretical results drawn from the rate of convergence theory. This in fact is due to different factors that arise from parallel implementation, such as the allocation of subsets to parallel paths, path cooperation, the assignment of processors to paths, and shared data access overheads. All these factors may decrease the factor of improvement from its theoretical counterpart.

On the other hand, the optimum ω obtained from the experiments is 1.62 for a 12 $\times$ 12 mesh size when different numbers of parallel paths up to the number of available processors are used. This optimum value is similar to the theoretical optimum ω, which is obtained from Eq. (28) as

$$\omega_b = \frac{2}{1 + \sqrt{1 - \rho^2 (B)}} = \frac{2}{1 + \sqrt{1 - \cos^2 \pi / 13}} = 1.614$$

The asynchronous behavior of the algorithm does not affect the value of the optimum ω when it is run in parallel as with some other algorithms. A method with such an effect is the AGS method implemented on a 12 $\times$ 12 mesh size with the overrelaxation factor ω. We call this method the asynchronous successive overrelaxation (ASOR) method. It is implemented to solve the two-dimensional Dirichlet problem using the five-point formula on a square mesh in the same manner as described for the AGS method except that each component is iterated according to the formula of Eq. (13).

Obviously, to solve the 12 $\times$ 12 mesh, that is, to evaluate 144 components of the mesh, it is required to have an optimum ω as 1.62, as described for the PAOR method. The results of the ASOR algorithm are listed in Table 10, where $1.3 \leq \omega < 2$ and the number of paths is 1, 2, 3, and 4. When $\omega = 1.0$ the method reduces to the AGS method. Thus this value of ω is included in Table 10 to compare the effectiveness of the acceleration factor.

From these results we observe the following points:

1. The timing results decrease as the number of paths increases. This is correct up to a point close to the optimum value of ω, and then the timing results increase as the number of paths increases.
2. As a result from point 1, the optimum ω decreases as the number of paths increases.
3. The range of convergence decreases when the number of parallel paths increases.

The exact optimum value of ω for any number of paths is obtained, however, and these values are listed in Table 11 with the timing results and speedup ratios. From Table 11 we see that no speedup is achieved for this method when it is implemented on k paths, $k = 2, 3, 4$, so that the timing results are better when this algorithm is run sequentially, that is, when $k = 1$.

As the nature of the SOR method is known, the ASOR method is implemented with the critical sections to read and update the global data structure that holds the final results. It uses the new values of the components in the subset as soon as they are available within this subset, and the update of the global data of all components within a set is performed within a critical section. Hence when k increases the dependent components are in different subsets carried out by

Table 10 ASOR Method to Solve the Two-Dimensional Dirichlet Problem

Mesh size (N × N)	No. of paths	ω	Time (s)	No. of iterations	No. of waiting cycles to access a critical section
12 × 12	1	1.0	148.590	162	—
		1.3	81.800	89	
		1.4	65.280	71	
		1.5	48.750	53	
		1.6	31.300	34	
		1.7	38.810	42	
		1.8	55.330	60	
		1.9	134.760	146	
	2	1.0	90.730	189	
		1.3	56.370	117	285,84
		1.4	50.050	102	50,95
		1.5	41.790	87	222,67
		1.6	38.160	78	17,65
		1.7	41.270	84	22,75
		1.8	156.020	320	146,255
		1.9	1131.610	No convergence	—
	3	1.0	67.930	199	—
		1.3	43.500	131	1451,12,544
		1.4	37.520	113	1204,13,474
		1.5	33.470	97	1939,17,909
		1.6	36.420	106	2111,26,986
		1.7	336.360	No convergence	
	4	1.0	54.960	211	—
		1.3	37.160	143	2264,11,1019,1429
		1.4	32.770	126	1984,893,13,1250
		1.5	28.900	111	1730,7,775,1153
		1.6	250.740	959	17096,33,6582,10279
		1.7	No convergence		

different processors running at different speeds. The probability of obtaining the new values of the dependent components is therefore small, which means that the components are forced to iterate on the old available values. This can generate a significant delay that causes degradation in the problem's running time. Therefore no speedup is achieved since the penalty paid for critical section access is much more than that for parallelism in the algorithm.

To illustrate this degradation, it was considered worthwhile to examine the time spent to access the critical sections (CS) made by the algorithm and also the time that the processors spent waiting for each other because the critical section

Table 11   Optimum ω for the ASOR Method

No. of paths	Optimum ω	Time (s)	No. of iterations	Speedup	Waiting cycle to access critical section
1	1.62	25.880	28	1	—
2	1.62	32.930	68	0.79	66
3	1.55	30.930	93	0.84	453
4	1.51	28.970	111	0.89	939

was being used by another processor. However, the time required to access a critical section in the Neptune system (i.e., the $ENTER/$EXIT construct) is ~400 μs. Since in the implementation of the ASOR method there are critical sections, one is at the commencement of each path to read all components, which is called just once, and the other critical section is to update the components of the subset of that path and to read in all the mesh components for the next cycle. The second critical section is called as many times as the path iterates on its components. Therefore the number of accesses to critical sections is obtained by adding 1 to the number of iterations obtained experimentally. The corresponding timing required for the waiting cycles made by the process to enter a critical section is ~200 μs. Therefore the time spent in the critical sections and for the processes to wait is calculated as

Time spent in all CS = number of accesses to CS × 400 μs and

Time spent for waiting cycles to access CS = number of waiting cycles × 200 μs.

We now calculate the time spent in the critical sections of the ASOR experimental results and the overhead measurements of these critical sections accesses. Such measurements are listed in Table 12. By static overhead we mean the actual accesses made by the algorithm, which can be obtained from the output results, and contention represents the time that the processes spend in waiting to access a critical section because it has been accessed by another process. The measurements in Table 12 are calculated from the results in Table 11.

The measurements in Table 12 declare that when the number of processes k increases, which results in an increase in the number of iterations, the critical sections accesses also increase. Therefore the contention among the processes increases, which yields an unwanted delay that reduces the value of the optimum ω.

Table 12   Critical Section Overheads in the ASOR Method

No. of paths	Optimum ω	Time (s)	No. of iterations	Time to access CS (s)	CS access overhead (%) Static overhead	Contention
1	1.62	25.880	28	0.012	0.04	—
2	1.62	32.930	68	0.028	0.08	0.04
3	1.55	30.930	93	0.038	0.12	0.3
4	1.51	28.970	111	0.045	0.16	0.65

It is also interesting to study the behavior of the acceleration factor ω for the Jacobi method, that is, a study of the JOR method given by Eqs. (7) and (8). However, for our model problem, that is, the two-dimensional Dirichlet problem, the Jacobi method cannot be accelerated by using ω since ω = 1.

Thus in this section we have solved the Dirichlet problem involving the Laplace equation on a grid of points using the five-point difference formula with the SOR method involving an overrelaxation parameter ω, which is described in more detail in previous sections.

This work included the study of asynchronous iterative methods. The problem to be solved was partitioned into k processes with $k \leq p$, the number of the available processors. The processes were not synchronized, and each process iterates on its subset using the new values of the components of the mesh when they are available. Hence the processes have a minimum of intercommunication through critical sections to allow processes to exchange their results to complete the solution of the problem. The purely asynchronous method described in this section is a typical asynchronous method with no critical sections at all. However, all asynchronous iterative methods show a great improvement over the synchronous methods that synchronize the processes at the end of each iteration step, which in turn degrades the performance of the algorithms.

Many experiments have been performed on the Neptune asynchronous system of four processors, and several asynchronous iterative methods have been implemented to solve a large system of equations drawn from the finite difference approximations, in particular the two-dimensional Dirichlet problem. These methods use the Jacobi method with full synchronization, the AJ and AGS methods with critical sections, and the PA method with no synchronization at all. However, the experimental results show good comparison between these methods, with the PA method the best because it exhibits full parallelism and has a linear speedup. The Jacobi method is considered the worst as it implements synchronizaton.

Two strategies of decomposing the problem were used, sequential and non-sequential allocaton of the rows of the mesh to each process. It was observed from the experimental results that sequential decomposition gave better results so that it was considered in all the other applications completed in this section.

The other factor obtained from experiments of the two-dimensional problem is the number of iterations, which remained the same for any number of processes in the Jacobi and PA methods and linearly increased with the number of processes in the AJ and AGS methods. In the one-dimensional problem the number of iterations decreases as the number of processes increases for the Jacobi, AJ, and PA methods, whereas the iterations increase when the number of processes increases in the AGS method.

Experiments were also performed on another asynchronous iterative method with an acceleration factor ω. The purely asynchronous overrelaxation method showed a great improvement over the PA method, and it has a linear speedup as well. In contrast, the AGS method with ω, that is, the ASOR method, did not gain speedup. Moreover, the optimum value of ω was decreased with the increase in the cooperating processes. This was shown to be related to the instability of the spectral radius of the matrix of the problem when the number of processes increases, and this is due to the existence of critical sections.

10   THE PARALLEL FOUR-POINT BLOCK ITERATIVE METHOD: EXPERIMENTAL RESULTS

After presenting the basic concepts of the four-point block iterative method in Section 8, we now implement the method in parallel, for which many different strategies are used to solve the model problem, the two-dimensional Dirichlet problem, by using Eqs. (48), (49), and (50). The strategies are mainly concerned with the way the problem to be solved is decomposed into many subsets that can be run in parallel. Clearly, different techniques may yield different results in running times and in speedup ratios. However, interesting results are obtained when the problem to be solved is programmed both synchronously and asynchronously.

We discuss in this section three strategies that we have developed; each strategy involves synchronous and asynchronous versions. The study includes comparisons between the synchronous and asynchronous versions in each strategy as well as comparisons between the results of different strategies. In all strategies the update of a component exactly follows the same scheme of the purely asynchronous algorithm discussed in Section 9, in which no critical section is used. The algorithms of these strategies were run for $\omega = 1.0$, that is, similar to the Gauss-Seidel point method, and for $\omega = \omega_{opt}$, that is, similar to the SOR point iterative method.

Table 13 Results of Strategy 1.1, the Synchronous Four-Point Block Method, Two Lines Taken at a Time (Natural Ordering)

Mesh size (N × N)	P	ω	Time (s)	No. of iterations	Speedup
12 × 12	1	1.0	89.00	89	1
	2	1.0	48.490	88	1.84
	3	1.0	32.880	88	2.71
	1	1.51	21.100	21	1
	2	1.51	11.140	20	1.89
	3	1.51	7.590	20	2.78
16 × 16	1	1.0	256.510	144	1
	2	1.0	138.690	143	1.85
	4	1.0	70.920	143	3.62
	1	1.60	47.860	27	1
	2	1.61	25.390	26	1.89
	4	1.60	13.150	26	3.64

Strategy 1

In the first strategy 1.1, the four-point block method is implemented so that a number of processes (paths) equal to the number of cooperating processors is generated. Each path works on a subset of lines $N_s = N/P$, where N is the number of rows in the mesh (divisible by P) and P is the number of cooperating processors. This means that P subsets are formed in which each subset contains N_s rows of the mesh. In this strategy N_s should also be a multiple of 2. Each processor then computes its own subset by taking up each successive two neighboring rows at a time so that each block on these two lines is evaluated, that is, the natural ordering. When the blocks on these two lines are all evaluated, the next two lines are taken and the algorithm proceeds as before until all the lines in the subset are evaluated.

This strategy is implemented synchronously, and in this program each processor evaluates its own subset in the manner already discussed and synchronizes itself after each iteration step. When all the processors are synchronized the convergence test is performed by one processor (usually the master processor), and if all the components of the mesh are obtained with the required accuracy the procedure terminates. Otherwise a new cycle is repeated and so on until all the components have converged. The results are listed in Table 13, where 12 × 12 and 16 × 16 mesh sizes were used in the experiments with ω = 1.0 and ω = optimum ω, which equals 1.51 in the 12 × 12 mesh and 1.60 in the 16 × 16 mesh.

Table 14 Results of Strategy 1.2, Synchronous Four-Point Block Method
(Red-Black Ordering)

Mesh size $(N \times N)$	P	ω	Time (s)	No. of iterations	Speedup
16×16	1	1.0	272.960	146	1
	2	1.0	150.880	146	1.81
	4	1.0	78.300	146	3.49
	1	1.60	52.240	28	1
	2	1.60	28.990	28	1.80
	4	1.60	15.010	28	3.48

The same synchronous strategy is again implemented in strategy 1.2, but the blocks on each two lines are evaluated in the red-black ordering, that is, blocks numbered 1, 3, 5, . . . , are taken first and then the even-numbered blocks 2, 4, 6, ... The results of this program are listed in Table 14, where the 16×16 mesh size only is used.

From the results of Tables 13 and 14 we note that the running time of strategy 1.2 is greater than that of strategy 1.1 and the speedup for the natural ordering of the latter is higher than the speedup of the red-black ordering of the former. We therefore choose the natural ordering scheme among the two implementations for the asynchronous version of this strategy.

The asynchronous version of this strategy was implemented in the same manner as in the synchronous version so that the blocks were taken in natural ordering. This scheme is programmed in strategy 1.3, where each subset of lines is allocated to a processor that runs asynchronously on its subset without waiting for other processors to complete their computations. In this case each processor iterates permanently on its subset as in the purely asynchronous algorithm (see Sect. 9) until this subset as well as the subsets that are carried out by other processors have converged. A flag is therefore assigned to each processor; the set of all flags is in the shared memory and can be accessed by all processors to check whether all the subsets have converged. The results of this implementation are listed in Table 15.

If we consider only the speedups obtained from strategies 1.1 and 1.3 we note that the asynchronous version is faster than the synchronous version for this algorithm strategy. Moreover, from Tables 13 and 15 we see that the qualitative trend of the asynchronous results is slightly better than that of the corresponding synchronous results because of the synchronization overheads, despite efficient implementation on the Neptune system. However, since each component is updated as soon as it is evaluated and is made available to all the other processors, which always obtain the most recent values of the components all the time,

Table 15   Strategy 1.3 Results, the Asynchronous Four-Point Block Method (Two Lines at a Time)

Mesh size (N × N)	P	ω	Time (s)	No. of iterations	Speedup
12 × 12	1	1.0	88.470	89	1
	2	1.0	46.340	90	1.91
	3	1.0	32.310	93	2.74
	1	1.51	20.950	21	1
	2	1.51	11.00	21	1.90
	3	1.52	7.450	22	2.81
16 × 16	1	1.0	254.920	144	1
	2	1.0	132.750	145	1.92
	4	1.0	68.530	145	3.72
	1	1.60	47.820	27	1
	2	1.61	24.990	27	1.91
	4	1.61	13.050	28	3.66

the overheads obtained from synchronizing the processes of each cycle may not be very high.

Now we compare the results of this asynchronous method for the explicit four-point block method when run with $\omega = 1.0$ and $\omega = \omega_{opt}$ and the purely asynchronous algorithm when run with $\omega = 1.0$ and $\omega = \omega_{opt}$. The timing results of the PA method are given in Table 16. The factor of improvement of this method over the point iterative method implemented in the purely asynchronous algorithm is given in Table 17 as γ, where

$$\gamma = \frac{\text{time for the point iterative method (PA)}}{\text{time for the four-point block method}}$$

From Table 17 we see that the factor is large for $\omega = 1.0$, which means that the four-point block method is about 1.5 times faster than the PA method. This is not true for $\omega = \omega_{opt}$, that is, when we have the SOR version, because this method does not show a very large factor over the PA point method. In fact, the four-point block method is more than $\sqrt{2}$ times faster than the PA point method, whereas the four-point block SOR method is faster than the PAOR point method by about $(\sqrt{2})^{1/2}$ times.

The running time results for the 16×16 mesh size of the PA method of Table 16 and that of the asynchronous four-point block method are illustrated in Figure 14. The speedup ratios of both methods are also illustrated in Figure 15.

Table 16   Results of the PA Point Method

Mesh size	P	ω	Time	No. of iterations	Speedup
12 × 12	1	1.0	136.800	162	1
	2		69.930	162	1.96
	3		47.210	163	2.90
	4		36.010	164	3.80
	1	1.62	26.270	28	1
	2		13.840	29	1.90
	3		8.980	28	2.93
	4		6.650	28	3.95
16 × 16	1	1.0	393.840	263	1
	2		207.010	265	1.90
	4	1.0	105.090	270	3.75
	1	1.70	55.600	37	1
	2	1.70	29.460	38	1.89
	4	1.70	15.130	38	3.70

Table 17   Improvement Factor Obtained from the Four-Point Block Method over the PA Point Method

Mesh size	P	ω	γ
12 × 12	1	1.0	1.55
		ω_{opt}	1.13
	2	1.0	1.51
		ω_{opt}	1.18
	3	1.0	1.46
		ω_{opt}	1.08
16 × 16	1	1.0	1.54
		ω_{opt}	1.16
	2	1.0	1.56
		ω_{opt}	1.22
	4	1.0	1.53
		opt	1.09

Strategy 2

In this strategy the mesh lines are grouped into subsets, and each subset contains N_s lines as in the first strategy. In this strategy the lines of a subset are taken one at a time; therefore the first two components of each block are evaluated when the line holding them is taken by a processor. When all components on this line

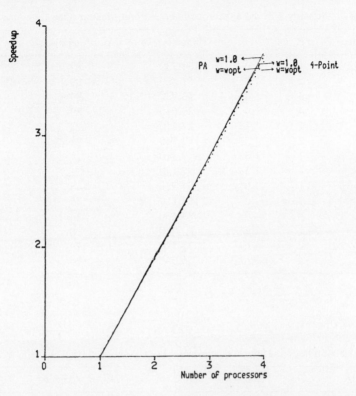

Figure 15   Speedup ratios for the PA and the asynchronous four-point block methods.

are evaluated, the next line that holds the other two components of each block is taken. The natural ordering scheme is used to evaluate the half-blocks on each line.

This strategy has also been implemented in both synchronous and asynchronous versions. The synchronous version was implemented in strategy 2.1 and the asynchronous version in strategy 2.2. The iteration cycles and the convergence test for both versions were implemented as in the first strategy. The results of strategy 2.1 and strategy 2.2 obtained from the Neptune system are listed in Tables 18 and 19, respectively.

From the results in Tables 18 and 19 we see that the synchronous and asynchronous versions of this strategy require approximately similar running times when only one processor is in use, whereas for two or more processors the asynchronous version requires slightly less time than the synchronous version. However, the speedup ratios for the asynchronous version when $\omega = 1.0$ are

Table 18   Strategy 2.2, the Synchronous Four-Point Method (One Line at a Time)

Mesh size	P	ω	Time (s)	No. of iterations	Speedup
12 × 12	1	1.0	104.730	89	1
	2	1.0	57.660	89	1.82
	3	1.0	38.930	89	2.69
	4	1.0	29.640	89	3.53
	1	1.52	27.210	23	1
	2	1.53	13.740	21	1.98
	3	1.53	9.310	21	2.92
	4	1.53	7.130	21	3.82
16 × 16	1	1.0	304.720	145	1
	2	1.0	166.110	145	1.83
	4	1.0	83.430	144	3.64
	1	1.61	62.920	30	—
	2	1.61	33.340	29	1.89
	4	1.61	16.340	28	3.85

Table 19   Strategy 2.2, the Asynchronous Four-Point Method (One Line at a Time)

Mesh size	P	ω	Time (s)	No. of iterations	Speedup
12 × 12	1	1.0	104.680	89	1
	2	1.0	56.090	91	1.87
	3	1.0	37.780	92	2.77
	4	1.0	28.310	92	3.70
	1	1.53	27.150	23	—
	2	1.53	14.200	23	1.91
	3	1.54	9.540	23	2.84
	4	1.54	7.140	23	3.80
16 × 16	1	1.0	303.480	145	1
	2	1.0	159.570	146	1.90
	4	1.0	80.140	146	3.79
	1	1.61	62.870	30	1
	2	1.62	33.620	31	1.87
	4	1.62	16.370	30	3.84

greater than those of the synchronous version; the converse is true for $\omega = \omega_{opt}$. Also, the optimum ω is slightly different for both versions.

By comparing the results of the asynchronous versions of the first and second strategies, as illustrated in Tables 15 and 19, respectively, we see that the strategy of taking up two lines at a time requires less time than that of taking up one line at a time. This is mainly related to the fact that in the strategy of taking up two lines at a time the whole block is evaluated at once and more recent values of components are available for use, but in the strategy of taking up one line at a time only half of the block at a time is evaluated and the second half is evaluated after the evaluation of the whole line is completed, and this naturally involves some extra time.

Generally speaking, in these two strategies it does not really matter whether the algorithm is synchronous as far as the timing results are concerned. This is because these algorithms were implemented so that each component is updated as soon as it is evaluated and then it is made available to all other processors, which means using only one shared array to hold the component's values. However, in the synchronous versions each processor allocates a subset of lines and the processors are synchronized at the end of each iteration cycle to ensure that the subsets that are carried out by other processors iterate on the new values of the components. This of course gives the correct approximation to the solution of the linear system of equations with a fixed number of iterations for any number of cooperating processors. Although the asynchronous versions are implemented in the same manner, except that each processor iterates on its subset and never waits for other subsets to be completed, no gain is obtained in time and this is explained by the first reason already mentioned.

Because the decomposition used in both strategies generates equal disjoint subsets to be carried out by different processors, the complexity of evaluating any component by any processor is the same. However, the computational complexity of a component update is only nine operations when the mesh size is 16×16; therefore, 16×9 operations are completed along each single line. If the subsets consist of N_P lines then $N_P \times 16 \times 9$ operations are performed by each processor. This is true for the synchronous and asynchronous algorithms, but synchronization at the end of each iteration cycle and consequently the generation of the parallel paths at every iteration cycle generate some overheads that reduce the timing results by approximately 1 s in these implementations.

Since the synchronous form makes the number of methods of access to a parallel path equal to the number of iterations required to satisfy the convergency test, this overhead generates a significant overhead on a MIMD computer. It follows that the use of the asynchronous version of these strategies may be better suited for the MIMD computer because asynchronous algorithms do not generate a significant amount of "parallel path access" overheads.

Strategy 3

The third strategy is completely different from the other two strategies in the manner of distributing the blocks of four points onto the processors. The asynchronous version is given in strategy 3.1, described here, following the work done by Barlow and Evans (1982). The row and column indices of the blocks of four points are stored in a shared list of two dimensions. The arrangement of the blocks in the list is represented by the red-black ordering. A shared index to the column of the list is used. Since the number of rows is stored in the first row of the list and the columns in the second row, updating the shared index by a critical section allows only one processor to evaluate the block that corresponds to that index. For the convergence test a list of the number of flags that is equal to the number of blocks in the linear system of equations is used so that the flag for the block under consideration is set to 1 if any single component of the block has not converged; otherwise the flag is set to 0. To iterate again each processor tests all the flags of all blocks, and if it finds any of them set to 1 then it iterates again, choosing any available block. Otherwise a global flag is set indicating that the convergence has been achieved for all components in the mesh. Since setting the global flag is performed by one processor that terminates its path afterward, the other processors terminate only when they attempt to take up another block. The results of this implementation are listed in Table 20.

To compare the asynchronous versions of the three strategies we list the timing results with the speedup ratios of the three strategies 1.3, 2.2, and 3.1 in Table 21, where a 16×16 mesh size is chosen. From Table 21 we note that the asynchronous algorithm of the third strategy achieves better speedup ratios than the other strategies when $\omega = 1.0$ and $\omega = \omega_{opt}$. These speedups are mostly linear, that is, of order P, where P is the number of processors in use. If we

Table 20 Strategy 3, the Asynchronous Four-Point Block Method (a List of Blocks)

Mesh size $(N \times N)$	P	ω	Time (s)	No. of Iterations	Speedup
16×16	1	1.0	271.830	146	1
	2		138.980	147	1.96
	3		93.100	148	2.92
	4		70.820	150	3.84
	1	1.61	52.120	28	1
	2		26.720	29	1.95
	3		17.880	30	2.92
	4		13.620	31	3.83

Table 21   Results of the Asynchronous Versions of the Three Strategies

Method	ω	P	Time	No. of iterations	Speedup
1. Strategy 1.3	1.0	1	254.920	144	1
(Two lines		2	132.750	145	1.92
at a time)		4	68.530	145	3.72
2. Strategy 2.2	1.0	1	303.480	145	1
(One line		2	159.570	146	1.90
at a time)		4	80.140	146	3.79
3. Strategy 3.1	1.0	1	271.830	146	1
(a list of blocks)		2	138.980	147	1.96
		3	93.100	148	2.92
		4	70.820	150	3.84
1. Strategy 1.3	1.60	1	47.820	27	1
	1.61	2	24.990	27	1.91
	1.61	4	13.050	28	3.66
2. Strategy 2.2	1.61	1	62.870	30	1
	1.62	2	33.620	31	1.87
	1.62	4	16.370	30	3.84
3. Strategy 3.1	1.61	1	52.120	28	1
		2	26.720	29	1.95
		3	17.880	30	2.92
		4	13.620	31	3.83

consider the timing results in Table 21, however, we see that the algorithm of the first strategy (i.e., the two-line strategy) requires less time than the other two strategies; meanwhile the algorithm of the second strategy (the one-line strategy) requires more time than the others.

If we consider the results of strategies 1.3 and 3.1, we see that the time in 1.3 is much less than that in 3.1, when $P = 1$, but the difference in the timing results of the two algorithms becomes smaller when P is greater than 1.

The difference in the running times for the three algorithms is because in taking up two lines at a time some time is saved since the two lines are evaluated together when each block on these two lines is carried out at a time. This is different from taking up one line at a time, as described earlier. On the other hand, evaluating the blocks in natural order, in which the blocks are stored in a list that is accessible by all processors, may require extra time to update the index to that list of blocks and to pick up a block of four points to evaluate. This last method seems to be better among the other methods for implementation on more than one processor, as is obvious from the speedup ratios of Table 21. For

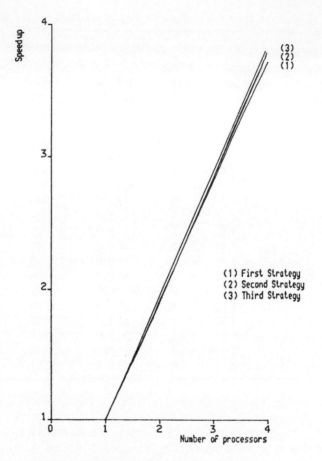

Figure 16   Speedup for the three strategies when $\omega = 1.0$

illustration, the speedup ratios for the three strategies are represented in Figures 16 and 17 when $\omega = 1.0$ and $\omega = \omega_{opt}$, respectively.

As in the first and second strategies, the synchronous version is also implemented in the third strategy. The implementation is such that a number of parallel paths equal to the number of blocks in the mesh is generated and carried out by any available number of processors. This means that each path picks up a block different from that picked up by the other paths, and this is ensured by updating a shared index to the list by a critical section. However, all the paths are synchronized at the end of each cycle of the iterate. When not all the blocks have converged, the paths are generated again when each path iterates on the available recent values. This procedure is continued until all the blocks have converged.

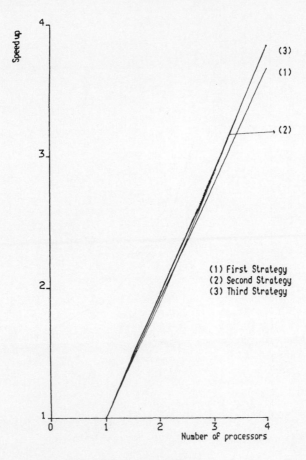

Figure 17   Speedup for the three strategies when $\omega = \omega_{opt}$.

Also, in this algorithm the blocks are arranged in red-black ordering in the list. The program for this implementation is strategy 3.2, the results of which are listed in Table 22.

If we compare the results of the asynchronous and synchronous versions of this strategy (Tables 20 and 22, respectively) we see that the time required by the asynchronous version is less than that for the synchronous one and also the asynchronous version indicates a very much higher speedup over the synchronous version. The timing results of the two versions are listed in Figure 18 for $\omega = 1.0$. This is different from the other two strategies, however, in which the synchronous and asynchronous versions have approximately equal timing results. The reason the two versions in the third strategy are different in their timing results is explained later in this chapter.

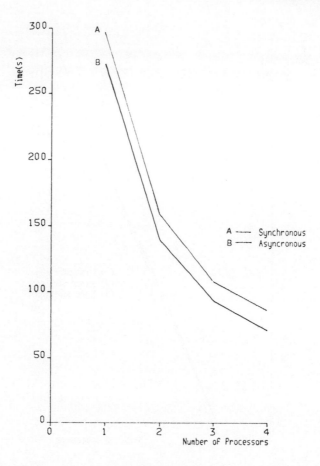

Figure 18 The synchronous and asynchronous versions of the third strategy when $\omega = 1.0$.

The synchronous version 3.2 has some limitations. Since the number of parallel paths generated is equal to the number of blocks of four points, a linear system of equations of a large size that has more than 75 blocks of four points cannot be used since at the time of the experiment the Neptune system can generate only 75 paths as a maximum limit, including the path generating the parallel paths. Thus for our model problem we run the problem for a 16 × 16 mesh size that contains 64 blocks. From another point of view this synchronous algorithm generates a large number of overheads, mainly parallel path overheads. As each component of the mesh requires nine arithmetic operations, 4 × 9 operations are required for a block of four points, which is actually the

Table 22   Strategy 3.2, the Synchronous Four-Point Block Method
(a List of Blocks)

Mesh size (N × N)	P	ω	Time (s)	No. of iterations	Speedup
16 × 16	1	1.0	296.770	146	1
	2		159.460	146	1.86
	3		108.410	146	2.74
	4		86.910	146	3.41
	1	1.61	56.970	28	1
	2	1.61	30.310	28	1.88
	3	1.61	20.680	28	2.75
	4	1.61	16.570	28	3.44

computation required for each single path. This amount of computation is considered small when compared with the cost paid for synchronizing all the paths. However, the synchronization cost is larger if the number of paths is greater than the number of available processors. This is because the paths that complete the job remain idle for a long time while waiting for the other paths to complete the job. The same situation occurs as many times as the algorithm iterates to achieve convergence. Clearly this generates a significant number of overheads that degrade the performance of the algorithm.

On the other hand, in the asynchronous version (3.1) the number of parallel paths generated is equal to the number of cooperating processors; each path is iterated to evaluate each block without synchronization. Therefore the asynchronous version is better than the synchronous in this strategy because it generates less overhead.

Finally, we compare the asynchronous version of the third strategy of the four-point block method with the explicit point method implemented in the purely asynchronous method, the results of which are listed in Table 16. Table 23 illustrates this comparison, where γ represents the ratio

$$\frac{\text{Time of the PA method}}{\text{Time of version 3.1}}$$

From Table 23 we note that when $\omega = 1.0$ the ratio γ becomes approximately 1.5, whereas the ratio is slightly greater than unity when $\omega = \omega_{opt}$. This matches the γ ratios of Table 17, which are shown to be $\sqrt{2}$ for $\omega = 1.0$ and $(\sqrt{2})^{1/2}$ for $\omega = \omega_{opt}$.

In Figure 19 the timing results of the point version of the PA method and the asynchronous versions of the three strategies of programs 1.3, 2.2, and 3.1 of the four-point method are illustrated.

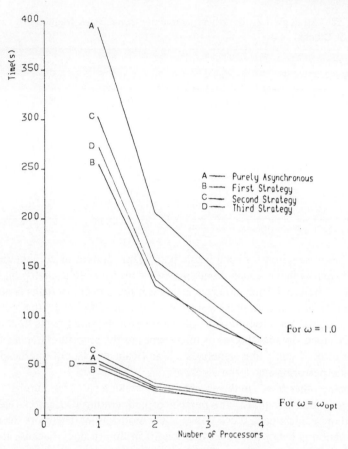

Figure 19 Timing results of the PA method and the asynchronous forms of the three strategies of the four-point block method.

Table 23 Improvement Factor Obtained from the Asynchronous Form (a List of Blocks) over the Point Method

Mesh size	P	ω	γ
16×16	1	1.0	1.45
	2	1.0	1.49
	4	1.0	1.48
	1	ω_{opt}	1.10
	2	ω_{opt}	1.10
	4	ω_{opt}	1.11

A 16×16 mesh size is used for $\omega = 1.0$ and $\omega = \omega_{opt}$. We see the difference in the timing results between the point method and the three implementations of the four-point block method, especially the first and third strategies, which take less time than the PA method. This time discrepancy is due to the computational work carried out by each processor, which in fact is different from one implementation to another. This amount of work is calculated and compared with the equivalent part of the synchronous version, which possesses significant overhead.

REFERENCES

Barlow, R. H., and Evans, D. J. (1982). Parallel algorithms for the iterative solution of partial differential equations. Computer Journal, Vol. 25, pp. 56–60.

Baudet, G. M. (1978a). The design and analysis of algorithms for asynchronous multiprocessors. Ph.D. Thesis, Carnegie-Mellon University, Dept. of Computer Science, April 1978 (Univ. Microfilm International, 300 N. Zeeb Rd., Ann Arbor, MI 48106, Ref. 78-15,195).

Baudet, G. M. (1978b). Asynchronous iterative methods for multiprocessors. J. ACM, Vol. 25, pp. 226–244.

Biggins, M. J. (1980). The numerical solution of elliptic partial differential equations by finite-difference methods. Ph.D. Thesis, Computer Studies Dept., Loughborough University.

Chazan, D., and Miranker, W. (1969). Chaotic relaxation. In: *Linear Algebra and Its Applications*, Vol. 2, pp. 199–222.

Dunbar, R. C. (1978). Analysis and design of parallel algorithms. Ph.D. Thesis, Computer Studies Department, Loughborough University.

Enslow, P. H., Ed. (1974). *Multiprocessors and Parallel Processing*. John Wiley and Sons, New York.

Evans, D. J., and Biggins, M. (1982). The solution of elliptic partial differential equations by a new block over-relaxation technique. International J. Comp. Math., Vol. 10, pp. 269–282.

Kung, H. T. (1976). Synchronized and asynchronous parallel algorithms for multiprocessors. In: J. Traub, Ed. Proc. of the Symposium on Complexity of Sequential and Parallel Numerical Algorithms, Academic Press, New York, pp. 49–82.

Kung, H. T. (1980). *The Structure of Parallel Algorithms*. Advances in Computers, Vol. 19. Academic Press, New York, pp. 65–112.

Neptune Programming Manual, 1982. Dept. of Comp. Stud., Loughborough University.

Parter, S. V. (1981). Block iterative methods. In: M. Schultz, Ed. *Elliptic Problem Solvers*. Academic Press, New York, pp. 375–382.

Rosenfeld, J. L., and Driscoll, G. C. (1969). Solution of the Dirichlet problem on a stimulated parallel processing system. Information Processing 68, North-Holland, Amsterdam, pp. 499–507.

Varga, R. S. (1962). *Matrix Iterative Analysis*. Prentice-Hall, Englewood Cliffs, N.J.

Young, D. (1954). Iterative methods for solving partial differential equations of elliptic type. Trans. Amer. Math. Soc., Vol. 76, pp. 92–111.

Young, D. (1971). *Iterative Solution of Large Linear Systems.* Academic Press, New York, 1971.

Young, D., and Hageman, L. H. (1981). *Applied Iterative Methods.* Academic Press, New York.

Yousif, N. Y. (1983). Parallel algorithms for asynchronous multiprocessors. Ph.D. Thesis, Loughborough University.

5
Parallel Execution of Ordinary Sequential Programs

Yoichi Muraoka and Hayato Yamana   *Waseda University, Tokyo, Japan*

1 INTRODUCTION

There are two approaches to implementing parallel processing. The first and obvious approach is to develop completely new programs based on new algorithms that are suitable for parallel processing. This approach seems very attractive because it is guaranteed to result in very good programs in terms of both execution speed and efficiency. However, it is generally very difficult, even for an expert in the field, to write a program for a parallel processing computer. There are several reasons for this:

1. Difficulty in exploiting parallelism: It is not an easy task for a programmer or a numerical analyst to divide a program into a set of tasks that are executable in parallel.
2. Difficulty in utilizing processors efficiently.

In writing a program for a parallel processing computer, we must keep the following two factors in mind:

1. The processor load balance: It is most desirable to balance the load of processors evenly so that the least execution time results.
2. The data communication time between processors: Processes should be allocated to processors so that the time to communicate data between processors is minimized.

To circumvent these problems an obvious answer is to let a compiler take care of these problems and free users from the burden.

Next let us look at the problem from a different point of view. As we know, most existing numerical algorithms are developed to minimize the computation time, that is, the number of operations required. To do so we must minimize not only the number of operations that appear explicitly in a program but also the number of of iterations involved in a relaxation process. As a very simple example let us look at a polynomial computation. The polynomial

$$p_n(x) = a_0 + a_1 x + a_2 x^2 + a_3 x^3 + \cdots + a_n x^n$$

requires $0(n^2)$ operations if we compute it as it is, but Horner's rule,

$$p_n'(x) = a_0 + x(a_1 + x\{a_2 + x[a_3 + \cdots + x(a_n - 1 + a_n x)\cdots)$$

allows us to minimize the number of operations required to $0(n)$. This is commonly used in an ordinary sequential program. If there are as many processors as required, however, the situation changes.

Because Horner's rule requires $0(n)$ steps regardless of the number of processors available, an original form of $p_n(x)$ now can be computed in $0 (\log_2 n)$ steps. In general, existing programs are written to minimize the number of operations on a sequential computer. They may not execute efficiently on a parallel processing computer, as discussed earlier. Of course we may ask people to start writing completely new programs suitable for a parallel processing computer based on new parallel numerical algorithms. Another approach that is attractive to users, however, is to let a compiler expose and exploit parallelism from an existing or ordinary program. To date, compilers for a vector computer have been developed and are being used extensively. These compilers are not powerful enough to apply to a parallel processing computer, however. In a vector computer parallel computation is done only in terms of vector data. Scalar operations cannot be computed in parallel even if they are independent of each other. Similarly, independent tasks cannot be computed in parallel. A parallel processing computer, on the other hand, has more freedom. Any independent operations may be computed simultaneously. Hence, although it is sufficient for a vector computer compiler to check only whether operations on elements of a vector can be performed simultaneously, a parallel processing computer compiler must do more. It must extract and expose more parallelism. In this chapter we introduce compiler algorithms and a parallel computer system to attain this goal.

Components of a usual program can be divided into the following three groups:

1. FOR loop
2. A block of assignment statements
3. IF statements

The descriptions here are given in reference to these groups.

2 FOR LOOPS

Obviously, the most promising candidate for parallel processing is a FOR loop. For example, in the loop

FOR I = 1 TO 100

A(I) = B(I) + C(I)

100 statements,

$$A(I) = B(I) + C(I) \qquad I = 1,2,\ldots , 100$$

can be computed simultaneously.

In general, statements in a program may be executed in any order other than the given order as long as they produce the same results as when they are executed in accordance with the given sequence. In principle, two statements S_1 and S_2 (originally written in a program in this order) may be computed simultaneously if the following two conditions hold:

1. S_2 does not use the result of S_1.
2. S_2 does not update a variable that is an input to S_1.

For example,

$$S_1: A = B + C$$
$$S_2: D = A + E$$

violates the first condition and

$$S_1: A = B + C$$
$$S_2: B = D + E$$

violates the second condition. Hence in both examples S_1 and S_2 may not be computed simultaneously. The necessity of the first condition is obvious. A statement may not be computed before its input data become available. The second condition assures that a variable is not updated before its value is used. This condition may be circumvented, however, by modifying S_2 as follows:

$$S_1: A = B + C$$
$$S_2: BT = D + E$$

In the statements that follow S_2, all occurrences of B must also be changed to BT. Hence only the first condition must be examined to check whether two statements may be computed simultaneously. In the case of loop

FOR I = 1 TO N

S(I);

the different iterations $S(h)$ and $S(k)$ ($h < k$) may be computed simultaneously if $S(k)$ does not use the result of $S(h)$. If this condition holds for all pairs of h and

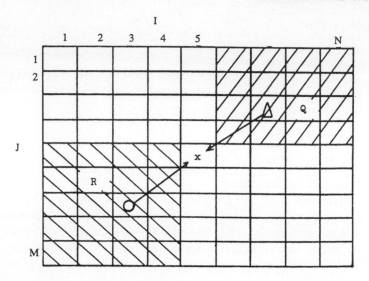

Figure 1   Condition of Parallel Computation in a loop.

k, then $S(I)$ may be compuated simultaneously for all values of I (I = 1,2,3, . . . , N).

    FOR I = 1 TO N
    FOR J = 1 TO M
    S (I,J);.

The sequential execution order of the loop may be explained as Figure 1. In this figure each square of the $N \times M$ two-dimensional plane represents the computation $S(k, h)$ for some $I = k$ and $J = h$. The computation of statements $S(I, J)$ proceeds from the leftmost column to the rightmost column, and in a column the computation proceeds from the top to the bottom sequentially. On the other hand, if the loop is executed for all values of I simultaneously, then we proceed to compute from the top row to the bottom row while we compute for all elements in each row simultaneously. For this execution to be valid we must make sure that the computation $S(k, h)$ (marked x in the figure) does not receive data to be updated by the computation in region R. Similarly, if the loop is computed for all values of J simultaneously, then we proceed to compute from the right to the left column while we perform computation in each row simultaneously. In this case we must make sure that the computation $S(k, h)$ does not receive data to be updated by computation in region Q. This is the basic algo-

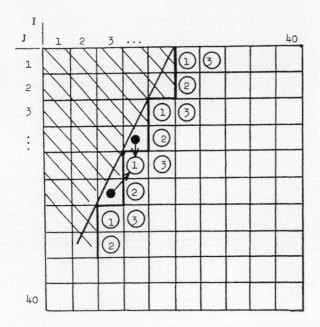

Figure 2   Wavefront.

rithm being used in compilers for a vector computer. As we mentioned earlier, we must go further and extract more parallelism from a program to fully utilize the power of a parallel processing computer. Let us now introduce several new techniques to attain this goal.

Wavefront Method.   Let us consider the loop

FOR I = 1 TO 40
FOR J = 1 TO 40
S(I,J): A(I,J) = A(I−1,J+2) + A(I,J−1);

The dependence relation between different iterations is illustrated in Figure 2. From this figure it is clear that the loop cannot be computed simultaneously for all values of I, nor can it be computed simultaneously for all values of J.

Now suppose that all $S(I, J)$ in the shaded area in the figure have been computed. Then at the next step those $S(I, J)$ marked 1 can be computed simultaneously, and so forth. We can see that a heavy zigzag line travels from left to right like a wavefront, indicating that all statements on the front can be computed simultaneously. Note that this kind of parallel computation is not adequate for a vector computer because only a column or a row sweep is appropriate for that type of the computer. Also, note that computation of the loop by this scheme takes approximately 120 steps, but if the loop is computed sequentially it takes $40 \times 40 = 1600$ steps.

This wavefront is such that all communications corresponding to points (I, J) next to the wavefront can be computed simultaneously.

Simultaneous Execution of Statements in a Loop.   Let us look at the loop

FOR $I = 1$ TO 40
BEGIN
S1: $A(I) = F(A(I-1),B(I-2))$
S2: $B(I) = G(A(I))$;
END

In this example S_1 and S_2 cannot be computed in parallel for all values of I. However, S_1 and S_2 can be computed simultaneously when I varies sequentially if the index expression in S_2 is skewed as follows:

FOR $I = 1$ to 40
BEGIN
S1: $A(I) = F(A(I-1),B(I-2))$;
S2: $B(I-1) = G(A(I-1))$;
END

Figure 3 illustrates the computation of the modified loop as well as the original loop.

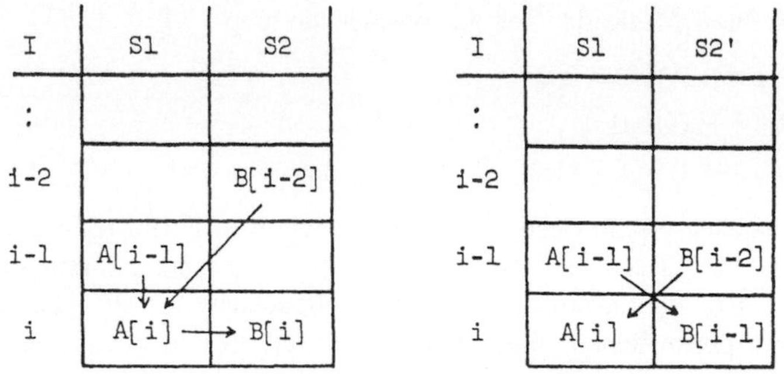

Figure 3   Simultaneous execution of body statements.

Partial Parallel Execution of a Loop.   Let us look at the loop

FOR I = 1 TO 100
BEGIN
S1: B(I) = A(2*I + 500);
S2: A(7*I + 35) = C(I)*C(I + 1);
END

Here, for $I < 95$, the computation of S_1 does not use a result of S_2. Hence we can compute S_1 simultaneously for all values of I ($0 < I < 96$) and then compute S_2 simultaneously. However, since to compute S_1 at $I = 100$ [$A(2 \times 100 + 500) = A(700)$] we need the result of S_2 at $I = 95$ [$A(7 \times 95 + 35) = A(700)$], S_2 must be computed before S_1. Therefore, for $I > 94$ the loop must be computed sequentially. In general, part of a loop may be computed in parallel as follows. Let us use the following simple loop for the sake of explanation:

FOR I = 1 TO n
BEGIN
S1: . . . = V(A1*I + C1) . . .
S2; V(A2*I + C2) = . . .
END

There are four possibilities for the relation between

$$F_1(I) = A_1 I + C_1$$

and

$$F_1(I) = A_2 I + C_2$$

Case 1

S_1 never uses the result of S_2. Computation of S_1 can then precede computation of S_2 for all values of I. Hence S_1 can be computed simultaneously for all values of I, and then similarly S_2 can be computed simultaneously for all values of I.

Case 2

The relation $F_1(h) = F_2(k)$, $h > k$, holds for all pairs (h, k). Let P be a set of these pairs. For pairs in P, the computation of S_2 must precede the computation of S_1. Hence the loop must be computed sequentially.

Case 3

For $h < k$ where $F_1(h) = F_2(k)$, we apply case 2. Otherwise we apply case 1.

Case 4

For $h > k$ where $F_1(h) = F_2(k)$, we apply case 2. Otherwise we apply case 1.

3 ASSIGNMENT STATEMENTS

Consider the assignment statement

$$X = A + B + C + D + E + F + G + H$$

and its syntactic tree. The tree is such that operations on the same level may be done in parallel. The height of a tree is the maximum level of the tree and indicates the number of steps required to compute an assignment statement in parallel. Note that there may be many different syntactic trees for an assignment statement, and among them the tree with the minimum height should be chosen to attain the minimum parallel computation time.

For the sake of argument let us assume that an assignment statement consists of additions, multiplications, and possibly parentheses. The associative, the commutative, and the distributive laws then hold. For example, the associative law allows one to compute the assignment statement $X = A + B + C + D$ as $X = [(A + B) + (C + D)]$ in two steps rather than as $X = \{[(A + B) + C] + D\}$, which requires three steps. Also, if the commutative law together with the associative law is applied, we may be able to further lower the tree height. For example, $X = [(A + B \times C) + D]$ requires three steps and $X = [B \times C + (A + D)]$ requires two. Now we turn our attention to the third law, that is, the distributive law, and see if it can help to lower the tree height. As we can readily see there are cases when distribution helps. For example, $X = A \times (B \times C \times D + E)$ requires four steps but its equivalent, $X = A \times B \times C \times D + A \times E$, which is obtained by distributing A over $B \times C \times D + E$, can be computed in three steps. However, distribution does not necessarily always speed the computations. For example, the undistributed form $X = A \times B \times (C + D)$ can be computed in fewer steps than the distributed form $X = A \times B \times C + A \times B \times D$. Hence nondiscriminative distribution is not the solution to the problem, and we need an algorithm to guide about us how and when to distribute and build a tree to attain the minimum computation time. We call this algorithm the distribution algorithm.

Given an assignment statement A, the distribution algorithm derives the assignment statement A^d by distributing multiplications over additions properly so that the height of A^d (denoted $h[A^d]$) is minimized. The algorithm works from the innermost parenthesis level to the outermost parenthesis level of an assignment statement and requires only one scan through the entire assignment statement. In what follows we assume that additions and multiplications require the same amount of time, that is, one unit of time. Small (lowercase) letters possibly with subscripts, denote single variables. The letter t, possibly with subscripts, denotes arbitrary arithmetic expressions, including single variables. We also write $h[A]$ for the height of a tree of A, and $T[A]$ for a tree of A. The logarithmic base in this chapter is 2, that is, $\log n = \log_2 n$.

The minimum height of tree using only the associative and commutative laws can be built as follows. Let us assume that either $A = \Sigma t_i$ or $A = \Pi(t_i)$ and that for each i a minimum height of tree $T[t_i]$ has been built. Then first we choose two trees, say $T[t_l]$ and $T[t_p]$, each with a height smaller than the height of any other tree. We combine these two trees and replace them by a new tree whose height is one higher than the maximum $\{h[t_q], h[t_p]\}$. This procedure is repeated from the innermost parenthesis level to the outermost parenthesis level, and at each level it is repeated until all trees are combined into one tree.

The effective length e of an arithmetic expression A is defined as $e[A] = 2^h[A]$. The height of a minimum height of tree can be obtained as follows.

Theorem 1

1. If

$$A = \prod_{i=1}^{p} a_i \quad \text{or} \quad A = \sum_{i=1}^{p} a_i,$$

then

$$h[A] = \log p$$

2. If

$$A = \sum_{i=1}^{p} t_i$$

then

$$h[A] = \log \left\{ \sum_{i=1}^{p} e[t_i] \right\}_2$$

3. If

$$A = \prod_{i=1}^{p} a_i^* \prod_{j=1}^{q} (tj)$$

then

$$h[A] = \log \left\{ \{p\}_2 + \sum_{j=1}^{q} e[t_j] \right\}_2$$

where $\{x\}_2$ is the smallest integer of power 2 which is greater than or equal to x.

Given an arithmetic expression A to obtain the height of a tree for a, theorem 1 is applied from the innermost parts of A to the outermost parts, recursively.

Now let us turn our attention to the main problem, that is, when and how one should distribute multiplications over additions. In an arithmetic expression there are four possible ways of parenthesis occurrence:

$P1$: $\cdots + (A) + \cdots$
P_2: $\cdots \# (t_1 \times t_2 \times \cdots \times t_n) \times (t_1' \times t_2' \times \cdots \times t_m') \# \cdots$
P_3: $\cdots \# a_1 \times a_2 \times \cdots a_n \times (A) \# \cdots$
P_4: $\cdots \# (t_1 + t_2 + \cdots + t_n) \times (t_1' + t_2' + \cdots + t_m') \# \cdots$

where $\#$ is either addition or multiplication or no operation.

From careful inspection we can conclude that distribution in case P_3 and partial distribution in case P_4 are the only cases that should be considered for lowering tree height. In cases P_1 and P_2 removal of the parentheses leads to a better result or at least gives the same tree height. Full distribution in case P_4 always increases tree height and should not be done. Also it should be clear that in any case the tree height of an arithmetic expression cannot be lower that that of a component term even after distribution is done. This assures that evaluation of distributions can be done locally. That is, if a distribution increases the tree height for a term, then that distribution should not be done because once the tree height is increased, it can never be remedied by further distributions.

In fact, there are two cases when distribution is beneficial. We explain these two cases informally by giving an example for each case. An example for the first case is as follows. Let $A = a(bcd + e)$. Then $h[A] = 4$. However, if we distribute a over $bcd + e$, then we get $A^d = abcd + ae$ and $h[A^d] = 3$. In this case we balance a tree by filling the ''holes'' in the tree because a balanced tree can accommodate the largest number of variables among trees of equal height. An example for the second case is as follows. Let $A = a(bc + d) + e$ and $A^d = abc + ad + e = t^d + e$. In this case, $h[A] = 4$ but $h[A^d] = h[t^d] = 3$. What happened here is that t is ''opened'' by distributing a over $bc + d$ and the ''space'' to place e is created.

At each level of parenthesis pair, for cases P_3 and P_4, instances of holes and spaces are checked and proper distribution is performed.

Next let us find out how much we can lower the height of a tree. First we examine the case in which only the associative and commutative laws are applied.

Theorem 2

$h[A]$ is less than or equal to $1 + 2d + \log n$, where n is the number of single variable occurrences in an arithmetic expression and d is the maximum parenthesis nesting level.

If an arithmetic expression is a polynomial of Horner's rule, then the second term becomes dominant and we obtain $h[A] = 1 + 2d$. Similarly, if an arithmetic expression is of a form $a + b + c + \cdots$, then we obtain $h[A] = 1 + \log n$.

How, then, does the application of the distributive law affect tree height? The following theorem gives an answer.

Theorem 3

By applying the associative, commutative, and distributive laws, the height of a tree can be reduced as much as $5 \log n$.

4 BACKSUBSTITUTION

The tree height reduction algorithm introduced in the previous section treats a single assignment statement, but it can be used for a set of assignment statements. Given a block of assignment statements we can rewrite the block with one assignment statement by substitution of assignment statements into one another. For example, the assignment statements

$$A = B + C$$
$$D = EF$$
$$G = A + C$$
$$h = A + GD$$

can be rewritten as

$$h = (B + C) + [(B + C) + C]EF$$

At this point the tree height reduction algorithm can be applied to the resultant statement. In this example five steps are required if each statement is computed independently, but only four steps are required to compute the backsubstituted statement. Backsubstitution plays a key role in speeding the computation of a sequential loop, that is, an iteration. To compute the iteration

$$y_i = f(y_i - 1)$$

or

FOR i = 1 TO n

$$Y(I) = F(Y(I-1));$$

we must usually repeat the computation sequentially for each index value. Assume that we are interested only in the value of y_n. Then instead of computing n statements sequentially, that is,

$$y_1 = f(y_0)$$
$$y_2 = f(y_1)$$
$$\vdots$$
$$y_n = f(y_{n-1})$$

we may obtain one statement for y_n by backsubstitution. For example, let

$$y_i = a_{i-1}y_{i-1}$$

Then

$$y_n = a_{n-1}y_{n-1}$$
$$= a_{n-1}(a_{n-2}y_{n-2})$$
$$\vdots$$
$$= a_{n-1}a_{n-2}\cdots a_2a_1a_0y_0$$

Then instead of computing each y_i repeatedly for $i = 1, 2, \ldots, n$, we compute the backsubstituted y_n directly after the height of a tree is reduced. Table 1 summarizes the results for some primitive yet typical iteration formulas. Note that on a vector computer an iteration is usually treated as sequential whereas on a parallel processing computer it can be computed in parallel as described here.

5   IF STATEMENT

A program written in a conventional language tends to exhibit little parallelism because of many control dependencies due to IF statements. To select an appropriate path that follows an IF statement, one must wait until the condition part of an IF statement is evaluated. To circumvent the problem and speed the computation, we propose a scheme to compute all possible alternatives that follow an IF statement in parallel even before the condition part of an IF statement is evaluated and select the right one when the condition part is evaluated. In this section a scheme called an unfolding scheme, which restructures a program to find all possible alternative paths following an IF statement, is proposed to expose parallelism.

First the unfolding scheme for codes including only forward branches is described. Then the scheme for the codes including backward branches as well as forward branches is described.

Table 1 The Computation Time for Reccurrence Equations

y_i	y_n^b	T_s	nT_s	T_p
ay_{i-1}	$a^n y_0$	1	n	$\lceil \log_2(n+1) \rceil$
$y_{i-1} + b$	$y_0 + \overbrace{b + \cdots + b}^{n}$ $\ \ n-1$	1	n	$\lceil \log_2(n+1) \rceil$
$a_{i-1}y_{i-1}$	$\prod_{k=0}^{n-1} a_k y_0$	1	n	$\lceil \log_2 n \rceil$
$y_{i-1} + a_{i-1}$	$\sum_{k=0}^{n-1} a_k + y_0$	1	n	$\lceil \log_2 n \rceil$
$ay_{i-1} + b$	$a^n y_0 + p_{n-1}(a)^*$	2	$2n$	$\approx 2\lceil \log_2(n+1) \rceil$
$ay_{i-1} + x_{i-1}$	$p'_n(a)^{**}$	2	$2n$	$\approx 2\lceil \log_2(n+1) \rceil$
$y_{i-1} + bx_{i-1}$	$\sum_{k=0}^{n-1} bx_k + y_0$	2	$2n$	$\approx \lceil \log_2 n + 1 \rceil$
$ay_{i-1} + bx_{i-1}$	$p''_n(a)^{***}$	3	$3n$	$\approx 2\lceil \log_2(n+1) \rceil$

$^*p_{n-1}(a) = ba^{n-1} + ba^{n-2} + \cdots + b$
$^{**}p_n'(a) = a^n y_0 + a^{n-1}x_0 + a^{n-2}x_1 + \cdots + ax_{n-2} + x_{n-1}$
$^{***}p_n''(a) = a^n y_0 + ba^{n-1}x_0 + ba^{n-2}x_1 + \cdots + bax_{n-2} + bx_{n-1}$
T_s: The time required to compute y_i in parallel, i.e., $h[y_i]$.
T_p: The time required to compute y_n^b in parallel, i.e., $h[y_n^b]$.

The first step to unfold a program is downward unfolding, which restructures a program to remove branches. We call this step "downward unfolding". To downward unfold we simply copy codes following the point at which control flow merges. An example is shown in Figure 4. In Figure 4a, control flows join at BB4 and BB8. As the result of downward unfolding the blocks following BB4 and BB8 are copied as shown in Figure 4b. Now the execution of all paths (we call them threads) is initiated in parallel. The executed path is then terminated as soon as it becomes clear that the execution of the path is not valid.

Next we propose the unfolding scheme for a program containing backward branches as well. A typical example is a sequential FOR loop. In many cases the number of iterations of a loop is unknown at the time of compiling. Thus it is impossible to unfold a loop into a set of threads corresponding to all possible paths. Moreover, the number of threads that can be initiated may be limited because of the limited number of processors. To solve these problems the following strategy can be applied.

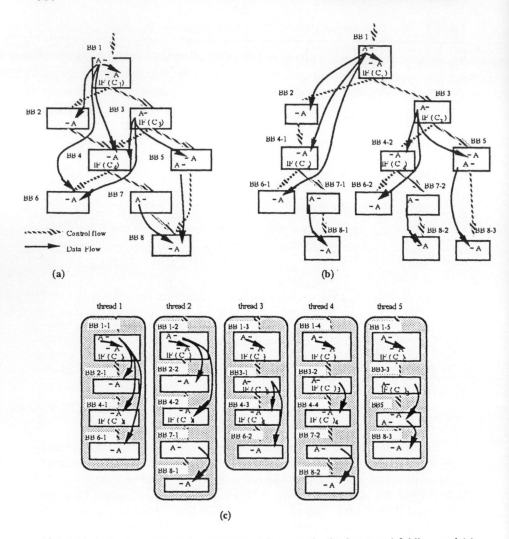

Figure 4 Creation of threads: (a) Original flow graph, (b) downward folding, and (c) upward folding.

1. A perfect IF-THEN-ELSE statement is not unfolded because only a small amount of parallelism is exposed if such a statement is unfolded.
2. A predefined number of threads is generated for an unfolded iteration. A set of threads is initiated again when the previous execution has finished to continue further iteration. A set of threads that has been unfolded is initiated

simultaneously when control flow reaches it. Each thread evaluates the predicate (the condition part of an IF statement) by itself whether it is valid or not.

The effect of this scheme was evaluated using typical linear algebraic codes. The result is shown in Figure 5. This figure shows the speedup ratio of the execution time using the unfolding scheme in comparison with that without using the scheme. Since the execution time depends on the selected path, the speedup ratio in the figure is the average of all paths. As shown in Figure 5, the average speedup ratio is 1.5 when the codes include only forward branches and is 2.7 when they include backward branches in addition to forward branches. These results show that the unfolding scheme is effective in speeding the execution of programs including branches. In Figure 5, KENO-BE(1), (2), and (3) are the names of programs being used in the experiment.

6 THE HARRAY SYSTEM

The Harray system being developed by the author's group is a parallel processing system that implements the aforementioned theories. The characteristics of the Harray system are as follows.

The main objective of the Harray system is to execute existing scientific calculation programs as fast as possible in parallel.

The Harray system is a shared-memory array type of machine with 1024 processors (PEs). These 1024 processors can work independently in parallel. The data flow control mechanism is used in a processor to execute a tree of an arithmetic expression in parallel (Fig. 6).

A parallel compiler is provided that restructures a program as described earlier. The compiler first divides a program into a set of blocks (called macroblocks). A block is basically a jump-free set of assignment statements or a FOR loop. The compiler then generates a flow graph (global control graph) that describes the dependence relation between blocks; that is, if a block b uses a result of block a, then b depends on a. Blocks are assigned to processors. If a block is a FOR loop that can be computed in parallel, multiple numbers of processors are allocated to the block. Execution of a block is initiated in a processor as soon as the execution of its preceding blocks has finished. This execution mechanism, which governs the execution of a program, is called the control data (CD) flow (Fig. 7).

The CD flow mechanism has been proposed to overcome the shortcomings of the data flow mechanism. Even though the data flow mechanism is suitable to expose the parallelism in a program naturally, it tends to consume a large amount of resources, such as the size of a memory unit for packet storage. Since a program is divided into macroblocks and data flow execution is done in terms

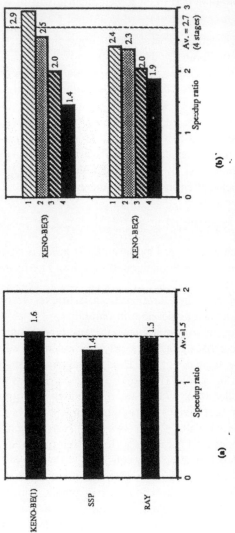

Figure 5 Speedup ratio of the execution time using an unfolded scheme in comparison with that without using it. Unfolded codes containing (a) only forward branches and (b) forward and backward branches.

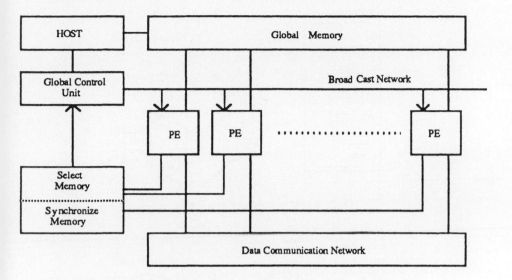

Figure 6 Organization of the Harray system.

of these macroblocks, the size of the memory unit for packet storage can be considerably reduced.

In the CD flow mechanism a FORTRAN program is divided into a set of macroblocks as described. A macroblock is a set of statements, including only one entry point and many exit points of control dependency. A macroblock is a unit whose execution is controlled by the control flow mechanism. Operations in a macroblock are executed by the data flow mechanism; the execution order of macroblocks is controlled by the control flow mechanism. The CD flow mechanism enables a macroblock-by-macroblock controllable execution. The execution of a macroblock is initiated as follows. The major factor limiting the parallelism in a program is dependency. Conceptually, every operation has two types of incoming dependencies: data dependency, which determines when the operation can be initiated, and control dependency from the operation that computes the predicate, that is, the condition part of an IF statement. To allow a maximum overlap of the execution of macroblocks we adopt a scheme called the preceding activation scheme to initiate the execution of a macroblock. A macroblock is initiated whenever idle processors exist even if the control does not reach it. The execution is suspended only when data from the preceding macroblocks are not yet available. It resumes execution as soon as they become available. The unfolding scheme is also adopted to lessen the effect of the control flow dependency.

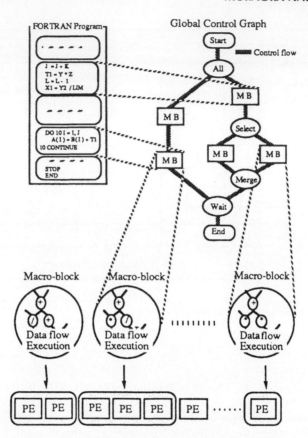

Figure 7   The controlled dataflow (CD flow) mechanism.

The Harray system consists of six kinds of units, as shown in Figure 6. The global control unit (GCU) is the central controller that controls the execution order of blocks assigned to processors according to the global control graph. The synchroniztion memory unit (SYM) synchronizes the completion of execution in the PEs. In the PEs, a block is executed in the data flow mechanism. Packets consisting of data and their identifier, called a tag, circulate in the PEs. Circulating packets are handled by units, such as a matching unit, an arithmetic unit, and a packet management unit.

Two facilities are provided for communication between PEs to enhance the communication bandwidth. one is the global memory (GM) unit and another is the data communication network (DCN). The GM is a shared-memory unit in which the PEs write data and other PEs read. PEs communicate directly through the DCN by sending packets. In general, structured data are stored in the GM and scalar data are sent directly as packets through the DCN.

A set of blocks that follows an IF statement is unfolded into multiple threads, and they are assigned to separate processors. These threads are executed in parallel.

A very large scale integrated chip for a processor is now being designed, and a compiler is under development.

7 CONCLUSION

In this chapter we presented algorithms to compute existing programs as fast as possible, and also the possible speedup. Furthermore, we presented an outline of the Harray system, which implements the algorithms. Because many parallel processing systems, such as Cedar and hypercube, are emerging, it is absolutely necessary to provide users with a parallel compiler that allows them to write programs with a language with which they are familiar. To this end we believe that the studies described in this chapter will be greatly appreciated.

REFERENCES

Muraoka, Y., and Yamana, H. (1989). Major research activities in parallel processing in Japan and an example—Harray. International Journal of High Speed Computing, Vol. 1, No. 1, pp. 185–206.

Yamana, H., Marushima, T., Hagiwara, T., and Muraoka, Y. (1988). System architecture of parallel processing system Harray. Proc 1988 International Conference on Supercomputing, Sant Marlo, pp. 112–125.

Yamana, H., Hagiwara, T., Kodata, J., and Muraoka, Y. (1989). A preceding activation scheme with graph unfolding for the parallel processing system Harray. Proc. of Supercomputing '89, Reno, pp. 89–98.

6
Parallel Simulation
of Discrete Systems

Richard M. Fujimoto *Georgia Institute of Technology, Atlanta, Georgia*

1 WHAT IS PARALLEL DISCRETE EVENT SIMULATION?

This chapter surveys the state of the art in executing discrete event simulation programs on a parallel computer. We first discuss fundamental concepts used in discrete event simulation and examine why this application is so difficult to parallelize. We then discuss several simulation strategies that have been proposed, as well as the ideas on which they are based. Finally, we survey recent performance results for these strategies and critique existing approaches to clarify their respective strengths and weaknesses.

A discrete event simulation is a computer model for a physical system in which the state of the system is viewed as changing at *discrete* points in simulated time. The simulation model "jumps" from one state to another, not unlike a cartoon in which the picture depicting the state of the cartoon's characters jumps from one frame to the next.

The simulation activity associated with producing the next "cartoon frame" is called an *event*. Events typically denote the occurrence of something "interesting" in the simulation model, for example, beginning the transmission of a message over a communication link in the simulation of a computer network or the destruction of a tank in simulated warfare. Each event has a time stamp associated with it to denote the point in simulated time at which the event occurs. The entire simulation computation consists of a sequence of event computations.

151

An event computation may do two things: (1) modify the state variables of the simulation to reflect the new state of the system after the event has occurred, and (2) schedule zero or more additional events into the simulated time future. Scheduling events into the simulated time future is one way in which cause-and-effect relationships are modeled by the simulator. For example, in the communication network simulation the event denoting the beginning of message transmission over a previously idle link might set a flag to indicate the link is now busy and schedule a new event (say) 5 s into the simulated future to denote when transmission will be complete (at which time the state of the link will again become idle if there are no other messages waiting to use it).

Parallel discrete event simulation (PDES) refers to the execution of a single discrete event simulation program on a parallel computer. PDES has attracted a considerable amount of interest in recent years. From a pragmatic standpoint this interest arises from the fact that large simulations in engineering, computer science, economics, and military applications (to mention a few) consume enormous amounts of time on sequential machines. It is not unusual for a single detailed simulation run to require hours or days of central processing unit (CPU) time on mainframe computers. Simulationists are frequently forced to simplify their simulation models to reduce the amount of time required to execute them.

Here we focus attention on discrete event simulations of *asynchronous* systems, that is, systems that are not synchronized by a global clock. In this case time-stepped methods based on the concurrent execution of events at the same time step are either inefficient because there are too few events occurring within a single time step or inaccurate because the fidelity of the simulation must be compromised by making the time step sufficiently large to include many events. Instead we exploit the concurrent execution of events at *different* points in simulated time. As we will see, this introduces interesting synchronization problems that are at the heart of the PDES problem.

Parallelization of discrete event simulation problems contrasts sharply with parallelization of *continuous* simulation models. The latter views the state of the system as changing continually over time and usually involves solution of a set of differential equations. Continuous simulations typically require a substantial amount of floating-point computations on large matrices of data, but discrete event simulations usually involve scalar computation on irregularly structured data. It is the irregular, data-dependent nature of the discrete event simulation problem that has made it one of the most difficult applications to parallelize on existing parallel computers.

1.1 Process-Oriented Simulation

Existing PDES strategies assume that a process-oriented simulation methodology is used. The system being modeled is assumed to be comprised of some number of *physical processes* that interact at various points in simulated time. The simulator is constructed as a set of *logical processes* LP_0, LP_1, . . . , LP_{N-1}, one per physical process. Interactions between physical processes are modeled by time-stamped event messages sent between the corresponding logical processes. Each logical process contains a portion of the state corresponding to the physical process it models. A crucial restriction of the algorithms discussed here is that logical processes may *not* have direct access to shared state variables (exceptions are described by Jones, 1986, and Fujimoto, 1989, however).

1.2 Examples

Consider a simulation of air traffic in the United States. Here the physical system consists of a collection of airports that interact by "sending" aircraft among each other. An actual air traffic system would also include radio transmission among aircraft and airports, but we ignore this aspect of the system to simplify the discussion.

Using the process-oriented simulation paradigm, each airport is modeled by a logical process and aircraft are modeled by event messages sent between airport processes. For example, the LAX process modeling the Los Angeles airport might send an event message with time stamp 10:00 to the SFO process (modeling the airport in San Francisco) to denote that an aircraft now departing from LAX will arrive at SFO at 10:00 A.M. Upon receiving this event message the SFO process then schedules new events for itself to denote the landing of the aircraft, routing the aircraft to a gate, and so on. Eventually the SFO process sends a message to another airport process to denote the arrival of the airplane at its next destination.

As a second example, consider the simulation of the communication system for a message-passing multicomputer, such as a hypercube-based machine like the Intel iPSC. Messages are routed through the network, hop by hop, using some routing algorithm. Simulations might be used to evaluate the performance of various network designs.

The communication network simulation also maps very naturally to logical processes that communicate by exchanging time-stamped event messages. A separate logical process can be defined to model each node in the hypercube, and event messages correspond to messages transmitted between hypercube nodes. Specifically one might use two types of event messages: an arrival event indicates that a message has been received (in its entirety) on one of the node's input links, and a departure event indicates that a message has been transmitted over one of the node's output links. To simplify this example we assume an unbounded number of buffers is available. As shown in Figure 1a, we assume each of the node's *output* ports contains a queue of message buffers to hold messages waiting to be transmitted on the corresponding link. We further assume that messages are transmitted over the output link in the order in which they were received; that is, a first-come–first-served protocol is used.

The operation of the simulator in modeling a message passing through a node of the cube is depicted in Figure 1. In Figure 1b an arrival event indicates that the message has been received at some input port at simulated time T. Parameters within the event indicate the port on which the message is arriving, the size of the message, the eventual destination, and so on. The simulation code responsible for processing this arrival event examines the destination address and selects the output port on which the message is to be forwarded. It allocates a buffer on the selected output port, for example, by incrementing a counter that indicates the number of allocated buffers on that port.

Suppose the selected output link is idle (this scenario is depicted in Fig. 1b). The hypercube node immediately begins sending the message to the neighboring node. The simulator models this fact by marking the link busy and scheduling a departure event S units of simulated time in the future, where S is the amount of time required to send the message over the link. S depends on the size of the message and the rate at which data are transmitted over the link. Later, when this departure event is processed, an arrival event with time stamp $T + S$ is scheduled at the logical process corresponding to the hypercube node that will receive the message.

On the other hand, suppose the output link is busy transmitting another message when this message is received at simulated time T. In this case (Fig. 1c) the original arrival event does *not* schedule any new events. Let us assume the message remains in the queue Q units of simulated time before the link begins transmiting the message. This implies that a departure event for the message *preceding* this one in the queue will occur at simulated time $T + Q$. This departure event will schedule a new departure event to denote the departure (end of transmission) of this message at simulated time $T + Q + S$. When this latter departure event is processed it will schedule an arrival event at simulated time $T + Q + S$ at another logical process. If no other messages remain in the queue at time $T + Q + S$, the link is marked idle and no new events are scheduled. Otherwise another event is scheduled into the simulated future to denote the departure of the next message that is transmitted over the link.

The simulator continues processing and scheduling new arrival and departure events in this way until the simulation is completed. Other simulation activities include generating new event messages and removal of messages that have reached their final destinations. Also, code is associated with these events to collect statistics (e.g., computing the average length of message queues) as the simulation proceeds.

2 PARALLEL EXECUTION

At first glance, discrete event simulation using the process-oriented methodology seems to be an ideal candidate for parallel processing. The examples just de-

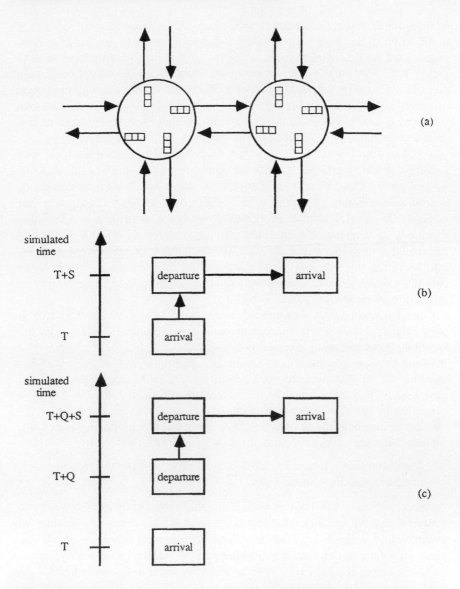

Figure 1   Simulator for a hypercube communication network. (a) Model for two nodes of the hypercube. (b) Scenario of events for a message being transmitted over a previously idle link. (c) Scenario of events when link is busy, necessitating queuing at the output port.

scribed, for example, are already mapped into parallel programs that consist of a collection of autonomous processes that communicate by exchanging messages. Further, one expects these examples to contain substantial amounts of parallelism. For instance, simulation of air travel along the East Coast could proceed concurrently with simulation of traffic in the West; similarly, simulation of messages in one part of the hypercube can proceed in parallel with traffic in other parts. Yet, effectively utilizing parallel computers to speed up large discrete event simulation problems has proven to be very challenging.

PDES is hard because the simulator must faithfully reproduce causal relationships in the system being modeled. Consider the air traffic simulator described earlier. Consider two simulator events, event E_{SFO} with time stamp 9:00 denoting the arrival of an aircraft at SFO at 9:00 A.M. and event E_{LAX} with time stamp 11:00 denoting another aircraft arriving at LAX at 11:00 A.M.. It is clear that the E_{LAX} computation cannot either directly or indirectly affect E_{SFO} because this would amount to a physical system in which the future affects the past! We informally call this rule that events cannot affect other events in the simulated past the *global causality constraint*. Violations of the causality constraint are referred to as *causality errors*.

In a sequential simulator adherence to global causality is straightforward: we need only process events in nondecreasing time stamp order. Operationally this is accomplished by sorting all pending events (those that have been scheduled but have not yet been processed) in a time stamp-ordered event list and by always selecting the smallest time-stamped pending event as the one to be processed next. Events may only schedule new events in the simulated time future, so adherence to global causality is guaranteed.

Let us now consider the parallel case. It can be shown that global causality is preserved if each logical process obeys the following local causality constraint:

> *Local causality constraint:* Each logical process must process events in nondecreasing time stamp order.

Adherence to this constraint is sufficient, although not always necessary, to guarantee that no causality errors occur. It is not always necessary because two events within a single LP may be independent of each other, in which case processing them out of time stamp sequence does not lead to causality errors.

Now comes the hard part. Operationally we must decide whether E_{SFO} at 9:00 can be executed concurrently with E_{LAX} at 11:00 without violating the local causality constraint. But how do we know whether E_{SFO} affects E_{LAX} without first performing the simulation for E_{SFO}? For example, upon arrival at SFO at 9:00, the aircraft might find that the airport is closed because of fog and may be rerouted to LAX! If this is the case, a new event E_{LAX2} appears at the LAX process for the rerouted aircraft that carries a time stamp of (say) 10:00 A.M. (assuming the flight time from SFO to LAX is 1 h). E_{LAX2} must be processed

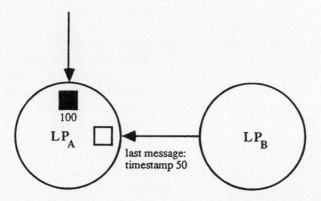

Figure 2 Process LP_A has a message waiting to be processed, but it must first wait to see if LP_B sends it a smaller time-stamped message.

before the original event (E_{LAX} at time 11:00) if we are to adhere to the local causality constraint. Therefore, if we execute E_{LAX} concurrently with E_{SFO} we violate the local causality constraint, compromising the correctness of the simulation.

This is the fundamental dilemma PDES strategies must address. The scenario in which E_{SFO} affects E_{LAX} can be very complex (e.g., the SFO-bound aircraft could be rerouted through several airports before finally arriving at LAX) and is critically dependent on the time stamp of events.

A similar problem can arise in the communication network simulator depicted in Figure 2. Let us assume messages are sent from one process to another in nondecreasing time stamp order. As shown in Figure 2, suppose logical process LP_A has received an arrival event message with time stamp 100 that is waiting to be processed. However, suppose no such message is pending from neighboring process LP_B and the last message received from LP_B contained a time stamp of 50. LP_A cannot process the pending event with time stamp 100 because a message with a time stamp smaller than 100 may be received later from LP_B, so LP_A must block. This blocking can lead to a deadlock situation: a cycle develops in which each process in the cycle is waiting to see if a neighboring process will send it a "small" time stamped message.

PDES is hard because the precedence constraints that dictate which computations must be executed before which others is, in general, quite complex and highly data dependent. This contrasts sharply with other areas in which parallel computation has had a great deal of success, such as vector operations on large matrices of data, where much is known about the structure of the computation at compile time. Thus it is not too surprising that a general solution to the parallel simulation problem has been elusive.

PDES mechanisms broadly fall into two categories: *conservative* and *optimistic*. *Conservative* approaches strictly *avoid* the possibility of any causality error ever occurring. These approaches rely on some strategy to determine when it is ''safe'' to process an event; that is, they must determine when all events that could affect the event in question have been processed. On the other hand, *optimistic* approaches use a *detection and recovery* approach: causality errors are detected, and a *rollback* mechanism is invoked to recover. We now describe some of the details and underlying concepts of several conservative and optimistic simulation mechanisms that have been proposed.

We assume that the simulation consists of N logical processes, LP_0 . . . LP_{N-1}. $Clock_i$ refers to the simulated time up to which LP_i has progressed: when an event is processed, the process clock is automatically advanced to the time stamp of that event. If LP_i may send a message to LP_j during the simulation, we say a *link* exists from LP_i to LP_j. In general, any number of links may exist from one logical process to another.

3   MACHINE ARCHITECTURES FOR PDES

Thus far most of the work in parallel discrete event simulation has focused on MIMD (multiple instruction–multiple data stream) computers. Most discrete event simulations contain a number of different types of events. Because it is highly desirable to be able to process different types of events concurrently, SIMD (single instruction–multiple data stream) computers have not been widely used, although they may be effective for simulations containing only a few different types of events.

Historically, parallel discrete event simulation has been identified as an application in which vectorization techniques using supercomputer hardware provide little benefit (Chandrak and Browne, 1983). This is because PDES problems typically do not contain floating-point operations on large matrices of data. For example, the communication network simulator described earlier does not use any matrix operations and may not require any floating-point operations at all, except perhaps for generating random numbers. Thus the high-speed (and high-cost) pipelined vector units of supercomputers, such as those manufactured by Cray and Fujitsu, provide little or no benefit. The parallel computers that are best suited for PDES problems are those that contain many very fast *scalar* processors. With the scalar performance of modern microprocessors rapidly approaching that of supercomputers, the suitability of expensive supercomputing hardware for this application is rather questionable.

Existing microprocessor-based parallel computers are either based on message passing (e.g., the Intel iPSC or NCube/10) or on shared memory (e.g.,

Sequent, BBN Butterfly, and Encore Multimax). Interestingly, PDES programs implemented on shared-memory machines usually use message-passing communications primitives that are built on top of shared memory; for example, sending a message is implemented by writing the contents of the message into a message buffer on another processor. A logical process's state variables cannot be directly shared by other logical processes, *even if the underlying hardware supports shared memory*. This is because each logical process is usually at a different point in simulated time. For example, the variables in LP_1 might correspond to the state of the process at time 100. This is of no value to LP_2 at time 200 because LP_2 is concerned only with the state of the system at time 200. As mentioned earlier, the logical process simulation approach used by all the algorithms proposed here prohibits the use of shared state variables.

Although shared memory cannot be used to directly reference state variables, it does have other benefits. In particular, message passing implemented using shared-memory multiprocessors is usually faster than communication in message-based machines, allowing finer grained parallelism to be exploited. Also, as is described later, various optimizations can be used if the underlying hardware supports shared memory. Further, some simulation algorithms utilize barrier synchronizations, so these are better suited to shared-memory machines. In general, however, message-based machines are about as well suited for most PDES algorithms (except those requiring barrier synchronizations) as shared-memory machines.

4 CONSERVATIVE MECHANISMS

Conservative approaches to PDES strictly *avoid* any possibility of violating the local causality constraint. As mentioned earlier, the basic problem that conservative mechanisms must solve is to determine when it is "safe" to process an event. More precisely, if a process contains an unprocessed event E_A with time stamp T_A (and no other with smaller time stamp) and that process can determine that it is impossible for it to later receive another event with time stamp smaller than T_A, then it can safely process E_A without fear of later violating the local causality constraint. Processes containing no "safe" events must block; this can lead to deadlock situations if appropriate precautions are not taken.

In our earlier example this means that the LAX process must first determine whether the rerouting of the SFO-bound aircraft will occur before it can process E_{LAX}. If in fact it turns out that no rerouting occurred, it may have been safe to process E_{LAX} concurrently with E_{SFO}. One cost of strictly avoiding any possibility of error is that certain opportunities for concurrent execution are lost.

In the preceding example, if event E_{SFO} carried a time stamp of 10:30 A.M. rather than 9:00, it would be impossible for E_{SFO} to affect E_{LAX} (with time stamp 11:00) because the minimum time delay for an aircraft to fly from SFO to LAX is 1 h. Thus such knowledge as the minimum time for events to propagate from one logical process to another is widely used in existing conservative algorithms.

4.1   A Simple Conservative Algorithm

We begin the discussion of conservative simulation mechanisms by first proposing a simple approach that ensures adherence to the local causality constraint. As we shall soon see, the principal problem with this scheme is that it is prone to deadlock. Much of the early work in devising conservative simulation algorithms was concerned with addressing these deadlock problems.

Using the process-oriented paradigm described earlier, we assume that the sequence of time stamps on messages sent from one process to another is nondecreasing and messages transmitted from one logical process to another are received in the same order in which they were sent. This guarantees that the time stamp of the last message received on an incoming link is a lower bound on the time stamp of any subsequent message that is received later. Messages arriving on each incoming link are stored in first-in–first-out order, which is also time-stamp order because of this restriction. Each link has an associated clock that is equal to the time stamp of the message at the front of that link's queue if the queue contains a message or the time stamp of the last received message if the queue is empty. The process repeatedly selects the link with the smallest clock and, if there is a message in that link's queue, processes it. If the selected queue is empty the process blocks. This protocol guarantees that each process processes events only in nondecreasing time-stamp order, thereby ensuring adherence to the local causality constraint.

As mentioned earlier, this scheme is prone to deadlock. All that is required is a cycle of empty queues to arise that contains very small link clock values. Each process in this cycle will block, waiting for a new message to be added to the empty queue, so the simulation is deadlocked. There may be messages waiting to be processed in other queues with larger link clock values, but because the process cannot guarantee adherence to local causality such messages cannot be safely processed. In general, if there are relatively few unprocessed event messages compared to the number of links in the network or if the unprocessed events are not uniformly distributed among the message queues, deadlock may occur very frequently.

4.2   Null Messages

This simulation mechanism can be augmented to avoid the possibility of deadlock. Chandy and Misra (1979) and Bryant (1977) developed a scheme in which processes send "dummy" or null messages among each other for this purpose. Null messages are used only for deadlock avoidance; they do not correspond to any simulation activity.

In effect, a null message sent from one process to another is a promise by the process sending the null message that it will not later send any actual event messages with a time stamp smaller than the time stamp on the null message. It is hoped that the receiving process can use this information to determine that it is now safe to process some other pending event message that it has already received. A null message denotes a *lack* of simulator events.

How does one determine the time stamp of null messages? The clock value of each incoming link provides a lower bound on the time stamp of the next unprocessed event that will be removed from that link's buffer. When coupled with knowledge of the simulation performed by the process (e.g., a minimum time-stamp increment of any event passing through the process), this incoming bound can be used to determine a lower bound on the time stamp of the next *outgoing* message on each output link. Each process sends a null message on each of its output links after processing a received event message (null or real). The receiver of the null message can then compute new bounds on its outgoing links, send this information to its neighbors, and so on. This mechanism avoids deadlock as long as no cycles exist in which the collective time stamp increment around this cycle is zero. A necessary and sufficient condition for deadlock using this scheme is that a cycle of links must exist with the same link clock time (Peacock et al., 1979).

In the aforementioned approach processes send null messages regardless of whether the receiving process is waiting for this information. Thus some null messages may be sent even though they serve no useful purpose. An alternative approach is to have processes request lower bound information when they need an updated link clock value (Reynolds, 1982; Misra, 1986; Bain and Scott, 1988). This helps to reduce the amount of null message traffic but may incur additional latency before the requested information is obtained.

4.3   Deadlock Detection and Recovery

Chandy and Misra (1981) also developed an alternative approach to parallel simulation that eliminates the use of null messages. Rather than sending null messages, the simulation is allowed to deadlock. A separate mechanism is used to detect deadlock situations, and yet another mechanism is used to break the deadlock.

Deadlock detection mechanisms are described by Dijkstra and Scholten (1980) and Misra (1986). These algorithms view the distributed simulation as a "diffusing computation." Once a deadlock occurs all processes are said to be disengaged. The deadlock recovery mechanism (described later) is used to determine a set of messages that are safe to process. A central controller sends messages to processes that hold these messages to instruct them to begin processing them. When these messages are processed new messages are sent to other processes, which in turn spawn yet other messages, and so on.

A disengaged processes that receives an event message and resumes computation is said to become engaged. As the computation spreads a tree of engaged processes is formed, with the central controller at the root. Once a process is (1) blocked and (2) is a leaf node in the "engagement" tree, it removes itself from the tree (it could later become engaged again if it receives new messages that enable it to process other event messages). Thus the computation can be viewed as a tree of processes that expands as processes become busy and contracts as leaf processes become idle. Deadlock occurs when the engagement tree is reduced to only the controller node because at this point all processes are again disengaged.

The deadlock recovery phase of the computation must determine a set of events that are safe to process. The smallest time-stamped message in the entire simulation is always safe to process, so it is guaranteed that one such event can always be found. In addition, one may use a distributed computation to compute lower bound information (not unlike the distributed computation using null messages already described) to enlarge the set of safe messages.

Unlike the deadlock avoidance approach, the deadlock detection and recovery mechanism does not prohibit cycles of zero time stamp increment, although performance may be poor if many such cycles exist. Deadlock detection and recovery is therefore more general than deadlock avoidance. Of course one can augment the deadlock avoidance approach to add a deadlock detection and recovery mechanism to make it a general simulation mechanism.

4.4 Other Conservative Techniques

Several researchers have proposed synchronous algorithms in which one iteratively determines which events are safe to process and then processes them (Ayani, 1989; Chandy and Sherman, 1989; Lubachevsky, 1989a). Barrier synchronizations are used to keep one iteration (or components of a single iteration) from interfering with each other. Because barrier synchronizations are necessary, these algorithms are best suited for shared-memory machines to keep the associated overheads to a minimum. This approach is similar to the deadlock detection and recovery mechanism in the sense that both approaches move through phases of (1) processing events and (2) performing some global synchronization function. Deadlock recovery is similar to the overhead function of the synchronous methods in that one attempts to determine which events are safe to process.

The feature that separates different synchronous approaches is principally the method used to determine which events are safe to process. Later we discuss ideas that have been introduced to streamline this process. A common thread that runs through many techniques is the minimum time stamp increment function used in the original deadlock avoidance approach. A simple extension of this concept leads to the *distance* between processes; distance provides a lower bound on the amount of simulated time that must expire for an unprocessed event in one process to propagate (and possibly affect) another process.

Another technique is to use a moving simulated time window to reduce the overhead associated with determining when it is safe to process an event (Lubachevsky, 1989a). The lower edge of the window is defined as the minimum time stamp of any unprocessed event. Only those unprocessed events whose time stamp reside within the window are eligible for processing. This window reduces the distance (as defined earlier) one must search in determining if an event with smaller time stamp will later be received. For example, if the window extends from simulated time 10 to time 20 and the application is such that each event processed by an LP generates a new event with a minimum time stamp increment of 8 units of simulated time, then each LP need only examine the unprocessed events in neighboring LPs to determine which events are safe to proceed. No unprocessed event two or more hops away can affect one in the 10–20 time window because such an event would have to have a time stamp earlier than the start of the window.

A third approach to optimize the execution of conservative algorithms is to exploit a property called *lookahead*. *Lookahead* refers to the ability to predict what will happen or, equally important for conservative methods, what will not happen, in the simulated time future based on knowledge of the application, events that have already been processed, and pending events waiting to be processed. Nonzero minimum time stamp increments are the most obvious form of lookahead and were essential for the deadlock avoidance approach to make progress. Because lookahead enhances one's ability to identify events that are safe to process, it is reasonable to expect that improving a process's lookahead ability is bound to improve performance.

In certain situations one can improve lookahead by precomputing portions of the computation for future events (Nicol, 1988). For example, in a queueing network simulation using first-come–first-served queues and no preemption, one can precompute the service time of jobs that have not yet been received. If the server process is idle, its clock has a value of 100, and the service time of the next job has been precomputed to be 50, then the lower bound on the time stamp of the next message it sends is 150 rather than 100. If the average service time is much larger than the minimum, then this provides a better lower bound on the time stamp of the next message. This in turn will (it is hoped) enlarge the set of safe events that may be processed in parallel. It should be noted, however, that this technique is not generally applicable to all simulation applications. Nevertheless, precomputation appears to be a useful technique when it can be applied.

Finally, Chandy and Sherman (1989a) propose a paradigm that combines mechanisms used in sequential simulations with conservative mechanisms. In a sequential simulation one often schedules an event (e.g., a job departure from a queueing network server) under the premise that this event will take place if no disruptive event (e.g., a job preemption) occurs first. Such events are referred to as *conditional events*.

All conservative approaches convert conditional events to definite events (events that are guaranteed to occur) before they can be processed. This is accomplished in sequential simulations by virtue of the fact that no events exist in the event list with a smaller time stamp than that of the conditional event when that event is processed. Like other conservative mechanisms, a protocol is required to determine when it is "safe" to process conditional events. Chandy and Sherman (1989a) propose both synchronous and asynchronous protocols to perform this task; these protocols use broadcasts to distribute "time of next event" information to avoid deadlock situations. The conditional knowledge approach to simulation arises from the unity theory of parallel programming (Chandy and Misra, 1988). An alternative approach, also based on unity, is described in Chandy and Sherman (1989b).

4.5 Conservative Performance

The degree to which processes can look ahead into the simulated future plays a critical role in the performance of conservative strategies. If a process can predict what events will occur (and, more importantly, what events cannot occur) L units of time into the simulated future, it is said to have lookahead L.

Consider the simulation of the hypercube-based communication network described earlier (Fig. 1). Recall that arrival and departure events are used to route messages through nodes of the hypercube and that each output port forwards messages in the order in which they arrive. This simulator has *poor* lookahead properties because each logical process must advance its simulated time clock to $T + Q + S$ (or $T + S$ if the link was idle) before it can generate a new arrival event with time stamp $T + Q + S$ ($T + S$). It has zero lookahead for generating new arrival events.

However, one can reprogram this simulator to have good lookahead properties. Utilizing the fact that messages are transmitted in first-come–first-served order, each arrival event can "look ahead" and predict the subsequent arrival event at the next node of the cube *without* the use of a departure event. This is quite easy if the link is idle when the message arrives because the time of arrival is simply $T + S$, and S can be computed from parameters in the message (e.g., the message length) and variables within the logical process (the rate at which the link transmits data). When the link is busy at time T, one need only maintain a variable with each link to indicate the time at which the link will become idle, assuming no new messages are sent on that link. If this variable holds the value T_{idle}, then a newly arriving message will utilize the link from time T_{idle} to $T_{idle} + S$; that is, the subsequent arrival event at the next node will occur at time $T_{idle} + S$. This is identical to the value $T + Q + S$ used before.

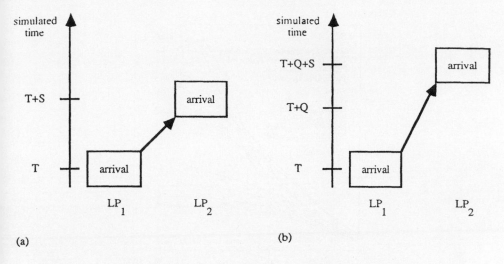

(a)   (b)

Figure 3   Revised simulator for a hypercube communication network that exploits lookahead. (a) Scenario of events for a message being transmitted over a previously idle link. (b) Scenario of events when link is busy.

Scenarios of events for the simulator when programmed in this way are shown in Figure 3. The key point is that the simulator is now looking ahead into the simulated feature to schedule the arrival events. The simulator has lookahead S (Fig. 3a) or $Q + S$ (Fig. 3b) for scheduling new arrival events.

Before continuing we should hasten to add that this approach is critically dependent on the fact that messages are processed by each output link in first-come–first-served order. Suppose it were the case that this simulator also included high-priority messages that *preempt* the sending of low-priority, that is, ordinary messages. In this case one could not reprogram the simulator to eliminate the departure event. The logical process must first advance to time $T + Q + S$ to determine that no preemption occurs, and only then can it generate the subsequent arrival event. If preemption occurs the departure event must be canceled and later rescheduled to reflect the new, projected departure time.

The performance of several simulators of the hypercube communication network are shown in Figure 4. Speedup over a sequential event list simulator is shown when the parallel simulator is executed on an eight-processor BBN Butterfly multiprocessor. The simulators use the deadlock avoidance strategy, but similar results were obtained using the deadlock detection and recovery algorithm. Performance of the simulator optimized to exploit lookahead (Fig. 3) as

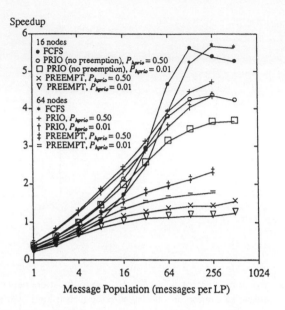

Figure 4 Performance of parallel simulators using the deadlock avoidance strategy (speedup of hypercube simulator; eight processors). As can be seen, only the simulators that have good lookahead properties can achieve good performance. P_{hprio} indicates the fraction of the message population with high priority; PRIO indicates prioritized messages are used but no preemption; and PREEMPT indicates both prioritized messages and preemption are used. FCFS indicates first-come–first-served queues are used, and the program is written to exploit lookahead.

well as the simulator using prioritized messages and preemption (which cannot be reprogrammed to exploit lookahead) are shown. In these simulators a fixed number of messages, referred to as the message population, move continually throughout the network. For the simulator with preemption it is assumed that 1% of the messages are high priority and the rest are ordinary (low-priority) messages. The time required to transmit a message is selected from a "shifted" exponential distribution with minimum of 0.1 and a mean of 1.0 units of simulated time. These speedup curves illustrate that although the simulator that exploits lookahead achieves good performance, the simulator than cannot performs very poorly. Further details of these experiments are described in Fujimoto (1988).

Historically, Reed et al. (1988) were among the first to report performance measurements for the deadlock avoidance and deadlock detection and recovery algorithms executing on a Sequent multiprocessor. They report very disappointing performance (the parallel simulators often run slower than the sequential simulators) for simulations of queueing networks, except in a few very restricted cases, such as feedforward networks that do not contain cycles. Reed et al. did not attempt to exploit lookahead, however. Fujimoto (1989a) reports more positive results using these algorithms on a BBN Butterfly multiprocessor. He demonstrated that performance is critically related to the degree to which logical processes can look ahead into the simulated time future. Fujimoto reproduced the poor performance reported by Reed and showed that reprogramming processes to exploit lookahead yielded dramatic improvements in performance for networks using first-come–first-served queues, as described earlier. Several experiments using synthetic work loads were also performed in which lookahead, computation granularity, time stamp increment function, message population, and the manner in which messages are routed through the network were varied. Depending primarily on lookahead, speedups ranged from no faster than sequential execution to performance approaching ideal (i.e., speedup N using N processors) using up to 16 processors.

Su and Seitz (1989) report some success in using variations of these algorithms to speed up logic simulations on an Intel iPSC multicomputer. Although speedups are relatively modest (8 using 64 processors and 10–20 using 128 processors are typical), they argue that better performance could be obtained on machines (e.g., shared-memory multiprocessors) with which the overhead of sending null messages can be substantially reduced. Reed et al. (1988), Fujimoto (1989a), and Wagner et al. (1989) exploit techniques using shared memory to improve the efficiency of these algorithms.

Wagner and Lazowska (1989), Lin and Lazowska (1989), and Nicol (1989) examined lookahead analytically. Lubachevsky (1989b) examined the performance and scalability of the bounded lag approach that uses synchronous execution, lookahead, and time windows to improve performance. Specifically he uses two forms of lookahead: minimum time stamp increments to allow *idle* logical processes to lookahead, and "opaque" periods that allow certain *busy* processes to do the same. The latter requires the exclusion of (for example) preemptive behavior. Lubachevsky argues that performance of this approach scales as the problem and machine size increase in proportion to within a factor of 0 (log N) of ideal, assuming adequate lookahead is available. He also demonstrates speedups as high as 16 on 25 processors of a Sequent Balance multiprocessor and over 1900 on a 16,384-processor connection machine for queueing network and Ising model (spinning atomic particles) simulations. Ayani (1989) and Chandy and Sherman (1989) also report some success in speeding up queueing network simulations using their approaches on Sequent and Intel iPSC systems, respectively. Speedup varies considerably depending on aspects of the simulation model. Ayani reports speedups as high as 5 on a 9-processor Sequent

Balance, and Chandy and Sherman report speedups as high as 9 on 24 iPSC processors and 7 using 12 processors.

Much has been learned concerning the performance of conservative mechanisms. As noted earlier, all conservative mechanisms rely on the ability to predict the future to ascertain which events may be safely processed. If no such capability existed, one could not guarantee the safety of any event other than the one containing the (globally) smallest time stamp, forcing sequential execution (ignoring the case of two events on distinct processes having identical time stamps).

To achieve good performance conservative mechanisms must be adept at predicting what will *not* happen because it is the fact that *no* smaller time-stamped event will later be received that is the firing condition that allows an event to be safely processed. Various forms of information are used to predict what will not occur, including the following:

1. The structure of the network of logical processes, that is, which processes can send messages to which other processes. This structure restricts the paths that ''dangerous'' events may use to reach other processes; sparsely connected networks present the best case for conservative mechanisms.
2. Received event messages. Assuming messages are transmitted in time stamp order, each received message excludes the possibility of later messages containing a smaller time stamp. Conservative mechanisms usually work best when there are many unprocessed events relative to the connectivity of the network (i.e., the number of processes) and these events are uniformly distributed among the links.
3. Knowledge of logical process behavior. Here the characteristic that one looks for is lookahead, that is, a *guaranteed invariance in behavior* in the physical system, regardless of any new events that might later occur. For example, new jobs arriving at a nonpreemptable queue do not affect the behavior of the job that is currently receiving service. Similarly, a minimum time stamp increment for some logical process is derived from the observation that the corresponding physical process does not perturb the system in any way up to some guaranteed time in the future, regardless of what happens next.

Any given simulation application may exhibit favorable or unfavorable characteristics for each of these properties. It appears that the inability to effectively exploit *any* of these aspects of behavior is fatal to existing conservative approaches. Conversely, most approaches can obtain good speedup if all these properties appear in a favorable manner. Depending on the specifics of the strategy in question, the inability to exploit one or more of these aspects may or may not be fatal. It remains to be seen to what extent applications that arise in practice exhibit these properties.

4.6   Critique of Conservation Mechanisms

Perhaps the most obvious drawback of conservative approaches is that they cannot fully exploit the parallelism available in the simulation application. If it is possible that event E_A *might* affect E_B either directly or indirectly, conservative approaches must execute E_A and E_B sequentially. If the simulation is such that E_A seldom affects E_B, these events could have been processed concurrently most of the time. In general, if the worst-case scenario for determining when it is safe to proceed is far from the typical scenario that arises in practice, the conservative approach is usually overly pessimistic and forces sequentiality when it is not necessary.

A related problem faced by conservative methods concerns the question of robustness; it has been observed that seemingly minor changes to the application may have a catastrophic effect on performance. For example, adding short, high-priority messages that interrupt "normal" processing in a computer network simulation may lead to severe performance degradations. This is problematic because experimenters often do not have advance knowledge of the full range of experiments that will be required, so it behooves them to invest substantial amounts of time parallelizing the application if an unforeseen addition to the model at some future date could invalidate all this work.

Critics of conservative methods also point out that many existing conservative techniques (the deadlock avoidance and deadlock detection and recovery mechanisms in particular) require static configurations: one cannot dynamically create new processes, and the interconnection among logical processes must also be statically defined. Techniques to circumvent this problem by creating "spare" processes at the start of the simulation and to define a fully connected network usually lead to excessive overheads; for example, broadcast communications may be required to determine when it is safe to proceed.

Many conservative schemes require knowledge concerning logical process behavior to achieve good performance. Such information as minimum time stamp increments or the guarantee that an event occurring at time T really has no effect on the behavior of certain other events may be difficult to derive for complex simulations. Users are ill-advised to give overly conservative estimates (e.g., a minimum time stamp increment of zero) because very poor performance may result.

Proponents of optimistic approaches argue that the user need not be concerned with the details of the synchronization mechanism to achieve good performance. Sequential simulation programs need not be concerned with the details of the implementation of the event list. Of course, certain guidelines that apply to *all* parallel programs must be followed when developing parallel simulation code, such as selecting an appropriate granularity and maximizing parallelism, but requiring the programmer also to be intimately familiar with the synchronization mechanism and to program the application to maximize its effectiveness often lead to "fragile" code that is difficult to modify and maintain.

5 OPTIMISTIC MECHANISMS

Conservative mechanisms strictly avoid any possibility of violating the local causality constraint, but optimistic approaches allow violations to occur but provide a mechanism for detecting and recovering from them. Thus rather than determining when it is safe to proceed, optimistic methods must determine (1) when an error has occurred and (2) how to recover. One advantage of this approach is that it allows the simulator to exploit parallelism in situations in which it is possible that causality errors *might* occur, but in fact do not.

Consider the example of the airport simulation discussed earlier. Recall that there were two pending events: E_{SFO} denoting an airplane arrival in San Francisco at 9:00 A.M. and E_{LAX} denoting an arrival in Los Angeles at 11:00 A.M.. An optimistic method would allow both these events to be processed concurrently. If the San Francisco-bound flight were not diverted and no other interfering events occurred, the simulation would proceed without violations of the local causality constraint. However, if the SFO flight were diverted to LAX as described earlier, E_{LAX} would have been processed incorrectly (i.e., it would not have taken into account the arrival of the diverted flight). All effects of erroneously processing E_{LAX} must be erased. This is disturbing because E_{LAX} may have incorrectly scheduled new events, which may have created still other events, and so on. All these events, as well as the original incorrectly processed event, must be rolled back.

5.1 The Time Warp Mechanism

The Time Warp mechanism, based on the virtual time paradigm (virtual time is synonymous with simulated time), provides a very elegant way out of this dilemma (Jefferson, 1985). Time Warp is by far the best known optimistic protocol. In Time Warp a causality error is detected whenever an event message is received that contains a time stamp smaller than that of the process's clock (i.e., the time stamp of the last processed message). The event causing rollback is called a *straggler*. Recovery is accomplished by undoing the effects of all events that have been processed prematurely by the process receiving the straggler (more precisely, those processed events that have time stamps larger than that of the straggler).

An event may do two things that must be rolled back: it may modify the state of the logical process, and/or it may send event messages to other processes. Rolling back the state is accomplished by periodically saving the process's state and restoring an old state vector on rollback. "Unsending" a previously sent message is accomplished by sending an *antimessage* that annihilates the original when it reaches its destination. If the annihilated positive message has already been processed, then that process must also be rolled back to undo the effect of processing the message. Recursively repeating this procedure allows all the effects of the erroneous computation to eventually be canceled. It can be shown that this mechanism always makes progress under some mild constraints.

As noted earlier, the smallest time-stamped, unprocessed event in the simulation is always safe to process. In Time Warp the time stamp on this event is called global virtual time (GVT). No event with a time stamp smaller than GVT is ever rolled back, so storage used by such events (e.g., saved states) can be discarded. Also, irrevocable operations (such as input/output I/O) cannot be committed until GVT sweeps past the simulated time at which the operation occurs. The process of reclaiming memory and committing irrevocable operations is referred to as *fossil collection*.

5.2 Variations on Time Warp

Numerous variations on the Time Warp mechanism just described have been proposed. We describe several of these techniques here.

Two optimizations have been proposed to "repair" the damage caused by an incorrect computation rather than completely repeat it. For instance, it may be the case that a straggler event does not sufficiently alter the computation of rolled-back events to change the (positive) messages generated by these events. The Time Warp mechanism uses *aggressive* cancellation: that is, whenever a process rolls back to time T, antimessages are immediately sent for all positive messages sent after simulated time T. In *lazy cancellation* (Gafni, 1988) processes do not immediately send the antimessages for any rolled-back computation. Instead they wait to see if the reexecution of the computation regenerates the *same* messages; if the same message is re-created there is no need to cancel the original message. An antimessage created at simulated time T is sent only after the process's clock sweeps past time T without regenerating the same message.

Although lazy cancellation may avoid some unnecessary rollbacks, it incurs some performance penalties that are not present using aggressive cancellation. Using lazy cancellation some additional overhead is required whenever an event is executed to determine if a matching antimessage already exists; one or more message comparisons are required if one is reexecuting previously rolled-back events. Also, lazy cancellation may allow erroneous computations to spread farther than they would under aggressive cancellation, so performance may be degraded if the simulator is forced to execute many incorrect computations. One can construct cases in which lazy cancellation executes a computation with N-fold parallelism N times *slower* than aggressive when N processors are used (Reiher et al., 1990).

On the other hand, lazy cancellation has the interesting property that it can allow the computations to be executed in less time than the critical path execution time (Berry, 1986). This is possible because computations with incorrect input may still generate correct results! Therefore, one may execute some computations prematurely and still generate the correct answer. For example, suppose the computation executes the statement $X = min(A, B)$ prematurely, before the proper value has been stored in the variable A. Further, suppose both the erroneous and correct values of A are larger than B. Then both the correct and incorrect computations produce the desired result. One can construct a case in which lazy cancellation can execute a sequential computation with N-fold speedup using N processors, but aggressive cancellation requires the critical path length execution time (Reiher et al., 1990).

A second optimization that attempts to repair incorrect computations rather than discarding and redoing them is called *lazy reevaluation* (West, 1988). It is somewhat similar to lazy cancellation but deals with state vectors rather than messages. Consider the case when the state of the process is the same after processing a straggler event message as it was before it executed. If no new messages arrived, then it is clear that the reexecution of rolled-back events will be identical to the original execution, so one need not reexecute them, but instead jump forward over these events. This requires a comparison of state vectors to see if the state has changed. The lazy reevaluation optimization is useful when a "read-only" or query event causes a rollback.

Time windows not unlike those proposed for conservative mechanisms have also been proposed for optimistic protocols (Sokol et al., 1988). As before, only events within the time window are eligible for processing. In optimistic methods the time window is used to prevent incorrect computations from propagating too far ahead into the simulated time future. Critics of this method point out, however, that such windows cannot distinguish good computations from bad, so they may impede the progress of correct computations. Further, incorrect computations that are far ahead in the future are already discriminated against by Time Warp's scheduling mechanism, which gives precedence to activities containing small time stamps. Finally, it is not clear how the size of the window should be determined.

Another technique with motivations similar to those of the time window technique is to send special control messages to stop the spread of the erroneous computation once an error (i.e., a straggler message) has been received (Madisetti et al., 1988). Receivers of the control message stop progressing forward until the error recovery procedure has been completed. This scheme requires one to determine which processes may have been "infected" by the erroneously processed straggler message, so knowledge of the real-time properties of the computation (e.g., the time to propagate incorrect computations and the time to transmit the special control messages) is required. Like the time window scheme, the disadvantage of this approach is that some correct computations may be unnecessarily frozen. Also, the overhead to implement this mechanism becomes excessive in certain applications because many control messages are required.

Finally, another Time Warp optimization called direct cancellation has been proposed that uses shared memory to streamline the cancellation of incorrect computations (Fujimoto, 1989b). Whenever an event E_1 schedules another event E_2, a pointer is left from E_1 to E_2. This pointer is used if it is later decided that E_2 should be canceled (using either lazy or aggressive cancellation). In contrast, conventional Time Warp systems must search to locate canceled messages. The advantages of this mechanism are twofold: it reduces the overheads associated with message cancellation, and it speedily tracks down erroneous computations to minimize the damage that is caused. Good performance has been reported on a version of Time Warp that uses direct cancellation.

5.3 Optimistic Performance

Several successes have been reported in using Time Warp to speed real-world simulation problems. Good speedups have been reported by researchers at the Jet Propulsion Laboratory (JPL) in simulations of battlefield scenarios (Wieland et al., 1989), communication networks (Presley et al., 1989), biologic systems (Ebling et al., 1989), and other physical phenomena (Hontalas et al., 1989). Typical speedups on the JPL Mark III hypercube (a 68020-based, message-based machine) are in the 10–20 range using 32 processors. Speedups as high as 37 using 100 processors of a BBN Butterfly are also reported.

Fujimoto has also demonstrated good performance using a separate, independently developed implementation of Time Warp for queueing network simulations and various synthetic work loads. This implementation uses direct cancellation to streamline the cancellation of incorrect computations. Speedup curves for the hypercube network simulator discussed earlier executing on a BBN Butterfly multiprocessor are shown in Figure 5. These curves show speedup as the number of processors is varied for both the simulators of networks that forward messages in first-come–first-served order as well as when high-priority messages preempt low-priority messages. The former simulator is programmed to exploit lookahead. Although the simulator that exploits lookahead achieves better performance, the simulator that cannot exploit it still achieves a respectable speedup. This contrasts sharply with conservative approaches. Speedups as high as 56.8 on a 64-processor system were obtained. Further details of these measurements are reported in Fujimoto (1989b).

A limited amount of work has been performed in deriving analytic models for Time Warp behavior. Models for two processors have been developed by Lavenberg and Muntz (1983) and Mitra and Mitrani (1984). Unfortunately, these models do not generalize to more processors.

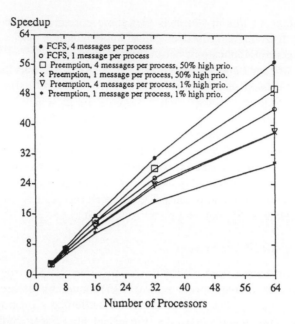

Figure 5   Performance of parallel simulators using Time Warp (speedup of Time Warp simulations; 256 logical processes). As can be seen, both simulators that have good lookahead properties and those that do not can achieve good performance. (Reprinted from Fujimoto, 1989b).

Lin and Lazowska (1989) derive a condition under which Time Warp will produce optimal performance, that is, corresponding to the critical path lower bound). They also identify situations in which Time Warp will outperform the Chandy-Misra algorithms. This work assumes that overheads for both Time Warp and the conservative mechanisms are negligible, so the results are largely applicable to large-grained events.

5.4   Critique of Optimistic Methods

A critical question faced by optimistic systems like Time Warp is whether the system will exhibit thrashing behavior when most of its time is spent executing incorrect computations and rolling them back. Thus far the experience of researchers at JPL, Georgia Tech, and Jade Simulations has been that such behavior is seldom encountered in practice and, when discovered, usually points to a correctable weakness in the implementation rather than any fundamental flaw in the algorithm. Critics argue, however, that no proof yet exists that Time Warp is stable.

An intuitive explanation of why empirical data suggest stable behavior is that erroneous computations can only be initiated when one processes a correct event prematurely; this premature event and subsequent erroneous computations must necessarily be in the simulated time future of the correct, straggler computation. Also, the farther the incorrect computation spreads the farther it moves into the simulated time future, lowering its priority for execution since preference is always given to computations containing smaller time stamps. Thus Time Warp systems tend to automatically slow the propagation of errors, allowing the error detection and correction mechanism to correct the mistake before too much damage has been done. A potentially more dangerous case is when the erroneous computation propagates with smaller time stamp increments than the correct one. It remains to be seen, however, to what extent this behavior can degrade performance or whether such pathologic situations arise in practice.

A more serious problem with the Time Warp mechanism is the need to periodically save the state of each logical process. Fujimoto (1989b) has demonstrated that state-saving overhead can seriously degrade the performance of many Time Warp programs, even if the state vector is only a few thousand bytes. The state-saving problem is further confounded by applications requiring dynamic memory allocation because one may have to traverse complex data structures to save the process's state. State-saving overhead limits the effectiveness of Time Warp to applications in which the amount of computation required to process an event can be made significantly larger than the cost of saving a state vector. This may be difficult to achieve for certain applications.

A more general solution is to use hardware support. Fujimoto (1988) has proposed a component called the *rollback chip* to eliminate state-saving overhead. When combined with conventional memory components, this component allows one to construct a memory board with state-saving and rollback capabilities. At any time one can instruct the hardware (by writing into a control register) to "mark" the state of the process as one to which it may later want to restore via rollback. A rollback operation restores the memory to a previously marked state. Read-and-write operations appear the same as read-and-write operations in conventional memory. The rollback chip acts as a special type of memory management unit that modifies memory addresses generated by the CPU so that areas of memory that were saved using a state saved are not overwritten, thereby avoiding excessive copying.

In addition to the rollback chip, a machine architecture based on Time Warp has also been proposed called the Virtual Time Machine (Fujimoto, 1989c). A shared-memory multiprocessor is proposed with a two-dimensional memory system (in contrast, conventional memory can be viewed as a one-dimensional array of words) that is addressed using a conventional word address as well as a virtual (i.e., simulated) time coordinate. For instance, one may request to read the contents of a variable at simulated time 100, and the memory system will arrange that the appropriate version of the variable is read. A Time Warp-like rollback mechanism is invoked when variables are accessed in a real-time order that is inconsistent with the desired virtual time order of operations.

Optimistic algorithms tend to use much more memory than their conservative counterparts. Although the space-time tradeoffs for optimistic systems are not yet understood, this appears to be an unavoidable aspect of optimism.

Finally, unlike conservative approaches, optimistic systems must be able to recover from arbitrary errors that can arise because such errors may be erased by a subsequent rollback. Erroneous computations may enter infinite loops, requiring the Time Warp executive to interact with the hardware's interrupt system. In certain languages pointers may be manipulated in arbitrary ways; Time Warp must be able to trap illegal pointer usages that result in running time errors and prevent incorrect computations from overwriting non–state-saved areas of memory. Although such problems are in principal not insurmountable, they may be difficult to circumvent in certain systems without appropriate hardware support. The alternative taken by most existing Time Warp systems is to leave the onerous task of analyzing incorrect execution sequences to the user.

6   IMPLEMENTATION ISSUES

6.1   Application Program Development

To the simulation programmer the central issues that must be addressed involve formulation of the simulation model, that is, decomposing the simulator into a collection of logical processes that communicate by exchanging event messages. Thus far PDES applications broadly fall into two categories: (1) simulators whose topology is a static network (e.g, communication network simulators, simulators of computer hardware, and assembly lines), and (2) simulators whose topology varies dynamically throughout the simulation. Examples of the latter case are situations in which simulation entities move across a physical domain and interact with other entities that are in physical proximity with it; in combat simulations, for example, troop units engage enemy troop units when they are sufficiently close.

Applications in the first category decompose very naturally to logical processes that communicate by exchanging messages due to the networklike structure of the system being modeled. Applications in the latter category are usually constructed by subdividing the physical domain into a grid structure, with a process assigned to each grid that keeps track of the entities (e.g., combat units) that reside in that grid. Grid processes exchange messages with neighboring grid processes as well as processes modeling the entities within that grid (or sometimes neighboring grids) to simulate interactions among the entities.

A challenging task for simulations that involve the physical proximity of interacting entities is management of the data structure that describes the state of the system without the use of shared variables (recall that the logical process paradigm excludes shared memory). For example, in a combat simulation each "grid sector process" contains state information, such as the number of units currently residing in that grid sector. A common activity performed by each combat unit is to scan its immediate environment to determine what other combat units are in the same or neighboring grid sectors and then act accordingly, for example, move into a neighboring sector to attack enemy units or retreat. The information indicating what resides in each grid sector must be shared among many combat units.

In the absence of shared state variables, the most natural approach to programming this simulation is to "emulate" shared memory by sending "read" and "write" event messages to access the shared information. However, this approach often leads to poor performance because message passing and synchronization delays are substantial, even if the underlying hardware supports shared memory. An alternative to this approach is to duplicate the shared information in the logical processes that need it. Because the shared state can be modified, a protocol is required to ensure coherency among the various copies of the shared state. This approach can be effective in achieving good performance; however, it significantly complicates the coding of the application, making it difficult to understand and maintain (Wieland and Jefferson, 1989).

Another important problem that must be addressed by the application program is constructing the model so that an appropriate computation granularity (i.e., computation per event) is used. If the granularity is too small, message passing and other overheads dominate, leading to poor performance. Sending a message typically requires several milliseconds on existing message-based multicomputers. Although communication costs are lower on shared-memory machines (e.g., the delay may be hundreds of microseconds), they are still sufficiently large that simulations with very few computations per event perform poorly. For Time Warp simulations the computation granularity must be significantly larger than state-saving overhead or that overhead will dominate.

Finally, the application programs should be written to maximize the degree to which they exploit lookahead. As we saw earlier, maximizing lookahead is crucial for conservative simulation algorithms; otherwise performance may be extremely poor. Although optimistic methods like Time Warp appear to be more resilient to poor lookahead, experience to date indicates that performance is significantly improved if lookahead is used.

6.2 Implementation of the Simulation Mechanism

Like all parallel programs, PDES applications must address such problems as load management and scheduling. For conservative algorithms these issues are essentially the same as for any parallel computation, so results in these problems are equally applicable here (e.g., see Eager et al., 1989). For optimistic algorithms, however, the problems are somewhat different.

Scheduling plays a critical role in simulators based on optimistic execution. At any given time correct computations are intermixed with incorrect ones. A poor scheduler may inadvertently favor incorrect computations over correct ones, leading to poor performance. The scheduling mechanism usually used in Time Warp is always to give priority to the smallest time-stamped process. In practice, experimenters have found that this simple approach performs well. It remains to be seen whether more sophisticated scheduling techniques can significantly improve upon this simple approach.

A second problem related to Time Warp involves the migration of logical processes from one processor to another. Logical processes in Time Warp contain much more state than conventional processes because they include a history of past state vectors and message queues. One would like to migrate only a small portion of this history information when migrating a process from one processor to another. Reiher and Jefferson (1990) have proposed an approach to do just this. The execution of a process over simulated time is divided into *phases;* for example, a process may consist of two phases, from time 0 to 100, and then from 100 to the remainder of the simulation. The latter phase can be viewed as a second process that is created at time 100 and can thus move to another processor without transporting the history from 0 to 100. This greatly reduces the overhead associated with migrating logical processes.

Numerous optimizations based on shared memory have been developed for both conservative and optimistic mechanisms. For deadlock avoidance using null messages, Fujimoto (1989a) proposes storing the time stamp of the null messages in a shared variable rather than explicitly sending messages. Also, deadlock detection is straightforward on a shared-memory machine because one can simply maintain a global counter to indicate the number of processes that are either running or ready to run; deadlock occurs when this counter becomes zero. Finally, for Time Warp the direct cancellation technique described earlier eliminates the need for explicit antimessages.

7 CONCLUSIONS

The purpose of this chapter is to provide insight into the problem of executing discrete event simulation programs on a parallel computer. We have surveyed existing approaches and analyzed the merits and drawbacks of various techniques. The state of the art in parallel discrete event simulation has advanced rapidly in recent years, and much more is now known about the behavior of proposed simulation mechanisms than a few years ago.

Optimistic methods like Time Warp offer the greatest hope as "general-purpose" simulation mechanisms, at least in simulating systems that contain some degree of parallelism. Significant successes have been achieved across a wide range of applications.

Conservative methods offer good potential for certain classes of problems. Significant successes have also been obtained, particularly when application-specific knowledge is applied to maximize the efficiency of the simulation mechanism. Conservative methods may find success in packaged simulation systems (e.g., logic simulators) in which the simulation code is optimized for the synchronization algorithm and users merely configure the provided simulation modules into specific systems.

An important application area that has not yet been adequately addressed by either optimistic of conservative simulation mechanisms is real-time applications. Theories of performance are not sufficiently developed to address this question, although some progress has been made.

ACKNOWLEDGMENTS

The author's research was supported in part by NSF Grant No. CCR-8902362. The author also thanks David Jefferson and members of the Time Warp group at the Jet Propulsion Laboratory for valuable discussions contributing to the development of this chapter.

REFERENCES

Ayani, R. (1989). A parallel simulation scheme based on the distance between objects. *Proceedings of the SCS Multiconference on Distributed Simulation*, 21(2):113–118.

Bain, W. L., and Scott, D. S. (1988). An algorithm for time synchronization in distributed discrete event simulation. *Proceedings of the SCS Multiconference on Distributed Simulation*, 19(3):30–33.

Berry, O. (1986). Performance evaluation of the Time Warp distributed simulation mechanism. Technical report, University of Southern California.

Bryant, R. E. (1977). Simulation of packet communication architecture computer systems. MIT-LCS-TR-188, Massachusetts Institute of Technology, Boston.

Chandak, A., and Browne, J. C. (1983). Vectorization of discrete event simulation. *Proceedings of the 1983 International Conference on Parallel Processing*, pp. 359–361.

Chandy, K. M., and Misra, J. (1979). Distributed simulation: A case study in design and verification of distributed programs. *IEEE Transactions on Software Engineering*, SE-5(5):440–452.

Chandy, K. M., and Misra, J. (1981). Asynchronous distributed simulation via a sequence of parallel computations. *Communications of the ACM*, 24(11):198–205.

Chandy, K. M., and Misra, J. (1988). *Parallel Program Design, A Foundation*. Addison-Wesley, Reading, Massachusetts.

Chandy, K. M., and Sherman, R. (1989a). The conditional event approach to distributed simulation. *Proceedings of the SCS Multiconference on Distributed Simulation*, 21(2):93–99.

Chandy, K. M., and Sherman, R. (1989b). Space, time, and simulation. *Proceedings of the SCS Multiconference on Distributed Simulation*, 21(2):53–57.

Dijkstra, E. W., and Scholten, C. S. (1980). Termination detection for diffusing computations. *Information Processing Letters*, 11(1):1–4.

Eager, D. L., Zahorjan, J., and Lazowska, E. D. (1989). Speedup versus efficiency in parallel systems. *IEEE Transaction on Computers*, 38(3):408–423.

Ebling, M., DiLorento, M., Presley, M., Wieland, F., and Jefferson, D. R. (1989). An ant foraging model implemented on the Time Warp Operating System. *Proceedings of the SCS Multiconference on Distributed Simulation*, 21(2):21–26.

Fujimoto, R. M. (1988). Lookahead in parallel discrete event simulation. *Proceedings of the 1988 International Conference on Parallel Processing*, pp. 34–41.

Fujimoto, R. M. (1989a). Performance measurements of distributed simulation strategies. *Transactions of the Society for Computer Simulation*, 6(2):89–132.

Fujimoto, R. M. (1989b). Time Warp on a shared memory multiprocessor. *Transactions of the Society for Computer Simulation*, 6(3):211–239.

Fujimoto, R. M. (1989c). The virtual time machine. *International Symposium on Parallel Algorithms and Architectures*, pp. 199–208.

Fujimoto, R. M., Tsai, J., and Gopalakrishnan, G. (1988). Design and performance of special purpose hardware for Time Warp. *Proceedings of the 15th Annual Symposium on Computer Architecture*, pp. 401–408.

Gafni, A. (1988). Rollback mechanisms for optimistic distributed simulation systems. *Proceedings of the SCS Multiconference on Distributed Simulation*, 19(3):61–67.

Hontalas, P., Beckman, B., DiLorento, M., Blume, L., Reiher, P., Sturdevant, K., Van Warren, L., Wedel, J., Wieland, F., and Jefferson, D. R. (1989). Performance of the colliding pucks simulation on the Time Warp Operating System. *Proceedings of the SCS Multiconference on Distributed Simulation*, 21(2):3–7.

Jefferson, D. R. (1985). Virtual time. *ACM Transactions on Programming Languages and Systems*, 7(3):404–425.

Jones, D. W. (1986). Concurrent simulation: An alternative to distributed simulation. *1986 Winter Simulation Conference Proceedings*, pp. 417–423.

Lavenberg, S., and Muntz, R. (1983). Performance analysis of a rollback method for distributed simulation. In: *Performance '83*. North-Holland, Amsterdam, pp. 117–132.

Lin, Y.-B., and Lazowska, E. (1989). Exploiting lookahead in distributed/parallel simulation. Technical report, Dept. of Computer Science, University of Washington, Seattle.

Lin, Y.-B., and Lazowska, E. (1989). Optimal performance of Time Warp simulation and a comparison with Chandy-Misra approach. Technical report, Dept. of Computer Science, University of Washington, Seattle.

Lubachevsky, B. D. (1989a). Efficient distributed event-driven simulations of multiple-loop networks. *Communications of the ACM*, 32(1):111–123.

Lubachevsky, B. D. (1989b). Scalability of the bounded lag distributed discrete event simulation. *Proceedings of the SCS Multiconference on Distributed Simulation*, 21(2):100–107.

Madisetti, V., Walrand, J., and Messerschmitt, D. (1988). Wolf: A rollback algorithm for optimistic distributed simulation systems. *1988 Winter Simulation Conference Proceedings*, pp. 296–305.

Misra, J. (1986). Distributed-discrete event simulation. *ACM Computing Surveys*, 18(1): 39–65.

Mitra, D., and Mitrani, I. (1984). Analysis and optimum performance of two message passing parallel processors synchronized by rollback. In: *Performance '84*. North-Holland, Amsterdam, pp. 35–50.

Nicol, D. M. (1988). Parallel discrete-event simulation of FCFS stochastic queueing networks. *Parallel Programming: Experiences with Applications, Languages and Systems*, 23(9):124–137 ACM SIGPLAN Notices.

Nicol, D. M. (1989). The cost of conservative synchronization in parallel discrete event simulations. Technical report, Department of Computer Science, College of William and Mary, Williamsburg, Virginia.

Peacock, J. K., Wong, J. W., and Manning, E. G. (1979). Distributed simulation using a network of processors. *Computer Networks*, 3(1):44–56.

Presley, M., Ebling, M., Wieland, F., and Jefferson, D. R. (1989). Benchmarking the Time Warp Operating System with a computer network simulation. *Proceedings of the SCS Multiconference on Distributed Simulation*, 21(2):8–13.

Reed, D. A., Malony, A. D., and McCredie, B. D. (1988). Parallel discrete event simulation using shared memory. *IEEE Transactions on Software Engineering*, 14(4):541–553.

Reiher, P., and Jefferson, D. R. (1990). Virtual time based dynamic load management in the Time Warp Operating System. *Proceedings of the SCS Multiconference on Distributed Simulation*, 22(1):103–111.

Reiher, P. L., Fujimoto, R. M., Bellenot, S., and Jefferson, D. R. (1990). Cancellation strategies in optimistic execution systems. *Proceedings of the SCS Multiconference on Distributed Simulation*, 22(1):112–121.

Reynolds, P. F., Jr. (1982). A shared resource algorithm for distributed simulation. *Proceedings of the 9th Annual Symposium on Computer Architecture*, 10(3):259–266.

Sokol, L. M., Briscoe, D. P., and Wieland, A. P. (1988). Mtw: A strategy of scheduling discrete simulation events for concurrent execution. *Proceedings of the SCS Multiconference on Distributed Simulation*, 19(3):34–42.

Su, W. K., and Seitz, C. L. (1989). Variants of the Chandy-Misra-Bryant distributed discrete-event simulation algorithm. *Proceedings of the SCS Multiconference on Distributed Simulation*, 21(2):38–43.

Wagner, D. B., and Lazowska, E. D. (1989). Parallel simulation of queueing networks: Limitations and potentials. *Proceedings of 1989 ACM SIGMETRICS and Performance '89*, 17(1):146–155.

Wagner, D. B., Lazowska, E. D., and Bershad, B. N. (1989). Techniques for efficient shared-memory parallel simulation. *Proceedings of the SCS Multiconference on Distributed Simulation,* 21(2):29–37.

West, D. (1988). Optimizing Time Warp: Lazy rollback and lazy reevaluation. Technical report, M.S. Thesis, University of Calgary, Ontario, Canada.

Wieland, F., and Jefferson, D. R. (1989). Case studies in serial and parallel simulation. *Proceedings of the 1989 International Conference on Parallel Processing,* 3:255–258.

Wieland, F., Hawley, L., Feinberg, A., DiLorento, M., Blume, L., Reiher, P., Beckman, B., Hontalas, P., Bellenot, S., and Jefferson, D. R. (1989). Distributed combat simulation and Time Warp: The model and its performance. *Proceedings of the SCS Multiconference on Distributed Simulation,* 21(2):14–20.

7

Finite Element Analysis on Concurrent Machines

Charbel Farhat   *University of Colorado at Boulder, Boulder, Colorado*

1   INTRODUCTION

New computer hardware with parallel processing capabilities is now commercially available and has the potential of revolutionizing scientific computing. Moving finite element applications to these computers faces significant obstacles that center on methods, algorithms, languages, and implementations. In this chapter we examine a general approach to the architectural design of the next generation of finite element software. We present several parallel computational strategies for both implicit and explicit static and dynamic computations and for I/O (input/output) manipulations. We describe their implementation on a large set of parallel environments and assess their adequacy for a given hardware architecture and a given problem. We address issues related to software portability. We report on performance results for a wide variety of multiprocessors, including the iPSC/1-32, Alliant FX/8, Cray-2/4, Cray Y-MP/8, and the CM-2 Connection Machine.

The realistic simulation of the nonlinear dynamics of complex structural systems remains beyond the feasible range of traditional computers. It has been the author's experience that the simulation of the transient response of a space station model with 100,000 degrees of freedom to various loading configurations consumes over 10 CPU h (central processing unit) on a Cray-2 supercomputer and that the simulation of the deployment of a space structure is even more computationally demanding, especially if the control-structure interaction problem is included. The aeroelastic response of a detailed wing-body configuration using a potential flow theory requires about 5 CPU h using the same supercomputer. To establish the transonic flutter boundary for a given set of aeroelastic parameters about 30 aeroelastic response analyses are required, which brings the total CPU time to 6 days. If the full Navier-Stokes equations are to be solved, it is estimated that the CPU time increases by two orders of magnitude. It is also clear that large amounts of data can be generated in a large-scale transient structural analysis or a large-scale computational fluid dynamic solution. These raw data must be interpreted, in real time if possible, to be understood.

Clearly, the true potential for execution improvement lies in massively parallel and/or parallel-vector supercomputing. The commercial supercomputer manufacturers of the last decade have extended their products into configurations that use a few vector processors coupled around a massive shared memory (Cary-2, Cray X-MP, and Cray Y-MP). Supercomputers with a larger number of vector processors are also under development (Cray-3). Concurrent multiprocessors with much finer granularity and a wide range of interconnection strategies are now appearing. Recently, massively parallel computers like the Connection Machine have demonstrated their potential to be the fastest supercomputers, a trend that may accelerate in the future (McBryan, 1988; Saati et al., 1990). The advent of advanced frame buffers and high-performance workstations now makes real-time visualization possible.

Moving finite element applications to concurrent processors faces significant obstacles that must be resolved as such machines become more and more available. The obstacles center on algorithms, languages, and education. In this chapter we emphasize the architectural, algorithmic, and portability aspects.

The various forms of parallel numerical algorithms that speed finite element computations are as numerous as the number of researchers working on the problem. Extensive lists of references on this topic may be found in the surveys of Noor (1987), White and Abel (1988), and Ortega et al. (1988). Throughout this chapter we present and discuss the adequacy of a set of parallel finite element computational strategies (mesh preprocessing, solution algorithms, and I/O manipulations) for a given parallel processor and a given structural and/or mechanical problem.

The remainder of this chapter is organized as follows. In Section 2 we begin with an overview of the present status of parallel computers that is pertinent to finite element computations. Through the examples of SIMD, MIMD, local memory and shared-memory multiprocessors we address the impact of hardware architecture on the design and implementation of parallel algorithms and parallel data structures. Next, in Section 3 we describe a portable parallel finite element software architecture based on the "divide and conquer" paradigm. Section 4 emphasizes linear and nonlinear static analysis, and Section 5 deals with dynamic analysis. Section 6 summarizes the experience of the author with massively parallel finite element transient computations on the Connection Machine. In Section 7 we discuss an approach to parallel I/O manipulations in finite element simulations, and in Section 8 we address the problem of "code portability." Section 9 illustrates the work described in this chapter with the solution of real-life problems on commercially available multiprocessors and reports the performance results measured. Finally, Section 10 concludes this chapter.

2 CHARACTERISTICS OF NEW PARALLEL COMPUTING ENVIRONMENTS

Several parallel computers have already been marketed commercially. Here we do not discuss these individually. Rather we focus on presenting an overview of their architecture and emphasize the impact of their hardware features on the design and implementation of parallel computational strategies for finite element computations. A review of some of the commercially available parallel systems can be found in Babb (1988), where programming examples are also provided.

Multiprocessors can be generally described by three essential elements: granularity, topology, and control.

1 *Granularity* relates to the number of processors and involves the size of these processors. A fine-grained multiprocessor features a large number of usually very small and simple processors. The Connection Machine (65,536 processors) is such a massively parallel supercomputer. NCube's 1024-node and iPSC's 128-node models are comparatively mdeium-grained machines. On the other hand, a coarse-grained multiprocessor is typically built by interconnecting a small number of large, powerful processors, usually vector processors. Alliant FX/8 (eight processors), IBM 3090 (six processors), Cray X-MP (four processors), Cray-2 (four processors), and Cray Y-MP (eight processors) are examples of such multiprocessors and supermultiprocessors. Granularity directly affects the parallel strategy. On a coarse-grained multiprocessor finite element computations can be parallelized at the subdomain level. On a fine-grained machine they are best parallelized at the element and sometimes at the degree of freedom level. When designing parallel algorithms for finite element computations on coarse-grained vector supermultiprocessors, one should preserve vectorization. This is because the potential speedup due to interconnecting a few processors cannot compete with the speedup due to the vector capabilities of a single processor. A good strategy for maximizing speedup consists of organizing the computations around two nested loops, where the outer loop is multiprocessed and the inner loop is vectorized.

2 *Topology* refers to the pattern in which the processors are connected and reflects how data flow. Currently available designs include the hypercube arrangement, a network of busses, and banyan networks. Usually the interconnection topology is related to the memory organization. For example, iPSC, NCube,

and the Connection Machine are local memory multiprocessors with a hypercube topology. On these systems a processor is assigned its own (local) memory and can access only this memory. Independent processors communicate by sending each other messages. The efficient solution of finite element simulations on these machines requires minimizing the interprocessor communication bandwidth. This requires the mapping of adjacent elements as far as possible onto directly connected processors, which can be a nontrivial problem. On the other hand, the processors on a shared-memory system, such as Alliant FX/8, are connected through a common memory bus and can access the same (global) large memory system. Adequate finite element parallel data structures are crucial for efficient computations on both shared and local memory multiprocessors. On a local memory machine one must introduce the concept of distributed data base and data structure. Each local memory is loaded only with the data relevant to the computational task assigned to its attached processor. For a system with thousands of processors the total amount of available memory can be very large, yet it is the storage capacity of each local memory that really matters. Different finite elements require different amounts of data to be stored. For each finite element in the mesh, a material and geometric nonlinear high-order shell element may require an amount of data storage two orders of magnitude higher than a simple linear truss element. Hence one may be able to assign one or several finite elements of a certain type to one processor but may fail in the attempt to assign one or several elements of another type to a similar processor. Also, in the case of MIMD (multiple instruction–multiple data) machines, such as iPSC and NCube, one must ensure that the compiled subroutines can be accommodated on the local memory. Consider the case in which a processor is mapped onto a submesh containing different types of elements. In this situation one must load into the processor's local memory all the element libraries for the types encountered in the assigned submesh. Generally one can overcome these problems by devising an intelligent partitioning scheme and a compact data structure. Careful data structures must also be designed for shared-memory multiprocessors to avoid potential serializations due to memory conflicts.

3   Finally, *control* describes the way the work is divided and synchronized. Of particular interest are the SIMD (single instruction–multiple data) and MIMD machines. The Cray Y-MP/8 (8 processors) and iPSC-32 (32 processors) are respectively a shared-memory MIMD supermultiprocessor and a local memory MIMD hypercube. They can simultaneously execute multiple instructions that can operate on multiple data. The Connection Machine is an SIMD system in which a single instruction is executed at a time, an instruction that can operate on multiple data. On an SIMD machine a single program typically executes on the front end and its parallel instructions are submitted to the processors. On an MIMD parallel processor separate program copies execute on separate processors.

Practically, local memory parallel processors are more difficult to program than shared-memory multiprocessors. However, it is believed that local memory systems are easier to scale to a large number of processors. Shared-memory multiprocessors are usually coarse grained because the bus to memory saturates and/or becomes prohibitively expensive above a few processors. Note that with SIMD machines one must devise special tricks to be able to process in parallel finite elements of different types since these do not involve the same instructions and only one instruction can be executed at a time.

Recommending one hardware architecture over another for parallel finite element computations is beyond the scope of this chapter. Our goal is to fully exploit the computational power of the fastest currently available multiprocessors of any architecture for solving large nonlinear finite element dynamic simulations.

3 A PORTABLE ARCHITECTURE FOR PARALLEL FINITE ELEMENT COMPUTATIONS

In this section we summarize a concurrent computer program architecture capable of handling linear and nonlinear and static and dynamic analyses. This software architecture is flexible enough to be implemented on all local memory and shared-memory SIMD and MIMD multiprocessors. It includes several special-purpose modules that constitute optimal algorithms for solving specific subproblems, such as domain decomposition, direct and iterative solution of systems of equations, solution of transient dynamics, and mesh refinement. These modules are designed to be consistent with the topology and data structure of the overall program architecture so that a maximum overall efficiency can be achieved during a complete finite element analysis.

3.1 DIVIDE AND CONQUER

In structural analysis the technique of dividing a large system into a system of substructures is very old and is still used extensively. For large aerospace structures its use is often motivated by the fact that different components are designed in parallel by different groups or companies. Therefore only the basic static and dynamic properties of the substructure need to be communicated between groups. This approach has also resulted in computational savings. If the design of one component is changed, only that substructure needs to be reanalyzed and the global system of substructures re-solved. In the case of limited nonlinear systems only the substructures that are nonlinear need be studied incrementally with time.

It is clear that this traditional substructuring approach can be used with parallel processors if the complete finite element system is subdivided so that each group of elements within a small domain is assigned to one processor. The data structure for such an approach is very simple. On local memory multiprocessors only the storage for the node geometry and element properties within the substructure need be stored within the random-access memory (RAM) of the processor assigned to that substructure. In addition, concurrent formation and reduction of the stiffness matrix for that region require no interprocessor communication. On shared-memory multiprocessors the substructure data are accessed only by the processor assigned to that substructure so that no memory conflict occurs. After the displacements are found the postprocessing of substructure stresses can be done in parallel.

3.2 Automatic Domain Decomposer

Here it becomes essential to distinguish between a substructure and a subdomain. By *subdomain* we mean a collection of elements of the global domain (mesh) that is assigned to a corresponding processor. A subdomain may thus represent anything from one element of a physical substructure to several substructures.

Domain decomposition is attractive in finite element computations on parallel architectures because it allows individual subdomain operations to be performed concurrently on separate processors. Given a number of available processors N_p, an arbitrary finite element domain is decomposed into N_p subdomains where each of the following computations can be carried out independently of similar computations for the other subdomains and hence performed in parallel:

Formation of element matrices
Assembly of global matrices
Partial factorization of the stiffness matrix (either in view of the solution of the
    equations of equilibrium with a direct method or for preconditioning these
    before the use of an iterative numerical scheme for that purpose)
State determination or evaluation of the generalized stresses

All these operations can be executed concurrently without synchronization or message passing between the processors. Since the time to complete a task is the time to complete the longest parallel subtask (which is the restriction of a task to a subdomain), an algorithm for domain decomposition is efficient only if it yields subdomains that require an equal amount of execution time for each of these operations. In other words, the algorithm must achieve a load balance among the processors.

Whatever numerical scheme or computational strategy is selected, the parallel solution for the generalized displacements **u** usually requires explicit synchronization on shared-memory multiprocessors and message passing on local memory multiprocessors. The reason is that the physical subdomains share physical information along their common boundaries that cannot be treated independently or asynchronously in each subdomain. This results in an overhead that, if not minimized, may ruin the sought-after speedup. Hence when partitioning an arbitrary finite element mesh, one must keep in mind the nature of the numerical parallel algorithm that will be selected. Parallel explicit computations require message passing only between neighboring subdomains. Therefore for explicit and explicitlike algorithms, such as an iterative solver for the linearized static problem or a time integration-explicit algorithm for the transient response analysis, an optimal decomposition is the one that minimizes the communication bandwidth of the problem, that is, the subdomain interconnectivity. For implicit computations an optimal decomposition is the one that minimizes the number of interface nodes.

These observations suggest that an automatic finite element domain decomposer must meet three basic requirements to be successful: (1) it must be able to handle irregular geometry and arbitrary discretization to be general purpose; (2) it must yield a set of balanced subdomains to ensure that the overall computational load is as evenly distributed as possible among the processors; (3) it must minimize the subdomain interconnectivity bandwidth when used with an explicit algorithm and/or the amount of interface nodes in presence of implicit computations to reduce the cost of synchronization and/or message passing between the processors. A simple and automatic finite element decomposer that meets these requirements and is completely transparent to the user has been described by the author (Farhat, 1988). It is applied to the discretized flexible aircraft shown in Figure 1a. Figure 1b displays the resulting subdomains for an eight-processor machine.

3.3 Parallel Setup of the Finite Element Simulation

Consider the automatic decomposition of the mesh shown in Figure 1a into a system of subdomains D_j. The mesh nodes, which are common to the subdomain interfaces, define a unique global interface denoted D_i. They are automatically identified by the program. The nodal point unknowns within the subdomains are numbered first, and the interface nodes are numbered last. The resulting stiffness matrix $\mathbf{K}$ has the following "arrow" pattern:

$$\mathbf{K} = \begin{bmatrix} \mathbf{K}_{11} & & & & & & \mathbf{K}_{si}(1) \\ & \mathbf{K}_{22} & & & & & \mathbf{K}_{si}(2) \\ & & \cdots & & & & \cdots \\ & & & \mathbf{K}_{jj} & \cdots & & \mathbf{K}_{si}(j) \\ & & & & & & \cdots \\ \mathbf{K}_{1i}^T & \mathbf{K}_{2i}^T & \cdots & \mathbf{K}_{ji}^T & \cdots & & \mathbf{K}_{ii} \end{bmatrix} \quad (1)$$

Each diagonal submatrix $\mathbf{K}_{jj}$ represents the local stiffness of a subdomain D_j. An off-diagonal submatrix $\mathbf{K}_{ji}$ denotes the coupling stiffness between D_j and the

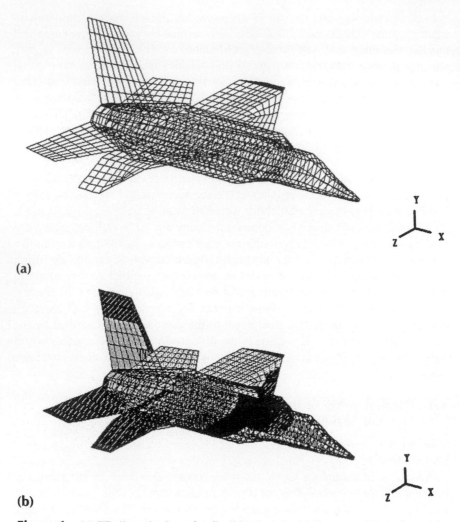

(a)

(b)

Figure 1 (a) FE discretization of a flexible aircraft. (b) Decomposition of an eight-processor machine and implicit parallel computations.

interface D_i. Block $\mathbf{K}_{ii}$ is the stiffness associated with D_i. Consequently the vector of unknown responses $\mathbf{u}$ is partitioned into subvectors $\mathbf{u}_j$ that correspond to the degrees of freedom lying in region D_j; similarly, $\mathbf{f}_j$ denotes the part of the loading vector $\mathbf{f}$ associated with D_j. If a consistent mass matrix $\mathbf{M}$ is used for the dynamic analysis, $\mathbf{M}$ has the same pattern as $\mathbf{K}$. If a lumped mass matrix is used,

M has the same pattern as **f**. Each processor p_j is assigned to a subdomain D_j. It forms and stores in parallel with the other processors the local stiffness $\mathbf{K}_{jj}$, its corresponding coupling term $\mathbf{K}_{ji}$, the mass $\mathbf{M}_j$, and the prescribed loading $\mathbf{f}_j$. An eventual damping term **C** is treated in a fashion similar to that used for the mass and stiffness terms. As noted earlier, the setup of the problem in each subdomain D_j does not require interprocessor communication or processor synchronization.

Implicit algorithms usually require the assembly of the stiffness, mass, and damping matrices. Explicit algorithms require only the assembly of forces and residuals at the finite element nodes. Hence both computational strategies involve some kind and some amount of assembly operations. Inside each subdomain D_j the assembly of all of $\mathbf{K}_{jj}$, $\mathbf{K}_{ji}$, $\mathbf{M}_j$, $\mathbf{C}_{jj}$, $\mathbf{C}_{ji}$, and $\mathbf{f}_j$ requires no interprocessor communication on a local memory multiprocessor and does not induce any memory conflict on a shared-memory multiprocessor. On the other hand, the assembly on local memory multiprocessors of such interface quantities as $\mathbf{K}_{ii}$, $\mathbf{M}_i$, $\mathbf{C}_{ii}$, and $\mathbf{f}_i$ requires that the processors assigned to adjacent subdomains communicate to add their local contributions. On a shared-memory multiprocessor the assembly of interface quantities in implicit algorithms and the assembly of nodal forces and residuals in explicit algorithms induce memory conflicts that may serialize the algorithms (see, for example, Berger et al., 1982). To resolve this problem the automatic finite element decomposer is augmented with a coloring algorithm that can be applied to the entire domain or to the interface only, depending on the computational strategy to be used. In essence this algorithm reorders the specified elements in groups of disconnected elements. The disjoint elements within each such group are processed in parallel without synchronization. This simple idea is illustrated in Figure 2, where it is applied to the domain of Figure 1a. Reordered elements belonging to one specific group (color) are displayed. More details on assembling computations on a shared-memory multiprocessor can be found in Berger et al. (1982), Farhat (1986), and Farhat and Crivelli (1989).

3.4 Parallel Compact Data Structures

Explicit algorithms operate with very simple data structures. Basically these algorithms operate at the element level so that no matrix need be formed or assembled. On the other hand, implicit computations often require the factorization of some matrix, for example the stiffness matrix **K**.

Since all the subdomains D_j are separated by the interface, they constitute a set of disconnected meshes that can be numbered independently. Each diagonal block $\mathbf{K}_{jj}$ is stored in skyline or profile form. The degrees of freedom in each subdomain are renumbered in parallel to minimize the local storage requirements (Hoit and Wilson, 1983). Note that the complexity for renumbering N_p disconnected subdomains D_j is much lower than for renumbering the entire domain D. Moreover, this phase is fully speeded by the multiprocessing capabilities.

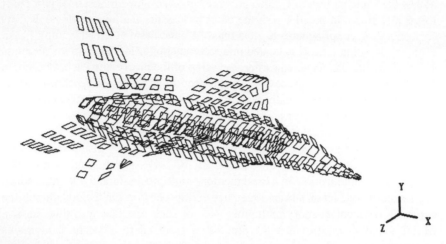

Figure 2   Graph coloring for element reordering.

The coupling blocks $\mathbf{K}_{ji}$ are usually very sparse and do not introduce fill-in if a suitable direct method is selected for the factorization of $\mathbf{K}$. Consequently only their nonzero values are stored columnwise in packed lists. In addition to $\mathbf{K}_{ij}$, $\mathbf{K}_{ji}$, and $\mathbf{F}_j$, each processor holds or is assigned to a set of columns of $\mathbf{K}_{ii}$, which is the stiffness associated with the subdomain D_i. These data structures and the processor assignments are summarized in Figure 3.

3.5   Postprocessing State Determination

The postprocessing phase, that is, the evaluation of field derivatives like stresses and fluxes, is parallelized subdomain by subdomain or element by element on both local memory and shared-memory architectures without communication or synchronization overhead. However, on local memory multiprocessors state determination in nonlinear analysis can face load-balancing problems, for example the number of elements, which may yield vary from subdomain to subdomain. Since the complexity of the computation of the internal forces depends on the constitutive equation that applies to the particular element in a specific subdomain, this implies that some processors would have to work less than others and hence would have to wait for each other. A load-balancing strategy for problems in which the nonlinearity is localized was described by Farhat and Wilson in 1987. It can be extended to the case in which the analyst can predict the regions in the finite element mesh that show only a linear behavior. The computational strategy is summarized here.

Let D be a given arbitrary finite element domain, and let $D^{(n)}$ be the known subset of D in which nonlinearities are localized and/or predicted. With these

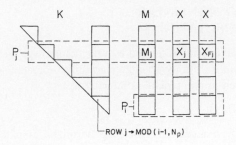

Figure 3 FE parallel data structures.

definitions $D^{(n)}$ may be a set of regions separated by linear substructures and hence need not be a single physical substructure. We shall assume that $D^{(n)}$ has somehow been marked during the mesh generation phase so that all its elements can be identified at any time by the software. In our program these elements are marked with a specific element attribute value. Also, we define $D^{(l)}$ as the collection of elements of D that are known (or predicted) to undergo only linear deformations. Hence we can write

$$D = D^{(l)} \cup D^{(n)}$$
$$D^{(l)} \cap D^{(n)} = 0$$

The automatic subdivision algorithm is extended to perform two passes on D. In the first pass the marked elements are skipped and $D^{(l)}$ is split into N_p subdomains. The second pass decomposes $D^{(n)}$. As a result D is subdivided into two sets of N_p subdomains each:

1. Linear subdomains (or predicted so): $D_j^{(1)}, j = 1, 2, \ldots, N_p$
2. Nonlinear subdomains (or predicted so): $D_k^{(n)}, k = N_p + 1, N_p + 2, \ldots, 2N_p$.

Renumbering is done as for the linear case, the linear subdomains being treated first. Each processor p_j is assigned the pair of subdomains $D_j^{(1)}$ and $D_{j+N_p}^{(n)}$ so that load balancing is achieved. Further details of the implementation of this strategy in local memory MIMD multiprocessors can be found in (Farhat, 1987).

On shared-memory multiprocessors load balance for state determination in nonlinear analysis is achieved simply by assigning the processors to elements in a self-scheduled manner; that is, elements are assigned to processors on the basis of first available,–first served. This is of course possible because each processor can access the large shared memory of the system.

4 LINEAR AND NONLINEAR STATIC COMPUTATIONS

In this section linear analysis is treated as a particular case of nonlinear analysis. Discrete equilibrium equations arising from finite element nonlinear formulations may be written in the general compact form

$$\mathbf{r}(\mathbf{u}, \mathbf{p}, \theta) = 0 \tag{2}$$

where $\mathbf{u}$ denotes the unknown vector of generalized displacements (rotations, temperatures, and so on) at the nodes of the discretized geometric domain, $\mathbf{p}$ denotes a set of control parameters, θ is a functional of past history of the generalized deformation gradients, and $\mathbf{r}$ denotes the residual vector of out-of-balance generalized forces (moments, fluxes, and others). Equation (2) covers all geometric nonlinearities, material nonlinearities, and several types of boundary condition nonlinearities.

The Newton-Raphson method and its numerous variants, collectively known as Newton-like methods, are the most popular class of methods for the solution of Eq. (2) on conventional computers. These methods can embed either a direct or an iterative algorithm for the solution of a linearized system of equations.

Direct solution techniques have been popular among engineers, mainly because of two advantages they possess over iterative schemes:

1. They are robust for ill-conditioned systems, which often arise in the analysis of flexible space structures.
2. Their execution time can be estimated for any given problem.

They can be sensitive to rounding off error (matrices with high condition number), but their main disadvantage is that they suffer from excessive storage requirements for large matrices so that an out-of-core solution is often required. However, they are still attractive for such supercomputers as Cray-2, in which the main memory can store up to 256 million double-precision words.

On the other hand, the use of iterative schemes for the solution at each step of the linearized system of equations has two desirable advantages:

1. It efficiently exploits the sparsity of the involved matrices and therefore requires less storage than direct schemes.
2. It provides a means of controlling the accuracy of the solution.

The preconditioned conjugate gradient (PCG) has emerged over the last decade as a favorite algorithm for solving large sparse systems of linearized equations on sequential, vector, and parallel computers (Carey and Jiang, 1984; Van Der Vorst, 1986; Law, 1986; Kowalik and Kumar, 1982; Benner and Montry, 1986). However, conventional Newton-like methods may behave poorly near bifurcation points and often fail to handle path-dependent problems, such as plastic flow, where the stiffness matrix may oscillate wildly as the solution changes by small amounts. For such problems explicit dynamic relaxation (DR) is a very robust iterative computational strategy.

Since the analyst would like to select the right strategy for the right problem, we have implemented within the same prototype parallel code Newton-like methods with iterative and direct solvers and explicit dynamic relaxation algorithms.

4.1 Newton-like Methods

Newton-like methods for solving nonlinear systems of equations having the general form $r(u, p, \theta) = 0$ are usually related to the following iteration scheme.

FOR $k = 0, 1, 2, \ldots$ until convergence DO:

Solve $\quad K(u_k) \, \delta u_{k+1} = -\delta f(u_k)$

Set $\quad u_{k+1} = u_k + \delta u_{k+1}$

where $K(u_k)$ is the stiffness matrix evaluated in the displacement state u_k, δu_{k+1} is the vector of unknown displacement increments, and $\delta f(u_k)$ is the vector of out-of-balance forces.

Next we examine both parallel direct and iterative methods for the solution at each iteration of the linearized system of equations. We emphasize only important parallel computation aspects and refer the reader to Farhat et al. (1987a, b), Farhat and Crivelli (1989), and Farhat and Wilson (1988) for implementation details on local and shared-memory multiprocessors.

Direct Method. A five-step direct algorithm for the solution at each iteration of the complete finite element system [see Eq. (1)] is given here in symbolic form:

1. Factor locally:

$$K_{jj} = L_j \, D_j \, L_j^T \qquad j = 1, N_p$$

2. Eliminate:

$$K_{ji}^T \, K_{jj}^{-1} \, K_{ji} \qquad j = 1, N_p$$

3. Update interface:

$$K_{ii} = K_{ii} - \sum_{j=1}^{j=N_p} K_{ji}^T \, K_{jj}{}^1 K_{ji}$$

and

$$\delta f_i = \delta f_i - \sum_{j=1}^{j=N_p} K_{ji}^T \, K_{jj}^{-1} \delta f_j$$

4. Solve interface:

$$K_{ii} \, \delta u_i = \delta f_i$$

5.   Backsolve locally:

$$\mathbf{K}_{jj}\, \delta\mathbf{u}_j = \delta\mathbf{f}_j - \mathbf{K}_{ji}\, o\mathbf{u}_i \qquad j = 1, N_p$$

Clearly, steps 1, 2, and 5 can be carried out concurrently on all N_p processors without interprocessor communication (local memory MIMD machines) or synchronization (shared-memory multiprocessors). Step 3 deserves special attention. On local memory MIMD multiprocessors some message passing is required because $\mathbf{K}_{ii}$ is scattered columnwise among the processors (Farhat, 1987). On shared-memory multiprocessors it can be shown that step 3 can be carried out in parallel without memory conflict (Farhat, 1989). Finally, step 4 is treated with a dedicated parallel profile equation solver (Farhat, 1988).

Remarks

The substructuring technique introduces a high level of parallelism, sometimes at the cost of additional floating-point computations, however. To be feasible this parallel algorithm requires the mesh partition to be such that the local bandwidth of each subdomain is much smaller than the bandwidth of the global problem. Otherwise it performs more floating-point operations than the serial algorithm. Therefore, if this condition is not satisfied, the parallel algorithm may be outperformed on a shared-memory multiprocessor by a global parallel direct solver like that applied to the interface problem (see Farhat, 1989 for details).

At each iteration, after the generalized displacements are found, state determination is carried out in parallel as indicated earlier.

On multiprocessors with a limited amount of memory, the interface final system of equations (Schur's complement) may not fit in the core. This can happen, for example, if the mesh size is small and the number of processors is relatively high. In this case the program switches to a parallel profile solver for the complete finite element system. Other possibilities for overcoming this difficulty include an out-of-core strategy or a semi-iterative solution scheme.

Semi-iterative Methods.   Equation (1) has inspired several preconditioners for the conjugate gradient method. Results from domain decomposition theory suggest that the interface equations are better conditioned than the entire system of equations (Bjorstad and Widlund, 1986). Consequently it is more efficient to apply PCG only to the system of interface equations:

$$(\mathbf{K}_{ii} - \sum_{j=1}^{j=N} \mathbf{K}_{ji}^T\, \mathbf{K}_{jj}^{-1}\, \mathbf{K}_{ji})\, \delta\mathbf{u}_i = \delta\mathbf{f}_i - \sum_{j=1}^{j=N_p} \mathbf{K}_{ji}^T\, \mathbf{K}_{jj}^{-1}\, \delta\mathbf{f}_j \qquad (3)$$

Note that the left side of Eq. (3) need not be formed and stored. During a conjugate gradient iteration it is used only multiplicatively. See Bjorstad and Widlund (1986) and Branble et al. (1984) for examples of preconditioners and their performance.

4.2 Explicit Dynamic Relaxation

Dynamic relaxation is a robust iterative method for solving highly nonlinear systems. Unlike the conjugate gradient method, it does not need to be embedded within a Newton outer loop. The algorithm solves the nonlinear discrete quasi-static finite element equations (4) by viewing them as the steady-state solution of the second-order pseudodynamic problem:

$$\mathbf{M\ddot{u}} + \mathbf{C\dot{u}} + \mathbf{S(u, p, \theta)} - \mathbf{f} = 0 \tag{4}$$

where $\mathbf{M}$ and $\mathbf{C}$ are fictitious mass and damping matrices constructed in a way that achieves computational efficiency when integrating Eq. (2) with the central difference scheme. $\mathbf{S(u, p, \theta)}$ denotes the internal forces. To preserve the explicit form of the central difference integrator $\mathbf{M}$ must be diagonal and $\mathbf{C}$ is chosen, for example, as $\mathbf{C} = c\mathbf{M}$. Parameter c and stepsize δt are selected to obtain the fastest convergence:

$$\delta t \le \frac{2}{(\omega_{max}^2 + \omega_{min}^2)^{1/2}}$$

$$c = \frac{2\omega_{min}}{(1 + \omega_{min}^2/\omega_{max}^2)^{1/2}}$$

where ω_{min} and ω_{max} are the lower and higher pseudofrequencies of the pseudodynamic problem, respectively. These quantities need not be computed exactly. Rough estimates are sufficient. Further details on $\mathbf{M}$, $\mathbf{C}$, c, and δt may be found in Papadrakakis (1981).

Dynamic relaxation may be slow in some cases. However, it presents two advantages: it is very robust, and it is fully vectorizable and parallelizable. The vector form of Eq. (4) clearly demonstrates the suitability of the algorithm for vector processors. By eliminating the need for matrices one saves not only storage but also the overhead associated with interprocessor communication that is usually required by algorithms manipulating two-dimensional arrays. The explicit nature of the central difference method allows computations on different vector subcomponents to be performed in parallel without processor synchronization and/or interprocessor communication, except for the evaluation of $\mathbf{S(u, p, \theta)}$, which requires special attention (Farhat et al., 1989). The internal generalized force vector is obtained by accumulating the contributions of several connected elements. If $\mathbf{u}_k$ denotes the generalized displacement associated with the kth degree of freedom, then

$$[\mathbf{S(u, p, \Theta)}]_k = \sum_{\mathbf{el}=1}^{\mathbf{el}=N} [\mathbf{S(u, p, \Theta)}]_k^{(\mathbf{el})}$$

where $[\mathbf{S(u, p, \theta)}]_k^{(\mathbf{el})}$ denotes the contribution of element $\mathbf{el}$ connected to the kth degree of freedom and is computed directly at the element level from a potential functional. On a local memory system the processors assigned to adjacent sub-domains communicate to exchange partial results at common interface nodes. On a shared-memory multiprocessor these partial results are accumulated via the coloring technique described earlier. See (Farhat et al., 1989) for implementation details.

5 EIGENVALUE AND TRANSIENT COMPUTATIONS

In this section we summarize parallel methods for the solution of the equations of equilibrium for a finite element system in motion:

$$\mathbf{M\ddot{u}} + \mathbf{C\dot{u}} + \mathbf{Ku} = \mathbf{f} \tag{5}$$

where $\mathbf{M}$, $\mathbf{C}$, and $\mathbf{K}$ are the real mass, damping, and stiffness $n \times n$ matrices and $\mathbf{f}$ is the prescribed external time-dependent load vector. First we focus on Rayleigh-Ritz solution schemes, and then we look at direct time integration algorithms. Detailed implementations on parallel processors for both strategies may be found in Farhat and Wilson (1986, 1987).

5.1 Rayleigh-Ritz Solution

For a very large n, say over 10,000, it is common to use a reduction technique before determining the solution of Eq. (5). In this approach the solution $\mathbf{u}(s, t)$ is assumed to be a linear combination of a set of independent vectors $\mathbf{v}_i(s)$:

$$\mathbf{u}(s,t) = \sum_{i=1}^{i=q} \alpha_i (t)\mathbf{v}_i(s) \tag{6}$$

where s represents the space variables, t is the time variable, and q is much smaller than n. If $\mathbf{V}$ is the $n \times q$ matrix whose columns are $\mathbf{v}_i$ and α is the vector of dimension $\mathbf{q}$ and with entries $\alpha_i(t)$, the matrix form of Eq. (6) is

$$\mathbf{u}(s, t) = \mathbf{V} \alpha \tag{7}$$

Premultiplying Eq. (5) by $\mathbf{V}^t$ and substituting Eq. (7) lead to reduced equations of motion:

$$(\mathbf{V}^t\mathbf{MV}) \ddot{\alpha} + (\mathbf{V}^t\mathbf{CV}) \dot{\alpha} + (\mathbf{V}^t\mathbf{KV}) \alpha = \mathbf{V}^t\mathbf{f} \tag{8}$$

These reduced equations of motion are of size $q \ll n$.

The Lanczos algorithm (Parlett, 1980) and other Lanczos-like algorithms (Wilson et al., 1983) generate a set of $\mathbf{M}$-orthonormal vectors $\mathbf{v}_i$, $i = 1, q$, at a fraction of the computational effort required for the calculation of exact mode shapes. The Lanczos vectors have the following properties:

$$\mathbf{V}^t \mathbf{M} \mathbf{V} = \mathbf{I}_q$$
$$(\mathbf{V}^t \mathbf{K} \mathbf{V})^{-1} = \mathbf{T}_q \tag{9}$$

where $\mathbf{I}_q$ is the $q \times q$ identity matrix and $\mathbf{T}_q$ is an $q \times q$ symmetric tridiagonal matrix.

Using the Lanczos vectors and assuming a Rayleigh damping matrix $\mathbf{C} = m\mathbf{M} + k\mathbf{K}$ transforms the original equations of miton (5) into the reduced tridiagonal system of second-order differential equations:

$$\mathbf{T}_q \, \ddot{\alpha} + (m\mathbf{T}_q + k\mathbf{I}_q) \, \dot{\alpha} + \mathbf{I}_q \alpha = \mathbf{T}_q \, \mathbf{V}^t \mathbf{f} \tag{10}$$

These reduced tridiagonal equations may be integrated directly, or the eigenvectors and eigenvalues of $\mathbf{T}_q$ may be calculated and the solution to Eq. (10) may be obtained by integrating the uncoupled equations. The latter approach is treated first. The direct integration of Eq. (10) is a particular case of the direct time integration of Eq. (5) and is treated next.

A Rayleigh-Ritz procedure based on a Lanczos-like algorithm (Wilson et al., 1983) is summarized here:

$\mathbf{K} = \mathbf{L}^t \, \mathbf{D} \mathbf{L}$ $n \times n$ system

$\mathbf{K}\bar{\mathbf{v}}_1 = \mathbf{f}$ solve for $\bar{\mathbf{v}}_1$

$b_1 = (\mathbf{v}_1^t \, \mathbf{M} \bar{\mathbf{v}}_1)^{1/2}$ **M** normalization

$\mathbf{v}_1 = \bar{\mathbf{v}}_1 \dfrac{1}{b_1}$

$\mathbf{K}\bar{\mathbf{v}}_i = \mathbf{M}\bar{\mathbf{v}}_{i-1}$ solve for $\mathbf{v}_i$

$a^{i-1} = \bar{\mathbf{v}}_i^t \, \mathbf{M}\mathbf{v}_{i-1}$ diagnoal of $\mathbf{T}_q$

$c^j = \mathbf{v}_j^t \, \mathbf{M}\bar{\mathbf{v}}_i$ compute for $j = 1, \ldots i-1$

$\hat{\mathbf{v}}_i = \bar{\mathbf{v}}_i - \displaystyle\sum_{j=1}^{j=i-1} c_j \, \mathbf{v}_j$ **M** orthogonalization

$b_i = (\hat{\mathbf{v}}_i^t \, M \, \hat{\mathbf{v}}_i)^{1/2}$ off-diagonal of $\mathbf{T}_q$

$\mathbf{v}_i = \hat{\mathbf{v}}_i \dfrac{1}{b_i}$ **M** normalization

$$[T_r] = \begin{bmatrix} a_1 & b_2 & 0 & . & . & . & 0 \\ b_2 & a_2 & b_3 & . & & . & . \\ 0 & b_3 & a_3 & . & & . & . \\ 0 & 0 & . & . & & . & 0 \\ . & . & . & b_{r-1} & a_{r-1} & b_r \\ 0 & . & . & . & b_r & a_r \end{bmatrix}$$

$$T_q Z = Z\Omega$$

$$[\omega^2] = \begin{bmatrix} \dfrac{1}{\Omega} \end{bmatrix}$$

$$V^* = VZ$$

This algorithm consists essentially of four phases:

1. Factorization of the stiffness matrix K
2. Generation of Lanczos vectors v and construction of T_l
3. Solution of the reduced eigenvalue problem:

$$\mathbf{T}_q \, Z = \mathbf{Z}\Omega$$

4. Computation of Ritz vectors and corresponding frequencies:

$$V^* = VZ$$

$$[\omega^2] = \begin{bmatrix} \dfrac{1}{\Omega} \end{bmatrix}$$

After these computations are performed, the uncoupled equations of motion are solved for $\alpha_i(t)$ and the solution to Eq. (5) is computed as $u = V^*\alpha = VZ\alpha$.

The parallel algorithms for finite element static analysis are used to factor K and generate the Lanczos vectors. The construction of $\mathbf{T}_q$ involves vector manipulations consisting of inner products and trivial scaling. On multiprocessors with local memory an inner product is first carried out locally and then the partial dot products are accumulated in $\log_2 N_p$ stages, following a binary tree. On a shared-memory multiprocessor the coloring technique described earlier is invoked to avoid memory conflicts during the accumulation phase.

As noted earlier, $\mathbf{T}_l$ is an $q \times q$ tridiagonal symmetric matrix and q is of the order of 100. On a local memory multiprocessor $\mathbf{T}_q$ can be duplicated concurrently, with a very low storage cost in each processor, as a set of 2 one-dimensional arrays storing the main and the upper diagonals, respectively.

The solution to the reduced eigenvalue problem $\mathbf{T}_q \, Z = \mathbf{Z}\Omega$ does not involve a significant amount of computational effort since the system is tridiagonal. However, it can be further reduced by the use of parallel processing. A simple algorithm for a concurrent extraction of all the eigenvalues of $\mathbf{T}_q$ was described by Farhat and Wilson in 1986. Basically, the spectrum of eigenvalues is divided into $N_p + 1$ subintervals $[s_p, e_p]$, each containing an almost equal number of eigenvalues (see Farhat and Wilson, 1987, for details on that splitting). The task of each processor p is to compute the eigenvalues of $\mathbf{T}_q$ that lie within its assigned interval $[s_p, e_p]$ as the roots of the polynomial $P(\lambda) = \det \mathbf{T}_q$ $- \lambda I$. A regulafalsi serach method together with the Sturm sequence property is used for this purpose. The corresponding eigenvectors are obtained via inverse shifted iteration. This procedure has two advantages:

1. No overlapping between the frequency subdomains can occur, and hence none of the eigenvalues or eigenvectors is computed twice (by two different processors).
2. All processors perform their task concurrently without communication or synchronization.

The parallel implementation of all four phases on local memory multiprocessors can be found in Farhat and Wilson (1986, 1987).

5.2   Direct Time Integration

Whether for integrating Eq. (5) or its reduced form [Eq. (10)] for linear or nonlinear problems, direct integration methods can be implicit or explicit. They involve algebraic computations that are similar to those of static analysis (e.g., factoring matrices and performing matrix-vector products and inner products). Implicit schemes are parallelized at the subdomain level, as for nonlinear static analysis, and explicit schemes are parallelized at the element level, as for dynamic relaxation. A prallel implementation of the Wilson $-\theta$ method on Intel's iPSC was described by the author in [1987]. Its conversion for shared-memory multiprocessors, such as Cray-2, is straightforward.

6   MASSIVELY PARALLEL FINITE ELEMENT COMPUTATIONS

The Connection Machine is one of the few massively parallel processors that are now commercially available. It consists of two parts: a front-end computer (VAX, Symbolics, or Sun), and a 64,000-K processor hypercube (65,536 single-bit processors). The front-end computer provides instruction sequencing and program development and has the ability to address any location in the hypercube distributed memory. The hypercube system provides number-crunching power. The Data Vault is the Connection Machine mass storage system. Each Data Vault unit is associated with one-eighth of a fully configured system. It stores its data in an array of 39 individual disk drives. With this disk-farming system the concept of performing parallel I/O is carried through: instead of regarding a file as a serial stream of bits, the file system of this parallel processor regards it as many streams of bits, which are read or written in parallel, one stream per processor. When eight Data Vaults operate in paralle, they offer a combined data transfer rate of 320 MB/S and hold up to 80 GB of data. The Connection Machine graphic display system, known as the Frame Buffer, also incorporates the concept of parallelism. It allows the user to visualize on a color monitor screen the data in the processors. The display can be updated as computations are performed.

Recently, Farhat et al. (1989b, 1990) investigated massively parallel transient finite element explicit computations on the Connection Machine. Preliminary results can be found in Farhat et al. (1989b) and more detailed information in Farhat et al. (1990). In general it has been found that this highly parallel processor can outperform vector supercomputers on explicit computations, but not on implications. Several features distinguish the Connection Machine from earlier hypercubes. On the hardware side we note the impressive number-crunching power and the fast parallel I/O capabilities. On the software side we note the virtual processor concept, which is somehow the dual of the well-known virtual memory concept. When this massively parallel processor is initialized for a run, the number of virtual processors may be specified. If this exceeds the number of available physical processors, then the local memory of each processor is split into a number of regions equal to the ratio between the number of virtual processors and the number of physical processors. Automatically, for every Paris (parallel instruction set) instruction, the processors are time sliced among the regions. If a physical processor is simulating N virtual processors, each Paris instruction is decoded only once for N executions. This results in an enhanced user performance.

Mesh decomposition and processor-to-element mapping are the two fundamental keys for efficient massively parallel finite element computations. A given finite element mesh is partitioned into 16-element subdomains that correspond to the 16-processor chips of the Connection Machine. This partitioning is carried out in a way that minimizes the number of nodes at the interface between the subdomains. As a result only those processors that are mapped onto finite elements at the periphery of a subdomain communicate with processors packaged on different chips. Moreover, this partitioning is such that the connectivity bandwidth of the resulting subdomains is large enough to allow an efficient use of the 12 interchip wires. The mapping algorithm attempts to reduce the distance that information must travel through the communication network. In essence it searches iteratively for an optimal mapping through a two-step minimization of the communication costs associated with a candidate mapping (see Farhat, 1989). We summarize here the basic conclusions reported in [1989, 1990]. The processor memory size of 64 Kbits penalizes high-order elements. Three-dimensional and high-order elements induce longer communication times. Mesh irregularities slow computation speed in many ways. The Data Vault is very effective at reducing I/O time. The Frame Buffer is ideal for real-time visualization. Finally, the virtual processor concept outperforms the substructuring technique on the Connection Machine.

7 EXPERIMENTS WITH PARALLEL I/O

Realistic finite element modeling of real engineering systems involves the handling of very large data spaces that can amount to several gigabytes of memory. To cope with this many programs in the general area of solid mechanics and structural analysis use out-of-core data base management systems. However, I/O traffic between the disk and the processor memory slows the computations significantly and can dominate the overall cost of the analysis.

In a typical finite element analysis nodal and element data are retrieved from a storage disk before their processing and then stored back on the same storage disk after their processing has been completed. Examples include the transfer of nodal point coordinates and of elemental mass and stiffness matrices in element-by-element computational procedures and of history response arrays in time-stepping algorithms for linear and nonlinear dynamics. Other examples include the movement into and out of core of blocks of an assembled stiffness or mass matrix in original or factored form and the output on disk of the final results of an analysis.

Here we briefly summarize two strategies that have been developed for speeding I/O manipulations in finite element analysis through parallel processing. The first strategy is tuned for local memory MIMD multiprocessors with parallel I/O capabilities. The second is targeted for shared-memory multiprocessors, such as Cray Y-MP and IBM 3090.

7.1 Parallel I/O on Local Memory MIMD Multiprocessors

It is very natural to extend the domain decomposition idea to achieve parallel I/O in finite element analysis. For example, on local memory multiprocessors it is tempting to imagine that in the same way that a processor is assigned its own memory, it could be attributed its own set of I/O devices (I/O controller, disk drive, and so on) and its own files. Each processor would then read or write the data for its subdomain from its own files and through its own data base, in parallel with the other processors. If assigning an I/O controller and/or a disk drive to each processor is impractical and/or impossible, as is probably the case for a fine-grained system, it is possible for a cluster of processors.

After a given finite element domain is decomposed it is grouped into regions R_i, each containing a cluster of subdomains $D^{R_i}_j$. A host processor p^h_i is uniquely mapped onto each region R_i. It is assigned the task of handling I/O manipulations associated with computations performed primarily in the cluster of subdomains within R_i. Host processor p^h_i directly transfers data from the p_j RAM to its attached disk, and vice versa. Each host processor p^h_i is loaded with the same program driver, which we call the listener, and the same copy of a data base manager

(DBM). The main task of the listener is to listen to processor p_j's requests for I/O and to activate the DBM accordingly. These requests may be as follows:

Receive data from p_j, and store it in disk using DBM.
Retrieve data from disk through DBM, and send it to p_j.
Retrieve data from disk through DBM, send it to another host processor p^h_j
together with the instruction for broadcasting it to a specified number of computational nodes that are directly connected to p^h_j; this particular operation implements the potential exchange of data between subdomains.

7.2 Parallel I/O on Shared-Memory MIMD Multiprocessors

Unlike the previous approach, a single executable version of a sequential DBM is stored in the global memory of the multiprocessor. Moreover, there is no need for a listener since all processors can directly access DBM, the I/O library, and the disks. However, the core of the computational routines must be slightly modified to distinguish between global variables, which are shared by all the defined processes, and local variables, which have a single name to ease programming but a distinct value for each process. The essence of the strategy consists of

Distinguishing between synchronous and asynchronous I/O requests
Distinguishing between shared and private data
Partitioning the data stream into a balanced number of contiguous subsets equal
to the number of calling processes

The design, implementation, and performance of a simple parallel I/O manager (PIOM) operating with this logic was reported by Farhat et al. (1989a). Here we illustrate its functioning with the example of a synchronous I/O request for a shared variable (SRSV). For each file related to an SRSV, PIOM consults an I/O table. If the request is for storing data the PIOM logic is as follows:

S1. It partitions the information into a number of contiguous subsets equal to the number of calling processes, each subset containing an equal amount of data.
S2. For each subset it computes a pointer to the location in the shared buffer where the subset data stream begins.
S3. For each calling process it creates a corresponding S I/O process. Each S I/O process is assigned a subset of the data with its pointer.
S4. It reports in the I/O table the total number of created S I/O processes. For each S I/O process it specifies the length of its assigned data and its destination on a hardware device.
S5. It fires the S I/O processes. Each S I/O process repartitions its assigned data into a number of records that is a multiple of the total number of

available processors on the machine and then calls the DBM independently of another S I/O process. The reason for the internal partioning will become clearer in the remarks that follow.

On the other hand, if the request is for retrieving data, the PIOM logic is as follows:

R1.   It retrieves the I/O table corresponding to the file. If the number of calling processes is equal to the number of processes registered in the table (the S processes that originally stored the file), the inverse logic to the "store" case is followed and the data are retrieved in parallel

R2.   If not, for each registered S I/O process it partitions its subset of information into a number of contiguous blocks of data equal to the number of calling processes, each block containing an equal amount of data.

R3.   For each block it computes a pointer to the location in the shared buffer where the block data stream begins.

R4.   For each registered S I/O process it creates a number of R I/O processes equal to the number of calling processes. Each R I/O process is assigned a block of the subset data with its pointer.

R5.   It fires the R I/O processes.

R6.   It follows with the next S process to be retrieved.

Clearly, step S5 and steps R2–R6 allow a file that was written in parallel using p processes to be read in parallel using p^* processes, where p^* is different from p. In this case the retrieval of the file is carried out in p waves, each of a degree of parallelism equal to p^*.

8   SOFTWARE PORTABILITY

No standard parallel language is currently available, even for a given class of multiprocessors. Consequently a researcher in the area of applied parallel processing who is lucky enough to have access to a wide variety of parallel computers may give up on experimenting with these after considering the amount of recoding that he or she would need to do to transport programs from one machine to another. The lack of a standard parallel language is also discouraging commercial finite element software developers from moving their products to parallel processors.

In the short term one can find a temporary solution in portable parallel constructs. The Force (Jordan, 1986) provides a FORTRAN-style parallel programming language for shared-memory MIMD multiprocessors utilizing an extensive set of parallel constructs. It offers two desirable advantages:

1.   It insulates programmers from process management, freeing them to concentrate on the synchronization issues of parallel programming.

2. It ensures the portability of the programmer's code to several different shared-memory multiprocessors. Basically, the same code is run on any machine in which The Force has been installed.

3. On Alliant computers the compiler recognizes inherent parallelism at the DO loop level without the need for the programmer to invoke any explicit parallel construct. This may facilitate the programmer's work. However, the code would not run on the Cray-2, for example, because software support for multitasking on this supermultiprocessor is at the library level, where the user makes calls to ask the system for multitasking functions. For synchronization it is commonly necessary to wait until all processes have terminated a given task or to make sure that at any given time only one process modifies a variable. The procedures for these synchronizations are machine dependent. The Force relieves the programmer from the burden of modifying the code to port it to a new shared-memory multiprocessor. Because only the Force constructs need to be reprogrammed from one multiprocessor to another, the precious code need not be modified. For example, if the desire is to request all processes to wait until the longest has terminated, the same Force construct "Barrier" is invoked on any multiprocessor. On the Cray-2 running under UNICOS, the Force preprocessor reads the simple barrier statement and generates the following complex FORTRAN code:

```
CALL LOCKON(BARLCK)
IF (FFNBAR.LT.(NP—1)) THEN
FFNBAR = FFNBAR + 1
CALL LOCKOFF(BARLCK)
CALL LOCKON(BARWIT)
ENDIF
IF (FFNBAR.EQ.(NP-1)) THEN
ENDIF
IF(FFNBAR.EQ.0) THEN
CALL LOCKOFF(BARLCK)
ELSE
FFNBAR = FFNBAR—1
CALL LOCKOFF(BARWIT)
ENDIF
```

which invokes the appropriate UNICOS multitasking software utilities.

The shared-memory version of our prototype parallel code has been successfully implemented on Encore Multimax, Sequent Balance, Alli-ant FX/8, and Cray-2 using the macros of The Force. Since we are not aware of any similar product for local memory machines, we had to write two other separate versions, one for Intel's iPSC and another for the Connection Machine.

9   REAL-WORLD EXAMPLES AND MEASURED PERFORMANCE

Here we demonstrate the validity of our approach to linear and nonlinear and static and dynamic finite element parallel computations and assess its performance on iPSC, Alliant FX/8, Cray-2, Cray Y-MP, and the Connection Machine. These machines scan the various trends in today's parallel processing technology. Whenever a MFLOP rate (million floating-point operations per second) is reported, it corresponds to the overall finite element simulation. A pair $+$, $*$ is counted as two flops. The speedup S_{N_p} is defined as the ratio T_1/T_{N_p}, where T_1 denotes the CPU time elapsed using only one processor and T_{N_p} denotes the CPU time elapsed using N_p processors. It is very important to note that T_1 measures the performance of a sequential version of the code that is different from the parallel version in the sense that it does not contain any of the synchronization calls and it does not include the preprocessing phases, such as domain decomposition and element coloring. Hence the results we report here account for the extra time spent in domain decomposition and element coloring.

9.1   Intel iPSC

The Intel Personal SuperComputer iPSC/1 was one of the first commercially available local memory parallel systems, with 32, 64, or 128 processors. Each processor is an Intel 80286 CPU with an Intel 80287 floating-point acclerator and has 513 KB of local memory. The interconnection network is a hypercube (or n-dimensional cube, $n = 5, 6, 7$). Approximately 275 KB of the 512 KB RAM on each processor is available to user application, the rest being consumed by the operating system. The latter provides a simple message-passing interface between the nodes (processors).

First we report our experience with solving finite element static problems on an iPSC-32/1. We seek the linear response of the flexible aircraft of Figure 1a to a prescribed aerodynamic loading. The mesh contains 1280 shell elements and 7680 degrees of freedom. We use the automatic decomposer of (1988) to partition the finite element mesh into 32 balanced subdomains. The interface subdomain has 2030 equations. The speedup of each of the phases of the static analysis is reported in Table 1. Clearly, relatively high speedups are achieved. Note, however, that the interface parallel solution is less efficient than the subdomain parallel solution.

Next we consider a modal analysis of the simplified space station model shown in Figure 4. The finite element mesh comprises 384 nodes, 1264 beam elements, and 2304 degrees of freedom. Since this is rather a small problem we use only 16 processors of the iPSC/1.

Table 1   Speedup of a Linear Static Analysis
on iPSC-32/1

Problem setup	31.4
Subdomain condensation	30.2
Interface solution	24.9
Subdomain solution	31.4
Stress computations	31.4
Overall	28.1

Figure 4   FE space station simplified model.

Table 2   Performance Results for Modal Analysis on iPSC-16/1[a]

Phase	Speedup	MFLOPS
Forming **K** and **M**	15	0.6
Factoring **K**	12	0.5
Generating Lanczos vectors	12	0.5
Extracting 200 frequencies	14	0.4
Computing 200 mode shapes	13	0.5

[a]Space station structural model, 2304 degrees of freedom, 200 modes.

The mesh is decomposed into 16 balanced subdomains, each containing approximately 79 elements. The size of the interface problem is 672 (112 nodes). The number of extracted modes is 200. The performance results for this eigenvalue analysis are reported in Table 2.

Our experience with the iPSC/1 is limited to linear finite element simulations because of the lack of true I/O facilities on the nodes (a processor cannot open or read/write in a file).

9.2   Alliant FX/8

The Alliant FX/8 is a shared-memory system consisting of up to eight processors. Each processor supports the M68020 instruction set augmented with instructions for supporting floating-point arithmetic, vector arithmetic, and concurrency.

Finite element computations on the Alliant FX/8 may be speeded through a proper combination of vectorization and concurrency. In the following we report separately each of the achieved speedups.

The nonlinear static analysis of NASA's COFS-I truss beam, shown in Figure 5, is performed twice on the Alliant FX/8, first using preconditioned conjugate gradient and then dynamic relaxation.

The finite element discretization of the longeron is shown in Figure 6. It contains 1971 beam elements and 2016 degrees of freedom.

The finite element mesh is first partitioned into nine sets of disconnected elements. Hence at each round of explicit computations over all the elements of the mesh, only nine synchronization points are required. The elements of each set are further partitioned into subsets, which are processed in parallel. Within each subset vectorization is achieved by processing the elements in blocks of 32 at a time. The measured speedups due to vectorization and to concurrency are reported spearately in Figures 7 and 8. Here the speedup due to vectorization is defined as the ratio of the elapsed time using p processors with vectorization turned off over the elapsed time using p processors with vectroization turned on.

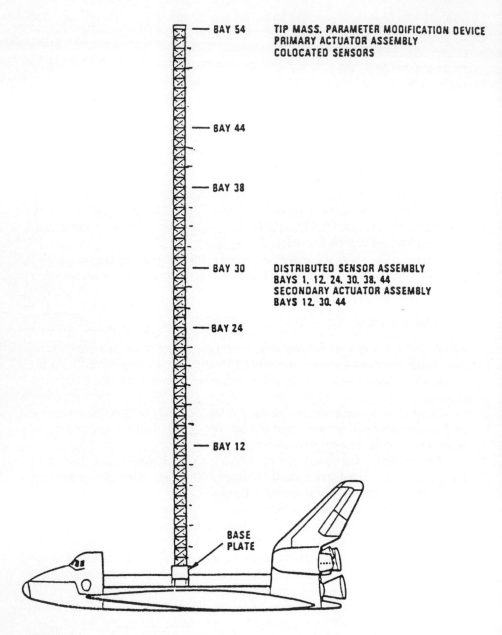

Figure 5   COFS-I longeron.

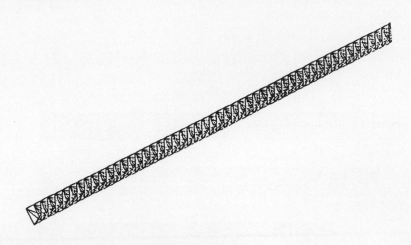

Figure 6   FE discretization of COFS-I longeron.

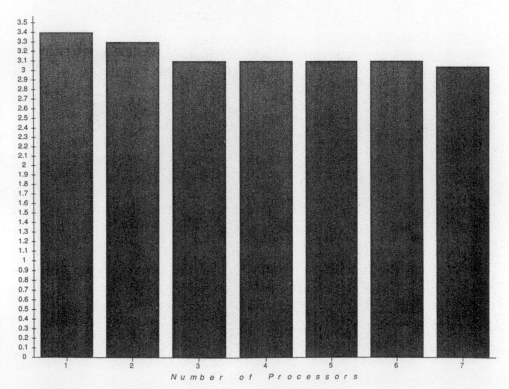

Figure 7   Speedup due to vectorization on Alliant FX/8.

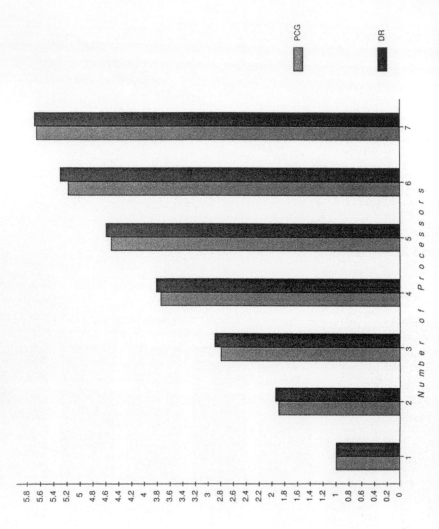

Figure 8 Speedup due to concurrency on Alliant FX/8 (PCG = preconditioned conjugate gradient; DR = dynamic relaxation).

For a given finite element problem the speedup due to vectorization slightly decreases when the number of processors increases because the length of a given vector within a process decreases. However, the total speedup due to the combination of concurrency and vectorization remains high.

9.3   Cray Y-MP

The Cray Y-MP supercomputer consists of up to eight super vector processors sharing a large memory of 256 Mwords. Finite element computations on this system can be speeded through both vectorization and parallelization.

Figure 9 displays a finite element mesh for the solid rocket booster (SRB). The discretized SRB model has 10,453 elements, 9206 nodes, and 54,870 degrees of freedom. The number of interface nodes corresponding to its subdivision into four subdomains is 165. After node renumbering the average profile bandwidth is 310.

We consider the static analysis of the SRB on a Cray Y-MP with eight processors. Following the first remark in Section 4, we select to perform the factorization of the stiffness matrix using a global parallel direct algorithm. For this purpose we have developed a parallel/vector version of the highly vectorized direct solver of Poole and Overman (1988). The measured performances for 1, 2, and 4 processors are listed in Table 3. (No results are available for the case $N_p = 8$ because the author could not arrange for dedicated time on the Cray Y-MP.)

Figure 9   FE discretization of the SRB.

Table 3   Performance Results for Parallel Solver on Cray Y-MP/8[a]

Number of processors	CPU time (s)	Speedup	MFLOPS
1	39	1	235
2	19.79	1.97	464
4	10	3.90	918
8	NA	NA	NA

9.4 CONNECTION MACHINE

The Connection Machine model 2 was briefly introduced in Section 6. Here we consider the transient response of a more detailed space station model to perturbations induced by shuttle docking. The finite element model for this analysis incorporates 7596 two-node beam elements, 572 eight-node brick elements, 24 three-node rigid elements, 9802 nodes, and 58,812 degress of freedom (Fig. 10). Given the size of this problem, we select to run it on an 8000 Connection Machine using the parallel central difference transient algorithm (1989). Table 4 summarizes the measured performances for computations and I/O manipulations. The latter correspond to dumping at each time step the computed displacements, velocities, accelerations, stresses, and strains onto the front end. The reported performances are scaled to the full 64,000 processor configuration (see 1988, for justifications). For this example, the Data Vault improves I/O by a factor of 1307!

Figure 10   FE discretization of a detailed space station model.

Table 4   Performance of a Transient Analysis on the Connection Machine[a]

Phase	CPU time (s)	MFLOPS (using C^*)
Mesh decomposition	3	—
Data loading in the CM2	41	—
Equation of motion solving	4,500	340
Computation time	2,500	665
Communication time	2,000	—
I/O through front end	18,300	—
I/O through Data Vault	14	—

[a]Detailed space station structural model, 58,812 degrees of freedom, 2000 time integration steps.

10   CONCLUSION

A portable software architecture for finite element parallel computations is presented. It is flexible enought to be ported on both shared-memory and local memory multiprocessors. It preserves vectorization and builds on top of it a layer of concurrency. Two kinds of parallelism are exploited:

1.   A *natural* parallelism, which is inherent to finite element formulations
2.   An *artificial* parallelism, which is introduced at the algebraic level

Parallel data structures that accommodate both kinds of parallelism are also described. With these data structures implicit static and dynamic computations are parallelized at the subdomain level. Explicit static and dynamic computations are parallelized at the element level using a coloring technique. Strategies for speeding I/O-bound problems are also described in this chapter. A finite-element prototype code based on the proposed software architecture has been implemented on iPSC-32, Alliant FX/8, Cray-2, and the Connection Machine. The performance of this code is assessed through several realistic finite element simulations. Very high speedups are demonstrated. It is hoped that the material presented in this chapter can serve as a model for porting commercial finite element software to parallel processors.

REFERENCES

R. G., Babb II, Ed. (1988). *Programming Parallel Processors*. Addison-Wesley, Reading, Massachusetts.

Benner, R., and Montry, G. (1986). Overview of preconditioned conjugate gradient (PCG) methods in concurrent finite element analysis. Internal Report from the Advanced Computer Science Project, Fluid the Thermal Sciences Department, Sandia

National Laboratories, Albuquerque, New Mexico 87185, NTIS Report No. SAND85-2727, 15 pp.

Berger, P., Brouaye, P., and Syre, J. (1982). A mesh coloring method for efficient MIMD processing in finite element problems. Proc. Int. Conf. Par. Proc., pp. 41–46.

Bjorstad, P. E., and Widlund, O. B. (1986). Iterative methods for the solution of elliptic problems on regions partitioned into substructures. SIAM. J. Numer. Anal, Vol. 23, No. 6, December.

Bramble, J. H., Pasciak, J. E., and Schatz, A. H. (1984). Preconditioners for interface problems on mesh domains. Dept. Math., Cornell Univ., Ithaca, New York.

Carey, G., and Jiang, B. (1984). Element-by-element preconditioned conjugate gradient algorithm for compressible flow. In: Liu, W., Belytschko, T. and Park, K. C. Eds. *Proceedings of the International Conference on Innovative Methods for Nonlinear Problems*. Pineridge Press International Limited, Swansea, UK, pp. 41–49.

Farhat, C. (1986). Multiprocessors in computational mechanics. Ph.D. Thesis, University of California at Berkeley.

Farhat, C. (1988). A simple and efficient automatic FEM domain decomposer. Computers & Structures, Vol. 28, No. 5, pp. 579–602.

Farhat, C. (1989a). On the mapping of massively parallel processors onto finite element graphs. Computers & Structures, Vol. 32, No. 2, pp. 347–354.

Farhat, C. (1989b). Which parallel finite element algorithm for which architecture and which problem. In: Grandhi, R. V., Stroud, W. J., and Venkayya, V. B., Eds. *Computational Structural Mechanics and Multidisciplinary Optimization*. ASME, New York, AD-Vol. 16, pp. 35–43.

Farhat, C., and Crivelli, L. (1989). A general approach to nonlinear FE computations on shared memory multiprocessors. Comp. Meth. Appl. Mech. Eng., Vol. 72, No. 2, pp. 153–172.

Farhat, C., and Wilson, E. (1986). Modal superposition analysis on concurrent multiprocessors. Engineering Computation, Vol. 3, No. 4, pp. 305–311.

Farhat, C., and Wilson, E. (1987). A new finite element concurrent computer program architecture. Int. J. Num. Meth. Eng., Vol. 24, pp. 1771–1792.

Farhat, C., Felippa, C., and Park, K. C. (1987a). Implementation aspects of concurrent finite element computations. In: Noor, A. K., Ed. *Parallel Computations and Their Impact on Mechanics*. ASME, New York, pp. 301–316.

Farhat, C., Wilson, E., and Powell, G. (1987b). Solution of finite element systems on concurrent processing computers. Engineering with Computers, Vol. 2, No. 3, pp. 157–165.

Farhat, C., Pramono, E., and Felippa, C. (1989a). Towards parallel I/O in finite element simulations. Int. J. Num. Meth. Eng., Vol. 28, No. 11, pp. 2541–2554.

Farhat, C., Sobh, N., and Park, K. C. (1989b). Dynamic finite element simulations on the connection machine. Int. J. High Speed Comp., Vol. 1, No. 2, pp. 289–302.

Farhat, C., Sobh, N., and Park, K. C. (1990). Transient finite element computations on 65,536 processors: The connection machine, Int. J. Num. Meth. Eng., Vol. 29.

Hoit, M., and Wilson, E. (1983). An equation numbering algorithm based on a minimum front criteria. Computers & Structures, Vol. 16, No. 1–4, pp. 225–239.

Jordan, H. F. (1986). Structuring parallel algorithms in an MIMD, shared memory environment. Parallel Computing, Vol. 3, No. 2, pp. 93–110, May.

Kowalik, J., and Kumar, S. (1982). An efficient parallel block conjugate gradient method for linear equations. Proc. 1982 Int. Conf. Par. Proc., pp. 47–52.

Law, K. (1986). A prallel finite element solution method. Computers & Structures, Vol. 23, No. 6, pp. 845–858.

McBryan, O. (1988). New architectures: Performance highlights and new algorithms. Parallel Computing, Vol. 7, No. 3, pp. 477–499.

Noor, A. K. (1987). Parallel processing in finite element structural analysis. In: Noor, A. K., Ed. *Parallel Computations and Their Impact on Mechanics*. American Society of Mechanical Engineers, New York, pp. 253–277.

Ortega, J., Voigt, R., and Romine, C. (1988). A bibliography on parallel and vector numerical algorithms. NASA Contractor Report 181764, ICASE Interim Report 6.

Papadrakakis, M. (1981). A method for the automated evaluation of the dynamic relaxation parameters. Computer Methods in Applied Mechanics and Engineering, Vol. 25, pp. 35–48.

Parlett, B. N. (1980). *The Symmetric Eigenvalue Problem*. Prentice-Hall, Englewood Cliffs, New Jersey.

Poole, E., and Overman, A. (1988). The solution of linear systems of equations with a structural analysis code on the NAS Cray-2. NASA CR 4159, December.

Saati, A., Biringen, S., and Farhat, C. (in press). Solving Navier-Stokes equations on a massively parallel processor: Beyond the one gigaflop performance. Int. J. Supercomp. Appl.

Van Der Vorst, H. A. (1986). The performance of FORTRAN implementations for preconditioned conjugate gradients on vector computers. Parallel Computing, Vol. 3, No. 1, pp. 49–58.

White, D. W., and Abel, J. F. (1988). Bibliography on finite elements and supercomputing. Communications in Applied Numerical Methods, Vol. 4, No. 2, pp. 279–294.

Wilson, E., Yuan, M., and Dickens, J. (1983). Dynamic analysis by direct super-position of Ritz vectors. Earthquake Eng. Struct. Dynam., Vol. 10, pp. 813–821.

8
Applications of Parallel Processing in Structural Engineering

Hsi-Ya Chang   *University of Miami, Coral Gables, Florida*

Senol Utku   *Duke University, Durham, North Carolina*

1 INTRODUCTION

Advanced computer architectures are centered around the concept of parallel processing. The basic idea behind parallel processing is that programs using parallel processors should run much faster than otherwise identical programs using only one processor. In the past decade many new parallel computers have emerged in research institutes as well as on the commercial market. Scientists and engineers are now facing the great challenge to meet the ever-increasing capability of the parallel computers, hardware, and software. They can now solve larger and more complex problems than hitherto. However, the parallel processing environment also introduces increased complexity in software design and algorithmic strategy. The development of parallel algorithms requires decomposing the solution of a problem into concurrently executable tasks (or processes), balancing computational loads in processors, arranging the task execution schedule, synchronizing operations, organizing data access and movement, planning communication among processors, and so on. It is believed that the main stumbling block to the use of parallel computers is the difficulty of formulating algorithms for them.

This chapter deals with the development of parallel algorithms to solve computationally intensive problems in the area of structural engineering. We are mainly interested in developing parallel algorithms for the response analysis of a given structural system of a finite element model. Today any complicated structure would easily have more than thousands of degrees of freedom in a finite element model. In the future the structures are likely to be modeled by more and more degrees of freedom because of the desire of accurately presenting the actual structure and optimizing its shape and strength. The response analysis of such structures under dynamic loads is notoriously time consuming and hence a prime candidate for the parallel processing environment.

The mode superposition method is generally used to solve the linear dynamic response analysis of a structural system. With this approach the problem is essentially reduced to computing the structure's natural frequencies and mode shapes, which are governed by the undamped free vibration equations. With a proper change of basis these equations may be transformed into a general eigenvalue problem. Since the solution time of an eigenvalue problem constitutes the major computational activity in a dynamic analysis, the parallel algorithm development in this chapter concentrates on algebraic eigenvalue problems. In addition, the reason for choosing eigenvalue problems over linear equations is that the solution of linear equations is much better understood and many efficient parallel simultaneous equation algorithms are already documented (Utku and Salama, 1986; Utku et al., 1989), although it is also computationally intensive.

Since there are so many different parallel computer architectures and intercommunication networks, compounded by a variety of problem decomposition techniques, algorithm strategies, and task scheduling arrangements, a unified theory of designing parallel processing procedures has not been developed. A parallel algorithm may perform very well on one machine but not on others. Despite the abundance of parallel algorithms in the literature, researchers are constantly searching for parallel algorithms with better efficiencies and speedups. Engineers, lacking sufficient mathematics background and complicated algorithmic experience, often do not have the confidence to use parallel algorithms for real applications. This chapter reviews the potential parallelism of some eigenvalue algorithms and discusses the concept of converting sequential algorithms into parallel procedures. Several parallel algorithms are given, some in detail and some briefly. The purpose is to illustrate the feasibility of developing parallel stratagems for engineering applications.

In the next section the basic issues of parallel computers are addressed briefly. A specific sublcass of parallel computers is defined as the theoretical machine for parallel implementation. Section 3 discusses the general considerations of stratagem design and problem decomposition. The criterion for assessing the performance of parallel stratagems is also discussed. Section 4 describes the engineering problem and reviews some existing solution methods. Section 5 presents several parallel stratagems for solving eigenvalue problems. Conclusions are summarized in the final section.

2   MACHINE AND ENVIRONMENT

State-of-the-art parallel computer systems may be characterized in three classes: pipelined computers, array processors, and multiprocessor systems (Hockney and Jesshope, 1981; Hwang and Briggs, 1984). In accordance with Flynn's classification (Flynn, 1966), most multiprocessor systems may be classified in the MIMD (multiple instruction–multiple data stream) category. From the viewpoint of designing parallel algorithms, multiprocessor systems offer the most challenge.

Another important characteristic of a computer system is the way processors communicate with each other. In general, the interconnections between processors can be of one of the following three kinds (Browne, 1984): (1) individual processors are connected to a common memory by a single or multiple data bus, (2) individual processors are connected to multiple memories by a network of connections, and (3) individual processors have their own memories, with interconnections directly among processors. Data access in shared memory is slower than that in local independent memory. On the other hand, information needs to be transferred among processors through interconnection network in machines with private memory only. Parallel computers have many possible designs, which differ primarily in the number and nature of the nodes and their interconnection topology.

Since there are many possibilities for parallel computer architectures, to attain some definite conclusion it is necessary to make assumptions with regard to the parallel machines and the mode of operations of these machines. Only a specific subclass of parallel machines is considered in this chapter, and for convenience it is referred to as concurrent computer. The concurrent computer is the ensemble of N identical uniprocessors placed at the nodes of some communication network. The uniprocessors are referred to as processing elements. Each processing element is an independent machine with its own central processing unit (CPU), communications control, and local memory, that is, a message-passing type of processor. This type of parallel machine is already available commercially, such as the NCube and the iPSC-VX vector concurrent computer of Intel.

We assume that the concurrent computer is working in irregular systolic mode (Utku et al., 1984); that is, all processing elements compute and then communicate with each other in cyclic unison by starting their computations simultaneously and their communications simultaneously in each cycle, which may or may not be of the same duration.

3   STRATEGEM AND PROBLEM DECOMPOSITION

When the algorithm for a given problem is prescribed, the solution possibilities created by the hardware are called stratagems. The term has been used before for uniprocessors (Wilkinson, 1965). The algorithm defines the solution process of the problem without ambiguity in terms of some conveniently selected operations and data chunks called primitives of the algorithm. Each primitive operation and relevant data may be called a task. The application of the algorithm in terms of tasks constitutes a stratagem. Since the task size and where it will be processed are variable, many stratagems correspond to a given algorithm.

3.1   Problem Decomposition

To express problem processing in terms of the tasks, each of which is a normal load to any of the processing elements, is called problem decomposition. Since each task involves some input, some output, and some computation, the decomposition involves not only the operations but also the data. The operations, data, and tasks related to problem decomposition are discussed here.

Operations.   These constitute the computational part of tasks. An algorithm always prescribes the sequential order of the serial tasks, whereas the order in which the concurrent task is to be executed is not prescribed. For the concurrent computer the identification of the concurrent tasks and the association of the concurrent tasks with the processing elements are very important. The pairing of the concurrent tasks with the processing elements should be done to minimize data transmission from processing element to processing element. The concurrent tasks with the same operations but with different data are preferable since to seed the processing elements with the same code by broadcasting is more economical than seeding the processing elements with different codes. Let P denote the set of operations of the decomposition.

Data.   The distribution of data among the processing elements is in general a function of the operations portion of the tasks. However, when the concurrent task operations are identical, the data distributions are those that ensure an even work load among processing elements and minimize the communication costs. Usually by reordering the data and/or increasing the task size one may ensure even work among processing elements. Let D denote the set of data items required and produced by the processing elements. Note that the data items that are not produced by the processing elements are the problem input and the data items produced by the processing element but not used by them are the problem output.

Tasks.   Let T denote the set of tasks corresponding to a stratagem. Suppose the ith task t_i is the execution of the jth operation p_j, requires kth data item d_k, and produces lth data item d_l. Then $(p_j, d_k, d_l)_i$ is the description of the ith task. Let n_i denote the size of set T; that is, the total number of tasks in the stratagem n_t is referred to as the granularity of the stratagem. Let n_P denote the size of set P, that is, the total number of distinct operations in the stratagem. The ratio n_P/n_t is referred to as the heterogeneity of the stratagem. One may in short form refer to the three-tuple $(j, k, l)_i$ as the description of the ith task, with the understanding that the first index is the operation label, the second is the input data item label, and the third is the output data item label of the ith task. When operations and data item sets are available and labeled, the decomposition of the stratagem is uniquely identified by n_t number of three-tuples: $(j, k, l)_i$, $i = 1, 2, \ldots, n_t$. Note that by changing the granularity and the initial data assignment policy for a given problem and algorithm, one may devise many decompositions and therefore many stratagems.

3.2 Speedup and Efficiency

The idea behind parallel processing is that programs using N parallel processors should run much faster than otherwise identical programs using only one processor. Because there are many stratagems to solve a problem on concurrent machines, specific criteria should be established to evaluate performance. Universally speedup is the accepted criterion to assess performance. The speedup of a parallel algorithm is defined as the ratio of the execution time of the original sequential algorithm using one processor to the execution time of the parallel algorithm using N processors. Ideally the speedup should approach N, although experience and theory show that the actual speedup is often much smaller. Efficiency is defined as the ratio of speedup to N, which indicates the effectiveness of the processors being used. Factors that affect the speedup of a parallel stratagem include the complexity of the computations, load balancing, the complexity of communications, the relative cost of communications over computations, and the granularity and heterogeneity of the stratagem. For the purpose of this chapter speedup and efficiency are estimated in terms of problem and machine parameters.

4 THE PROBLEM AND THE METHODS

The equations of motion in a multiple degree of freedom system are usually in the form of a set of second-order coupled ordinary differential equations in which the dependent variables are the scalars or coordinates describing the motion and the independent variable is the time. The equations governing linear dynamic responses of structures may be expressed as

$$\mathbf{M\ddot{u} + C\dot{u} + Ku = p} \tag{1}$$

where $\mathbf{K}$, $\mathbf{M}$, and $\mathbf{C}$ are the stiffness matrix, mass matrix, and damping matrix of the structure, respectively. For an n degrees of freedom structure these matrices are all of order n. Vector $\mathbf{u}$ is a list of nodal displacements, and $\mathbf{p}$ is the excitation force vector, both of order n.

If the structure is discrete $\mathbf{K}$ and $\mathbf{M}$ are easy to assemble when the problem is linear. For a continuous solid behaving linearly, finite element methods can be used to form $\mathbf{K}$ and $\mathbf{M}$. The damping matrix $\mathbf{C}$ may be obtained by using such methods as the proportional damping method or the modal damping method. Excluding some special cases these may be characterized as

$$\mathbf{K} = \bar{\mathbf{K}} = \mathbf{K}^T$$
$$\mathbf{M} = \bar{\mathbf{M}} = \mathbf{M}^T \qquad (2)$$
$$\mathbf{C} = \mathbf{C}^T$$

where the bar is used to indicate a complex conjugate; therefore $\mathbf{K}$ and $\mathbf{M}$ are real matrices. $\mathbf{K}$ may be positive or positive definite and $\mathbf{M}$ is positive definite. The superscript T represents a transpose; that is, $\mathbf{K}$, $\mathbf{M}$, and $\mathbf{C}$ are symmetric matrices.

In general the method used to solve the linear dynamic responses of structures is the mode superposition method (Bathe, 1984). Response analysis by the mode superposition method requires natural vibration modes and frequencies. The motion of a linear dynamic system can always be expressed as a linear combination of its vibration modes and the participation factors, which represent the amount each mode contributes. In other words, the nodal point displacements of the system can be obtained by superposing suitable amplitudes of the modes of vibration. Each natural mode vibrates with a corresponding frequency. The natural modes and frequencies of a system are the solutions of the governing equations of the system for undamped free vibrations. With a change of basis the undamped free vibration equations may be transformed into an algebraic eigenvalue problem. The resulting eigenvectors are the natural modes of the system, and the eigenvalues are the squares of the corresponding frequencies. This information is then used to define a reduced set of uncoupled equations of motion for the system. The uncoupled equations are then easily solved using a Duhamel integral (Paz, 1985).

Linear dynamic response analysis thus includes three procedures: (1) assemblage of matrices $\mathbf{K}$, $\mathbf{M}$, and $\mathbf{C}$ by the finite element method, (2) computation of vibration modes and frequencies, and (3) solution of the nodal displacements at different time steps from the uncoupled governing equations in the eigenspace or subspace. Among these three the computation of vibration modes and frequencies dominates the solution time for a large structure. The calculation of one vibration mode and frequency requires time of order n^3. The solution time increases as the number of required vibration modes and frequencies increases. In most engineering applications, however, the largest magnitude of natural frequency and mode is not as important as the smallest magnitude. Furthermore, for most types of loading the contribution of various modes is generally greatest for the lowest frequencies and tends to decrease for the higher frequencies. Consequently analysts are usually interested in only a narrow band of frequencies.

The natural modes and frequencies may be obtained from the free vibration equations with damping neglected:

$$\mathbf{M}\ddot{\mathbf{u}} + \mathbf{K}\mathbf{u} = 0 \qquad (3)$$

The solution of this equation can be postulated to be the form (Clough and Penzien, 1975)

$$\mathbf{u} = \mathbf{x} \sin \omega(t - t_0) \tag{4}$$

where $\mathbf{x}$ is a vector of order n, t the time variable, t_0 the time constant, and ω a constant representing the frequency of vibration of the motion.

Substituting Eq. (4) into (3) we obtain the general algebraic eigenvalue problem

$$\mathbf{Kx} = \lambda \mathbf{Mx} \tag{5}$$

where

$$\lambda = \omega^2$$

The eigenvector $\mathbf{x}_i$ is called the ith mode shape, and ω_i is the corresponding frequency of vibration. There are n eigenvalues and corresponding eigenvectors satisfying Eq. (5). The ith eigenpair is denoted $(\lambda_i, \mathbf{x}_i)$. The eigenvalues are ordered according to their magnitudes:

$$0 \leq \lambda_1 \leq \lambda_2 \ldots \leq \lambda_n \tag{6}$$

where the number of zero eigenvalues is equal to the number of rigid body modes of the structure.

If $\mathbf{M}$ is an identity matrix in Eq. (5), the general eigenvalue problem is reduced to

$$\mathbf{Kx} = \lambda \mathbf{x} \tag{7}$$

which is a special eigenvalue problem, the most common eigenvalue problem that is encountered in general scientific analysis. Many standard solution algorithms are available for special eigenvalue problems. Unfortunately, $\mathbf{M}$ is seldom an identity matrix. To utilize the existing solution algorithms it is in our interest to reduce the general eigenvalue problem to a special one.

If $\mathbf{M}$ is a nonsingular diagonal matrix, as in a lumped mass analysis, it is advantageous first to reduce the general eigenvalue problem to a special one by a transformation matrix, which is symbolically the square root of the matrix $\mathbf{M}$. When $\mathbf{M}$ is a banded real symmetric positive definite matrix, as in consistent mass analysis, the general eigenvalue problem may be reduced to a special one by decomposition techniques, such as the Cholesky method. However, the procedure may be costly and the property of a banded matrix is not preserved, although the symmetric property is preserved because the transformations involved are congruent transformations.

In the solution of special eigenvalue problems it is always beneficial to reduce the nth order matrix by similarity transformations into upper Hessenberg form first and then apply the preferred iteration process. When the matrix is full initially this takes an effort of order n^3; however, in the ensuing iteration process the cost of each iteration may drop from n^3 to n^2 or from n^3 to n if the matrix is symmetric. In the case of real symmetric matrices the Hessenberg form becomes tridiagonal.

Many algorithms are available for solving eigenproblems. For examples, Householder tridiagonalization followed by the QR method (with shift) (Strang, 1986), the power method with shift (Wilkinson, 1965), simultaneous iteration (subspace iteration) (Rutishauser, 1970; Jennings, 1977; Golub and Van Loan, 1983; Bathe, 1984), divide and conquer (Cuppen, 1981), determinant search (Bathe and Wilson, 1972), the preconditioned conjugate gradient method (Nguyen and Arora, 1986), the Lanczos method (Lanczos, 1950; Paige, 1971; Cullum and Willoughby, 1983), and others. The iterative nature of the eigensolvers, coupled with the fact that large eigensystems are being solved, tends to make them one of the computationally intensive modules of an overall engineering analysis package.

In recent years many new parallel architectures have been developed, and consequently a variety of parallel algorithms have been designed for them (Sameh, 1977; Kuck, 1977; Heller, 1978). In the literature a considerable number of parallel algorithms are related to eigenvalue problems because of their importance in many applications. Some of the algorithms are listed here.

Kuck and Sameh (1972) described the parallel implementation of three standard matrix special eigenvalue computation methods (Jacobi, Householder, and QR methods) on the ILLIAC IV computer. Dongarra and Sorrensen (1987) reported a parallel algorithm for a symmetric eigenvalue problem based on the divide and conquer method. Hung (1974) and Barlow et al. (1983) investigated parallel implementation of a bisection algorithm. Ispen and Jessup (1987) provided an algorithm to solve a symmetric tridiagonal eigenproblem on hypercube. Lo et al. (1986) showed the results of a multiprocessor algorithm for a symmetric tridiagonal eigenproblem on Alliant FX/8 and Cray X-MP/48. Ma et al. (1988) discussed a parallel hybrid algorithm for the generalized eigenproblem involving Raleigh quotient iteration (Szyld, 1988). Bostic and Fulton (1987) described a parallel Lanczos single-vector procedure for a generalized eigenvalue problem on FLEX/32 multicomputer. Bennighof (1988) presented a substructuring algorithm for a large eigenproblem. Regarding structural engineering applications, Goehlich et al. (1989) illustrated a parallel algorithm for solving static finite element problems, Chien and Sun (1989) developed parallel techniques for the finite element analysis of a nonlinear truss structure, and Noor and Peters (1989) presented a computational procedure for the nonlinear dynamic analysis of unsymmetric structures on vector multiprocessor systems.

The next section presents parallel algorithms for Householder tridiagonalization, eigenvalue extraction from a tridiagonal form, and simultaneous iteration for generalized eigenvalue problem. Some algorithms are accompanied by detailed pseudocodes; others are discussed only briefly.

5 SEQUENTIAL ALGORITHM AND PARALLEL STRATAGEM

The basic procedures for designing parallel algorithms are (1) select and restructure a proper existing algorithm or discover others, (2) balance the computational load among processors, and (3) minimize the communication among processors. These procedures depend heavily on the nature of the problem. One reasonable sequence for designing a parallel algorithm consists of two steps: (1) break the serial procedure into different parts, each with its own distinct algebraic operations, and (2) connect these parts together to integrate into a complete package. This is the strategy in this section.

5.1 Householder Method

In the Householder method, $n - 2$ successive similarity transformations are performed using Hermitian unitary elementary matrices (in this case, real symmetric and orthogonal) to reduce the matrix of order n into the final tridiagonal form (Ortega, 1967). Each such transformation annihilates all elements in a particular row and the corresponding column, except the tridiagonal elements. These zeroed elements remain zero in the succeeding transformations. Let $\mathbf{P}_k$ denote the kth such transformation matrix and $\mathbf{A}_{k+1}$ the resultant after kth transformation. Thus,

$$\mathbf{A}_{k+1} = \mathbf{P}_k \mathbf{A}_k \mathbf{P}_k \tag{8}$$

where

$$\mathbf{P}_k = \begin{bmatrix} \mathbf{I}' & \mathbf{O} \\ \mathbf{O} & \mathbf{P}_k' \end{bmatrix}$$

and

$$\mathbf{P}_k' = \mathbf{I} - 2\mathbf{w}\mathbf{w}^H \qquad \mathbf{w}^H\mathbf{w} = 1$$

The identity matrix $\mathbf{I}'$ is of order k. Because similarity transformations are used the final tridiagonalized matrix $\mathbf{A}_{n-2}$ has the same eigenvalues as the original matrix. An effective procedure may be written as

$$\mathbf{A}_{k+1} = \mathbf{A}_k - \mathbf{v}\mathbf{z}^T - \mathbf{z}\mathbf{v}^T \tag{9}$$

where

$$\mathbf{v} = 2r\mathbf{w}$$

and

$$\mathbf{z} = \mathbf{u} - \mathbf{w}^T\mathbf{u}\mathbf{w} \qquad \mathbf{u} = \left(\frac{1}{r}\right)\mathbf{A}_k\mathbf{w}$$

The procedure for the choice of **w** and r is detailed in the algorithm as shown in Table 1. In the algorithm, only the upper triangle of **A** is considered because all $\mathbf{A}_j$ are symmetric. The algorithm transforms the input matrix by row order and starts from the first row downward. Unfortunately, the sequence of transformations is inherently serial. Each transformation acts on elements that are the result of previous transformations. Therefore the strategy is to exploit potential concurrency within each transformation by distributing the computational load to as many processors as possible without incurring excessive communication among processors.

Let N be the number of processors in the parallel computer. It is assumed that $n >> N$. Also let k, $k = 1, \ldots, n - 2$, denote one of the $n - 2$ successive transformations. At the kth transformation one is involved with the $(n - k) \times (n - k)$ submatrix formed from the first $n - k$ rows and columns of the matrix. It is also found that the most time consuming step during the kth transformation appears to be the product of an $(n - k) \times (n - k)$ matrix and an $n - k$ vector. Therefore the natural way to apply the Householder algorithm on a parallel computer is to distribute rows of the matrix **A** among processors. However, if one stored groups of rows in processors starting from the top of the matrix sequentially, as the tridiagonalization propagates the processors containing upper rows would be idle while the processors containing lower rows were still active. A better way is to distribute upper, middle, and lower rows among processors to balance the computational load. This requires that n be an integer multiple of N. If it is not, one may increase n by adding zero rows and columns with identity diagonals to the matrix **A**. Thus $n = Ns$ where s is an integer. Each of the N processors is assumed to bear an identification number I. The processors are represented by PE#I, $I = 1, \ldots, N$. The distribution of data into processors may be stated as follows: assign rows $(I + iN)$ to PE#I, where I varies from 1 to N and i from 0 to $s - 1$. Therefore each processor is required to hold s rows of **A**. This kind of data distribution is referred to as scattered row decomposition (Fox, 1985).

Assume that there is a local matrix **B** in each processor; then **B** is an $s \times n$ matrix. To establish the relation between the elements of **A** and the elements of **B** in each processor, let

$$M(i) = \text{MOD}\ (i-1, N) + 1$$

$$I(i) = \text{INT} \left[\frac{i-1}{N} \right] + 1 \tag{10}$$

$$D(i) = (i-1)\ N + \rho$$

where MOD gives the remainder after dividing the first argument by the second. This allows us to identify the processor holding element a_{ij} as PE#$(M(i))$.

Table 1 Serial Householder Algorithm in Pseudocode Format

Step	Operation	Comment
1	For $k = 1, \ldots, n - 2$ do steps 2–8	
2	Set $k_1 \leftarrow k + 1$	
3	$$T \leftarrow \left[\sum_{i = k_1}^{n} (a_{ki}^2) \right]^{1/2}$$	
4	If $a_{k,k_1} < 0$ then $T \leftarrow -T$	$R = 2r^2$, $\mathbf{w} = \dfrac{1}{2r} \mathbf{v}$
	Set $a_{k,k_1} \leftarrow a_{k,k_1} + T$	
	Set $R \leftarrow a_{k,k_1} \times T$	
	Set $v_i \leftarrow a_{ki}$ for $i = k_1, \ldots, n$	
	Set $a_{k,k_1} \leftarrow -T$	
5	For $i = k_1, \ldots, n$	$\mathbf{u} = \dfrac{1}{R} \mathbf{A}_k \mathbf{v} = \dfrac{1}{r} \mathbf{A}_k \mathbf{w}$
	set $u_i \leftarrow \sum_{j=i}^{n} (a_{ij} \times v_j)$	
	For $i = k + 2, \ldots, n$	
	set $u_i \leftarrow u_i + \sum_{j=k_1}^{i-1} (a_{ji} \times v_j)$	
	For $i = k_1, \ldots, n$	
	set $u_i \leftarrow \dfrac{1}{R} u_i$	
6	Set $C \leftarrow \sum_{i = k_1}^{n} (v_i \times u_i)$	$C = \mathbf{v}^T \mathbf{u}$
7	For $i = k_1, \ldots, n$	$\mathbf{z} = \mathbf{u} - \dfrac{1}{2R} \mathbf{v}^T \mathbf{u} \mathbf{v}$
	set $z_i \leftarrow u_i - \dfrac{C}{2R} \times v_i$	
8	For $i = k_1, \ldots, n$, $j = i, \ldots, n$	$\mathbf{A}_{k+1} = \mathbf{A}_k - \mathbf{v}\mathbf{z}^T - \mathbf{z}\mathbf{v}^T$
	Set $a_{ij} \leftarrow a_{ij} - v_i z_j - z_i v_j$	

Function INT represents integer division; thus element a_{ij} may be found as $b_{I(i)j}$ in PE#($M(i)$). On the other hand, elements of $\mathbf{B}$ in processor ρ may be mapped back into $\mathbf{A}$ through function D. For example, element b_{ij} in processor ρ actually represents element $a_{D(i)j}$. The parallel algorithm for the kth transformation is described here.

We start from calculating the Euclidean norm of the last $n - k$ elements of the kth column of $\mathbf{A}$. Each processor calculates

$$T^\rho = \sum_{\substack{i=1 \\ D(i)>k}}^{s} (b_{ik})^2 \tag{11}$$

where T^ρ is the partial sum in processor ρ. The sums of squares of elements in each processor must now be combined into one sum. In general this can be done in $\log_2 N$ steps. The simplest way is to send each T^ρ to a single processor for addition. The Euclidean norm is then obtained in processor PE#$M(k)$ as

$$T = \left[\sum_{\rho=1}^{N} T^\rho \right]^{1/2} \tag{12}$$

The scalar R and vector $\mathbf{v}$ in step (4) of Table 1 are calculated by PE#$M(k)$ and then broadcast to all other processors. Next the vector $\mathbf{u}$ is formed concurrently in all processors by

$$u_{D(i)} = \frac{1}{R} \sum_{j=k_1}^{n} b_{ij} v_j \quad for \quad \begin{cases} i = 1, \ldots, s \\ D(i) > k \end{cases} \tag{13}$$

In this equation each b_{ij} is actually an element of the lower right $(n - k) \times (n - k)$ square of $\mathbf{A}$. This step shows that we could have considered an upper (or lower) triangular portion like that in the sequential procedure; however this would require much communication among processors.

To calculate constant C in step (6) of Table 1, the partial sums are first calculated in all processors by

$$C^\rho = \sum_{\substack{i=1 \\ D(i)>k}}^{s} v_{D(i)} u_{D(i)} \tag{14}$$

Again, partial sums are passed to PE#$M(k)$ to compute C from

$$C = \sum_{\rho=1}^{N} C^\rho \tag{15}$$

Also, the constant Y is obtained in the same processor by

$$Y = \frac{C}{2R} \tag{16}$$

and then broadcast to all other processors. The concurrent calculation of vector $\mathbf{z}$ in step (7) of Table 1 can now be initiated in every processor by

$$z_{D(i)} = u_{D(i)} - Y \cdot v_{D(i)} \qquad \begin{cases} i = 1, \ldots, s \\ D(i) > k \end{cases} \tag{17}$$

To complete the kth transformation the last $n - k$ elements of z must be first broadcast to all processors. With this the upper triangular portion of the transformed $(n \times k) \times (n - k)$ submatrix is computed in all processors concurrently. The assigned task to each processor is

$$b_{ij} = b_{ij} - v_{D(i)}z_j - z_{D(i)}v_j \qquad j = D(i) \ldots n \qquad \begin{cases} i = 1, \ldots, s \\ D(i) > k \end{cases} \tag{18}$$

The lower triangular portion of the transformed submatrix may be obtained by proper data transmission among processors. The pseudocode of the parallel stratagem just described is given in Table 2. Initial loading and final unloading are not included in the table. At the end of the tridiagonalization each processor except PE#N contains s numbers of diagonal and s numbers of superdiagonal elements. The processor PE#N contains s numbers of diagonal elements and $s - 1$ numbers of superdiagonal elements. In each processor $b_{i,D(i)}$, $i = 1, \ldots, s$ corresponds to $a_{D(i),D(i)}$ and $b_{i,D(i)+1}$, $i = 1, \ldots, s$ corresponds to $a_{D(i),D(i)+1}$ of A_{n-2}.

The computation time of Table 2 may be estimated in terms of floating-point arithmetic operation times. The communication time depends on the topology of the interconnection network. For example, assume q_t as the time to transmit one number from a processor to its nearest neighbor. If this algorithm is implemented on a ring-type interconnection network and a processor can send or receive one number at a time, then transmitting one number from one processor to all other processors takes time $(N/2)q_t$. Transmitting an array of numbers of length l from one processor to all other processors requires $(N/2 + 2l - 2)q_t$. The estimation of these times is not shown here.

The speedup g of this parallel stratagem on a concurrent machine is defined as

$$g = \frac{T_s}{T_c} = \frac{\text{execution time of Table 1 with 1 PE}}{\text{execution time of Table 2 with } N \text{ PEs}} \tag{19}$$

and the efficiency e is defined as

$$e = \frac{g}{N} \tag{20}$$

Chang et al. (1988a) showed that this parallel Householder stratagem may be efficiently implemented on a concurrent machine provided that $s \gg N$ and the ratio of the transmission time from one processor to any other processor to the floating-point arithmic operation time is small enough. Also, the specialization

Table 2 Pseudocode for Parallel Householder Algorithm Using Scattered Row Decomposition

Step	Step from Table 1	Operation	Processor
1	1	Set $k \leftarrow 0$	All
2	2	Set $k \leftarrow k + 1$ and $k_1 \leftarrow k + 1$	All
3	3	Compute T^p by Eq. (11)	All
4	3	Transmit T^p to PE#($M(k)$)	All
5	3	Compute T by Eq. (12)	$M\phi(k)$
6	4	If $b_{I(k),k_1} < 0$ then $T \leftarrow -T$	$M(k)$
		Set $b_{I(k),k_1} \leftarrow b_{I(k),k_1} + T$	
		Set $R \leftarrow b_{I(k),k_1} \times T$	
		Set $v_i \leftarrow b_{I(k),i}$ for $i = k_1, \ldots, n$	
		Set $b_{I(k),k_1} \leftarrow -T$	
7	5	Transmit R and the last $n - k$ components of **v** to all other processors	$M(k)$
8	5	Compute parts of **u** by Eq. (13)	All
9	6	Compute C^p by Eq. (14)	All
10	6	Transmit C^p to PE#($M(k)$)	All
11	6	Compute C and Y by Eqs. (15) and (16)	$M(k)$
12	7	Transmit Y to all other processors	$M(k)$
13	7	Compute parts of **z** by Eq. (17)	All
14	7	Transmit parts of **z** to all other processors	All
15	8	Compute updated parts of **B** by Eq. (18)	All
16	8	Update the components of the lower triangle of the transformed $(n - k) \times (n - k)$ submatrix	All
17		If $k < n - 2$ then GO TO STEP 2 else STOP	All

of this stratagem on a hypothetical ring type of processing array demonstrated satisfactory speedup. For a matrix of order 10,000 to be tridiagonalized, the speedup is about 6.6 if the concurrent machine has 10 processors and 44 if 100 processors. It is noted that (1) these numbers are based on the condition that the time to transmit one number from one processor to its nearest neighbor is nearly equal to one typical floating-point operation time, and (2) they would probably increase should the machine have a more sophisticated interconnection network.

Another decomposition scheme, scattered square decomposition (Fox, 1985), can also be used for the parallel implementation of the Householder tridiagonalization. Scattered square decomposition assumes the number of processors is $N = D^2$, where D is an integer and greater than one. Also $n = sD$ and s is an integer. Assume the processors are placed on the nodes of a $D \times D$ square grid. Each processor is identified as PE#(I, J), where $I = 1, \ldots, D$ and $J = 1, \ldots, D$. In other words, I is the label of the row and J is the label of the column of the node where the processor is placed. We then assign the matrix elements $a_{((i-1)D+I,(j-1)D+J)}$ for $i = 1, \ldots, s$ and $j = 1, \ldots, s$ to PE#(I, J). Based on this decomposition scheme Chang et al. (1988b) designed a parallel stratagem for concurrent computers with mesh-type interconnections.

5.2. Eigenvalue Extraction

In the special eigenvalue problem the original matrix $\mathbf{A}$ is first reduced to a tridiagonal real symmetric matrix $\mathbf{T}$ by the Givens (1953) or Householder method. Because similarity transformations are used in either of the two methods, the tridiagonal form possesses the same eigenvalues as the original matrix. To extract the eigenvalues from the tridiagonal form is much easier than that from the original matrix especially when the order of the matrix n is large.

After tridiagonalization the new eigenvalue problem becomes

$$\mathbf{Tx} = \lambda\mathbf{x} \tag{21}$$

noting that there are only $(2n - 1)$ independent elements in the tridiagonal matrix $\mathbf{T}$.

Many powerful eigenvalue extraction schemes may be applied to the problem of Eq. (21). One of the schemes is the method of bisection (in conjunction with the Strum sequence property). This method involves a series of iterations in which an eigenvalue is bracketed by upper and lower bounds and the interval between the bounds is successively halved with each iteration. To locate any individual eigenvalue the Sturm sequence property is used to determine the number of eigenvalues inside a certain range. The number of eigenvalues smaller than λ^{o}, a shift value, is equal to the number of negative main diagonal elements of the diagonal matrix $\mathbf{D}$ on the right side of the identity

$$\mathbf{T} - \lambda^{o}\mathbf{I} = \mathbf{LDL}^{T} \tag{22}$$

where $\mathbf{L}$ is the unit lower factored matrix corresponding to $\mathbf{T} - \lambda^{o}\mathbf{I}$ and $\mathbf{I}$ is an identity matrix.

Because $\mathbf{T}$ is a tridiagonal real symmetric matrix, the diagonal elements of the diagonal matrix $\mathbf{D}$ can be calculated by the formula

$$
\begin{aligned}
d_1 &= t_{11} - s \\
d_i &= (t_{ii} - s) - \frac{t_{i-1,i}}{d_{i-1}} \qquad i = 2, \ldots, n
\end{aligned}
\tag{23}
$$

The starting upper and lower bounds of the range that contains all eigenvalues may be determined by the Gerschgorin's theorem (Gourlay and Watson, 1973). If only a subset of all eigenvalues is interested, the bounds are prescribed.

In a typical serial algorithm the factorization of Eq. (23) is the most time consuming process, especially when the order of $\mathbf{T}$ is large. This observation provides a basis for the parallel implementation of the bisection method. Recall that N is the number of processors in a concurrent machine. In the parallel implementation N shifts are properly chosen at each iteration and N factorizations are carried out in parallel by N processors. At the start one may merely conveniently divide the eigenvalue band of interest into $N + 1$ segments and each processor takes one of the N internal end points as a shift. However, the determining of the shifts from the second iteration on becomes complicated. The technique of determining N shifts is based on two considerations: minimizing the waste of shifts and a simple rule of assigning shifts.

The parallel algorithm requires that every processor store a bounds table that includes the upper and lower bounds of every eigenvalue in the band of interest. Each processor modifies its bounds table at the end of every iteration. It may be arranged such that all processors work independently in determining the shifts they should choose, in factorizing the shifted matrix into $\mathbf{LDL}^T$ triple factors, and in modifying the bounds of all eigenvalues. The only time they need to communicate in an iteration is when the numbers of eigenvalues less than the selected shifts are calculated in all processors. Since the parallel algorithm is developed based on the method of bisection, it is referred to as the parallel N-section method (Chang et al., 1988c).

The pseudocode for the parallel N-section method is given in Table 3. In the table, l denotes the lower limit and u the upper limit of the eigenvalue band of interest; m is the number of eigenvalues within the prescribed interval. The label of each eigenvalue is denoted L_i. Therefore the eigenvalues of interest are λ_{Li}, for $i = 1, \ldots, m$. Index i of L_i is referred to as the relative label of the eigenvalue. Two vectors $\boldsymbol{\lambda}^l$ and $\boldsymbol{\lambda}^u$, both of order m, are used for the lower and upper bounds, respectively, such that λ_i^l and λ_i^u represent the lower and upper bounds of the ith eigenvalue. The bounds table also includes two other vectors $\mathbf{s}^l$ and $\mathbf{s}^u$, both of order m, such that s_i^l and s_i^u are the numbers of eigenvalues less than the lower and upper bounds of the ith eigenvalue, respectively.

The N shifts of the first iteration step are selected as the abscissas of the N equally spaced points between l and u. The calculation of these points is repeated in all processors. Assume the labels of processors are denoted ρ, $\rho = 1, \ldots, N$. The ρth processor is then assigned to carry out the factorization of $(\mathbf{T} - \lambda_\rho^0 \mathbf{I})$. The N factorizations are therefore performed in parallel by all processors. The number of the eigenvalue less than the selected shift in each processor, shown as $s(\lambda_\rho^0)$, is stored in a local vector $\boldsymbol{\tau}$ such that $\tau_\rho = s(\lambda_\rho^0)$. The first iteration step in every processor ends by modifying the bounds table and transmitting τ_ρ to all other processors.

At this point the original band of interest has been divided into $N + 1$ segments. Only those segments containing at least one uncalculated eigenvalue are our concern. Consequently we need also to keep track of these nonnull segments as well as the uncalculated eigenvalues. The idea is to select proper shifts from the nonnull segments and avoid any unnecessary shift. The scheme employed in Table 3 starts from the nonnull segment closest to l. The number of eigenvalues in this segment is counted and an equal number of shifts is assigned to this segment. The shifts are calculated as the abscissas of the equally spaced points between the lower and upper bounds of that segment. The procedure continues to the second nonnull segment and stops whenever N shifts are distributed. When N is larger than the number of uncalculated eigenvalues, remedy is provided in step 2 of Table 3 to avoid a waste of shifts, which would result in idling processors. After selecting and calculating the shifts the remaining work in this step is already described in the first iteration step. That is, calculate $s(\lambda_p^0)$ in each processor, and transmit it to all other processors and then modify the bounds table. The procedure continues until all eigenvalues are computed within the desired accuracy ε.

In Table 3 ω_i and υ_i are used as the relative label of the first eigenvalue and the number of shifts in the ith nonnull segment at a particular iteration step, respectively. The scalar η is the number of the nonnull segments and x the number of current uncalculated eigenvalues. All the steps listed in Table 3 are carried out by all processors, starting simultaneously and finishing at the same time. That is, all processors are used all of the time. Initial loading of the diagonal and superdiagonal elements of $\mathbf{T}$ into all processors is not considered. The only communication takes place at step 5 of Table 3. From the nature of this method it may be seen that it works best when the number of eigenvalues in the band of interest m is much less than n and $N << n$. The required communication time is expected to be small relative to the computation time as long as $m <<$ n and $N << n$. Take a large-scale problem as an example. Suppose the order of $\mathbf{T}$ is 10,000 and there are 100 eigenvalues inside the band of interest. The problem is to be solved on a concurrent machine with 100 processors and a ring-type interconnection network. The speedup is about 94 if the time to transmit one number from one processor to its nearest neighbor is nearly equal to one typical floating-point operation time. It is also worthwhile noting that this method can be applied to a general real symmetric matrix directly without tridiagonalizing the matrix first.

5.3 Simultaneous Iterations

Linear dynamic problems are generally associated with general eigenvalue problems. When converting the general eigenvalue problem to a special eigenvalue problem can be done without incurring costly operations, it is advantageous to do so before solving the problem. However, the conversion usually involves a huge computation for a matrix with large order. In these cases the solution directly from the general eigenvalue problem is preferred.

Table 3   Pseudocode for Parallel N-Section Method

1. Initialize the bounds table.

$$\left.\begin{array}{l} \lambda_i^l \leftarrow l \\ \lambda_i^u \leftarrow u \\ s_i^l \ \leftarrow s(l) \\ s_i^u \ \leftarrow s(u) \end{array}\right\} i = 1, \ldots, m$$

$$\lambda_0^o(0) \leftarrow l$$
$$\lambda_{N+1}^o \leftarrow u$$
$$\lambda_0^u \leftarrow -\infty$$
$$x \leftarrow m$$

2. Count the number of segments and assign shifts to segments.

$\eta \leftarrow 0, T \leftarrow 0$
For $i = 0, \ldots, m - 1$
    If $\lambda_{i+1}^u = \lambda_i^u$ GOTO 10
    If $\lambda_{i+1}^u = 0$ GOTO 10
    $\eta \leftarrow \eta + 1$
    $\omega_\eta \leftarrow i + 1$
    $v_\eta \leftarrow s_{\omega_\eta}^u - s_{\omega_\eta}^l$
    $T \leftarrow T + v_\eta$
    IF $T \geq N$ then $v_\eta \leftarrow v_\eta + N - T$ GOTO 11
10 Next i
11 If $N > x$ then
    $a \leftarrow$ INT $(N - x, \eta)$
    $b \leftarrow$ MOD $(N - x, \eta)$
    If $b = 0$ then
    $v_i \leftarrow v_i + a, \ i = 1, \ldots, \eta$
    Else
    $v_i \leftarrow v_i + a + 1, \ i = 1, \ldots, b$
    $v_i \leftarrow v_i + a, \ i = b + 1, \ldots, \eta$

3. Determine the shifts.

$k \leftarrow 0$
For $i = 1, \ldots, n$
    $c \leftarrow \dfrac{\lambda_{\omega_i}^u - \lambda_{\omega_i}^l}{v_i + 1}$
For $j = 1, \ldots, v_i$
    $k \leftarrow k + 1$
    $\lambda_k^o \leftarrow \lambda_{\omega_i} + c \cdot j$
Next j
Next i

(*continued*)

Table 3 Pseudocode for Parallel N-Section Method (Continued)

---

4. Calculate $s(\lambda_p^o)$ and set $\tau_p = s(\lambda_p^o)$.

5. Transmit τ_p to all other processors.

6. Modify the bounds table.

$\tau_o \leftarrow s_{\omega_1}^l$, $\tau_{N+1} \leftarrow s_{\omega_n}^u$

For $i = 1, \ldots, N + 1$

For $j = \tau_{i-1} - s(l) + 1$ to $\tau_i - s(l)$

If $\lambda_j^l < \lambda_{i-1}^o$ set $\begin{cases} \lambda_j^l \leftarrow \lambda_{i-1}^o \\ s_j^l \leftarrow \tau_{i-1} \end{cases}$

If $\lambda_j^u < \lambda_i^o$ set $\begin{cases} \lambda_j^u \leftarrow \lambda_i^o \\ s_j^u \leftarrow \tau_i \end{cases}$

Next j

Next i

7. Check required accuracy.

For $i = 1, \ldots, m$

If $\lambda_i^u = 0$ GOTO 12

If $\lambda_i^u - \lambda_i^l < \varepsilon$ set $\begin{cases} \lambda_{L_i} \leftarrow \dfrac{\lambda_i^u + \lambda_i^l}{2} \\ \lambda_i^u \leftarrow -\infty \\ x \leftarrow x - 1 \end{cases}$

12 Next i

8. If $x > 0$ return to step 2, else STOP.

This section addresses a strategy for the extraction of the eigenpairs of the general eigenvalue problem

$$\mathbf{Kx} = \lambda \mathbf{Mx} \tag{24}$$

where $\mathbf{K}$ and $\mathbf{M}$ are of large order n and half-bandwidth $b << n$. In structural dynamic problems the interest is usually in computing a set of eigenpairs within a prescribed range. Let m be the number of eigenpairs to be extracted. One may use a power method by iterating one vector with proper shift to obtain one eigenpair at a time. One may also use m iteration vectors $\mathbf{V}_k$ to obtain m eigenpairs altogether, in which case the recursive relation may be shown as

$$(\mathbf{K} - s\mathbf{M}) \mathbf{V}_k = \mathbf{MU}_{k-1} \tag{25}$$

and the transformation

$$\mathbf{U}_k = \mathbf{V}_k \mathbf{Q}_k \qquad k = 0, 1, \ldots \tag{26}$$

where s is a shift. Initially, for $k = 0$, $\mathbf{V}_0$ is any set of m linearly independent vectors. For suceeding steps $\mathbf{Q}_k$ is determined to satisfy the eigenvalue problem of order $m << n$:

$$\mathbf{K}_k\mathbf{Q}_k = \mathbf{M}_k\mathbf{Q}_k \text{ diag}(\mathbf{W}_k) \tag{27}$$

where

$$\mathbf{K}_k = \mathbf{V}_k^H(\mathbf{K} - s\mathbf{M})\mathbf{V}_k \tag{28}$$

$$\mathbf{M}_k = \mathbf{V}_k^H \mathbf{M}\mathbf{V}_k \tag{29}$$

where the superscript H denotes the Hermitian operation. Normalization of $\mathbf{Q}_k$ with respect to $\mathbf{M}_k$, that is $\mathbf{Q}_k^H\mathbf{M}_k\mathbf{Q}_k = \mathbf{I}$, is enforced so that

$$\mathbf{U}_k^H \mathbf{M}\mathbf{U}_k = \mathbf{I} \tag{30}$$

With these the m eigenpairs nearest to the shift may be obtained from $\mathbf{U}_k$ and diag($\mathbf{W}_k$). In short, the serial simultaneous iteration method stated here consists of four basic operations: (1) factorizing a banded symmetric matrix of large-order n into upper and lower tirangular factors, (2) forming the product of matrices that may be banded, rectangular, or square, (3) extracting all eigenpairs of m-order problem by a generalized Jacobi technique, and (4) solving for the unknowns of an algebraic set of equations by forward pass and then by backward pass. The parallel algorithm may therefore be constructed by considering the parallel implementation of these four operations.

To quantify the processing costs two architectural parameters q and μ are used. The computational cost is characterized by q, which is defined as the time required to perform a floating-point operation. One floating operation is assumed on average to consist of a multiplication followed by an addition. For simplification any single arithmetic operation (e.g., add, multiply, or square root) is counted as a floating-point operation. The cost of interprocessor communication is characterized by μ, which is defined as the ratio of the time required to transmit one floating-point number from one processor to its nearest neighbor to the time required to perform a floating-point operation. The cost of propagating data through a network of N processors depends on the transmission speed μq, the communication topology among processors, the number of data items to be transmitted, and whether all processors should receive the same data or each should receive different data. The speedup and efficiency expressions are to be specialized on a cosmic cube. In such computers the time required for transmitting γ floating-point numbers to N processors is $\gamma(\log_2 N)\mu q$ if the same γ data items are sent to all N processors and $(\gamma + N - 2)\mu q$ if different processors are sent different γ (Tuazon et al., 1985). In the following, operations on matrices of large order are reduced to operations on smaller partition of size s. It is assumed that $1 < s << b$. Further, by proper padding with zeros we assume

$$\text{MOD } (n, s) = 0$$
$$\text{MOD } (b, n) = 0 \tag{31}$$
$$\text{MOD } (b, s) = 0$$

and

$$b = 2Ns < < n$$
$$n_s = \frac{n}{s}$$
$$b_s = \frac{b}{s} = 2N \tag{32}$$
$$n_N = 1 + \text{INT } \frac{n_s - 1}{N}$$

Now some ideas for developing the parallel versions of the some of the basic operations listed previously are briefly introduced. The first basic operation is to factorize a banded symmetric matrix into triangular factors. Two stratagems for the parallel implementation of decomposing a banded Hermitian matrix of large order n into its triangular upper and lower factors $\mathbf{U}$ and $\mathbf{U}^H$ were given in Utku et al. (1986a). These stratagems use two different interpretations of the Cholesky factorization. Their performance is nearly equal. With the definitions and notations already outlined, the time required to factor an $n \times n$ matrix with half-bandwidth b using N processors is derived as $qns^2[2(1 + 3N/4) + (10 + 1/N) \mu/s]$, where the coefficient of μ is due to the interprocessor transmission of data.

The second basic operation is to form the product of matrices that may be banded, rectangular, or square. Only a scheme for performing $\mathbf{C} = \mathbf{A}^H\mathbf{B}$, where $\mathbf{A}$ and $\mathbf{B}$ are both rectangular, is given here.

Let $\mathbf{A}$ and $\mathbf{B}$ be $n \times m$ matrices, each of which is partitioned into blocks $\mathbf{A}_i$, $\mathbf{B}_j$, $i = 1, \ldots, n_s$, with block size $s \times m$. The inner product can be expressed as

$$\mathbf{C} = \sum_{l=1}^{N} \mathbf{C}_l \tag{33}$$

where

$$\mathbf{C}_l = \sum_{k=1}^{n_N} \mathbf{A}^H_{(k-1)N+l} \mathbf{B}_{(k-1)N+l} \tag{34}$$

or

$$\mathbf{C}_l = \sum_{k=1}^{n_N} \mathbf{A}^H_{(l-1)n_N+k} \mathbf{B}_{(l-1)n_N+k} \tag{35}$$

The choice between placing the operands $\mathbf{A}_i$ and $\mathbf{B}_i$ in the N processors according to $i = (k - 1)N + l$ of Eq. (34) or $i = (l - 1)n_N + k$ of Eq. (35) may be dictated by the step previous to the matrix multiplication. With N processors, N values of $\mathbf{C}_l$ can be computed in parallel. Each processor performs n_N serial floating-point operations on their operands. No interprocessor communication is needed, and all $\mathbf{C}_l$ are completed in time of order $n_N m^2 q (1 + s)$. The cascade sum of all $\mathbf{C}_l$ is performed next according to Eq. (33) in time proportional to $m^2 q (1 + \mu) \log_2 N$, during which the final results are collected in the host computer. The total cost of the inner matrix product is therefore $m^2 q [n_N (1 + s) + (1 + \mu) \log_2 N]$.

The third basic operation is to obtain all eigenpairs of the reduced m-order eigenvalue problem in Eq. (27) by the generalized Jacobi technique. In this technique both $\mathbf{K}_k$ and $\mathbf{M}_k$ are diagonalized simultaneously by a sequence of orthogonal transformations. Each transformation involves one pre- and postmultiplication with proper rotation matrix and results in changing two rows and two columns. More importantly, two off-diagonal symmetric elements of $\mathbf{K}_k$ and $\mathbf{M}_k$ are zeroed. Unfortunately, the zeroed elements do not remain zero in the succeeding transformation. Cyclic generalized Jacobi is used here, and no attempt is made to determine which off-diagonal element is the largest to reduce it to zero first. Instead, all off-diagonal elements are operated upon in row-major order sweep after sweep. There are $m(m - 1)/2$ orthogonal transformations in each sweep.

That each orthogonal transformation in the generalized Jacobi causes only two columns and rows of $\mathbf{K}_k$ and $\mathbf{M}_k$ to change suggests considerable concurrency in its parallel implementation. Denote the typical rotation matrix as $\mathbf{P}(i, j)$, which causes the i_th and j_th rows and columns of $\mathbf{K}_k$ and $\mathbf{M}_k$ to change. By storing data properly in all processors participating in the operation and selecting a set of $m/2$ nonoverlapping pairs of (i, j) indices, it is possible to perform $m/2$ parallel transformations on $\mathbf{K}_k$ and $\mathbf{M}_k$. The process is repeated $m - 1$ times, each time using a different set of nonoverlapping indices, in a sweep. Completion of all $(m - 1)m/2$ transformations constitutes a typical sweep that may be repeated until diagonalization is achieved. This technique requires rotating data among processors at each transformation, which is acceptable since $m << n$. A detailed account of this parallel Jacobi technique and a systematic row exchange scheme for selecting nonoverlapping rotation planes can be found in Utku et al. (1986b).

Utku et al. (1986b) also presented a parallel algorithm for forward and backward passes to complete the parallel simultaneous iteration stratagem, which demonstrates efficiencies ranging from 60 to 96% for selected representative problems.

6 CONCLUSIONS

In the area of computational mechanics structural dynamic analysis is one of the many notorious computationally intensive problems. It is believed that parallel processing is one of the most promising approaches to reduce the execution time of a large problem. This chapter has introduced some parallel stratagems for large-scale eigensolvers. The pseudocodes of parallel procedures for Householder tridiagonalization and N-section eigenvalue extraction are given. The concept of converting the simultaneous iteration method into parallel algorithms is also discussed. The performance of these algorithms, evaluated by the time complexity in terms of problem and machine parameters, indicates satisfactory efficiencies and speedups provided that, among others, (1) the problem size n should be much greater than N and (2) the relative cost of communications over computations is small. The former corresponds well to the structural dynamic problem with a large finite element model, and the latter is constantly improved as computer technology rapidly advances. The prospect of parallel processing in engineering applications is definitely encouraging.

REFERENCES

Barlow, R. H., Evans, D. J., and Shanehchi, J. (1983). Parallel multisection applied to the eigenvalue problem. Computer Journal, 26:6.

Bathe, K. J. (1984). *Finite Element Procedures in Engineering Analysis*. McGraw-Hill, New York.

Bathe, K. J., and Wilson, E. L. (1972). Large eigenvalue problems in dynamic analysis, ASCE J. Eng. Mech. Div., 98:1471.

Benninghof, J. K. (1988). An iterative substructuring algorithm for parallel solution of large eigenvalue problem. Symposium on Super Large Problems in Computational Mechanics, ONR, Mystic, Connecticut.

Bostic, S.W., and Fulton, R.E. (1987). Implementation of the Lanczos method for structural vibration analysis on a parallel computer. Computers & Structures, 25(3): 395.

Browne, J. C. (1984). Parallel architectures for computer system. Physics Today, May: 28.

Chang, H. Y., Utku, S., and Salama, M. (1988a). A parallel Householder tridiagonalization stratagem using scattered row decomposition. International Journal for Numerical Methods in Engineering, 26:857.

Chang, H. Y., Utku, S., Salama, M., and Rapp, D. (1988b). A parallel Householder tridiagonalization stratagem using scattered square decomposition. Parallel Computing, 6:297.

Chang, H. Y., Utku, S., and Salama, M. (1988c). Eigenvalue computation of large symmetric tridiagonal matrices on concurrent processors. Computers & Structures, 29(2):323.

Chien, L. S., and Sun, C. T. (1989). Parallel processing techniques for finite element analysis of nonlinear large truss structures. Computers & Structures 31(6):1023.

Clough, R. W., and Penzien, J. (1975). *Dynamics of Structures*. McGraw-Hill, New York.

Cullum, J., and Willoughby, R. A. (1983). Lanczos Algorithms for large symmetric eigenvalue computations. In: Abarbanel, S., Glowinski, R., Golub, G., Henrici P., and Kreiss, S., Eds. *Volume in Progress in Scientific Computing Series*. Birkhauser Boston, Boston.

Cuppen, J. J. M. (1981). A divide and conquer method for the symmetric tridiagonal eigenproblem. Numer. Math., 36:177.

Dongarra, J. J., and Sorrensen, D. C. (1987). A fully parallel algorithm for the symmetric eigenvalue problem. SIAM J. Sci. Stat. Comput., 8:s139.

Flynn, M. J. (1966). Very high-speed computing systems. Proc. IEEE, 15:1901–1909.

Fox, G. C. (1985). Square matrix decomposition—Symmetric, local, scattered. Internal report Hm-97, California Institute of Technology, Pasadena.

Givens, J. W. (1953). A method of computing eigenvalues and eigenvectors suggested by classical results on symmetric matrices. U.S. Nat. Bur. Standards Applied Mathematics Series, 29:117.

Goehlich, D., Komzsik, L. and Fulton, R. E. (1989). Application of a parallel equation solver to static FEM problems. Computers & Structures, 31(2):121.

Golub, G. H., and Van Loan, C. F. (1983). *Matrix Computation*. Johns Hopkins Press, Baltimore.

Gourlay, A. R., and Watson, G. A. (1973). *Computational Methods For Matrix Eigenproblems*. John Wiley and Sons, New York.

Heller, D. (1978). A survey of parallel algorithms in numerical linear algebra. SIAM Rev., 20(4):740.

Hockney, R. W., and Jesshope, C. R. (1981). *Parallel Computers*. Adam Hilger, Bristol, UK.

Hung, H. M. (1974). A parallel algorithm for symmetric tridiagonal eigenvalue problems. CAC Document No. 109, Center for Advanced Computation, University of Illinois at Urbana-Champaign.

Hwang, K., and Briggs, F. A. (1984). *Computer Architecture and Parallel Processing*. McGraw-Hill, New York.

Ispen, I. C. F., and Jessup, E. R. (1987). Solving the symmetric tridiagonal eigenvalue problem on the hypercube. Research Report 548, Dept. of Computer Science, Yale University, New Haven, Connecticut.

Jennings, A. (1977). *Matrix Computations for Engineers and Scientists*. John Wiley and Sons, New York.

Kuck, D. J. (1977). A survey of parallel machine organization and programming. ACM Comput. Surv., 9(1):29.

Kuck, D. J., and Sameh, A. H. (1972). Parallel computation of eigenvalues of real matrices, Vol. 2, IFIP Congress 1971, North-Holland, Amsterdam, pp. 1266–1272.

Lanczos, C. (1950). An iteration method for the solution of the eigenvalue problem of linear differential and integral operators. J. Res. Nat. Bur. Standards, B45:225.

Lo, S. S., Philippe, B., and Sameh, A. (1986). A multiprocessor algorithm for the symmetric tridiagonal eigenvalue problem. CSRD report no. 513.

Ma, S. C., Patrick, M. L., and Szyld, D. B. (1988). A parallel hybrid algorithm for the generalized eigenproblem. Proceedings of Third SIAM Conf. on Parallel Processing for Scientific Computing, December, Chicago, Illinois.

Nguyen, D. T., and Arora, J. S. (1986). An algorithm for solution of large eigenvalue problems. Computers & Structures, 24(4):645.

Noor, A. K., and Peters, J. M. (1989). A partitioning strategy for efficient nonlinear finite element dynamic analysis on multiprocessor computers. Computers & Structures, 31(5):795.

Ortega, J. (1967). The Givens-Householder method for symmetric matrices. In: Ralston, R. A., and Wilf, H. S., Eds. *Mathematical Methods for Digital Computer*. John Wiley and Sons, New York.

Paige, C. C. (1971). The computation of eigenvalues and eigenvectors of very large sparse matrices. Ph.D. Thesis, University of London, UK.

Paz, M. (1985). *Structural Dynamics,* 2nd Ed. Van Nostrand–Reinhold, New York.

Pissanetsky, S. (1984). *Sparse Matrix Technology*. Academic Press, New York.

Rutishauser, H. (1970). Simultaneous iteration method for symmetric matrices. Numer. Math., 16:205.

Sameh, A. H. (1977). Numerical parallel algorithms—A survey. In: Kuck, D. H., and Sameh, A. H., Eds. *High Speed Computer and Algorithm Organization*. Academic Press, New York, pp. 207–228.

Strang, G. (1986). *Introduction to Applied Mathematics*. Wellesley Cambridge Press, Wellesley, Massachusetts.

Szyld, D. B. (1988). Criteria for combining inverse and Raleigh quotient iteration. SIAM J. Numer. Anal., 25(6):1369.

Tuazon, J., Peterson, J. Pniel, M., and Liberman, D. (1985). CALTECH/JPL MARK II hypercube concurrent processor. Proceedings of 1985 International Conference on Parallel Processing, IEEE, St. Charles, Illinois, August, pp. 666–673.

Utku, S., and Salama (1986). Parallel solution of closely coupled systems. International Journal for Numerical Methods in Engineering, 23:2177.

Utku, S., Melosh, R. J., and Salama, M. (1984). A mathematical characterization of concurrent processing: Machines, jobs, and stratagems. In: Hodge, C. S., Ed. *Proceeding of Third Conference on Computing in Civil Engineering*. ASCE, San Diego. pp. 258–284.

Utku, S., Salama, M., and Melosh, R. J. (1986a). Concurrent Cholesky factorization of positive definite banded Hermitian matrices. International Journal for Numerical Methods in Engineering, 23:2137.

Utku, S., Chang, H. Y., Salama, M., and Rapp, D. (1986b). Simultaneous iterations algorithm for general eigenvalue problem on parallel processors. Proceedings of 1986 International Conference on Parallel Processing, IEEE, St. Charles, Illinois, August pp. 59–66.

Utku, S., Salama, M., and Melosh, R. J. (1989). A family of permutations for concurrent factorization of block tridiagonal matrices. IEEE Transactions on Computers, 38(6): 812.

Wilkinson, J. H. (1965). *The Algebraic Eigenvalue Problem*. Clarendon Press, Oxford.

9

Parallel Computations in Solid Mechanics

C. T. Sun   *Purdue University, West Lafayette, Indiana*

L. S. Chien   *Air Force Institute of Technology, Wright-Patterson Air Force Base, Ohio*

1   INTRODUCTION

The development of parallel computational methods, including parallel numerical algorithms and parallel processing techniques, has become a necessity and has generated new challenges in structural mechanics. Classic structural formulating methods and numerical methods are generally unable to exploit multiprocessors and powerful hardware. As a result, efficient parallel computational methods can be created by reformulating the original existing methods or by discovering new methods. Research in developing parallel computational methods for structural mechanics applications covers a broad spectrum of different considerations. At one end of the spectrum, transforming existing sequential programs using a parallel compiler speeds solutions of old mechanics problems on a parallel computing system and, at the other end, discloses new numerical simulations suitable for multiprocessor computation. The latter research motivation is perhaps the most important result of exploiting the power of parallel computation. The present research extends from developing parallel processing techniques for the finite element method to discovering inherent parallel numerical algorithms that make them attractive bases for parallel computation.

Parallel processing architecture and concurrent computing software were studied by various computer research organizations in the early age of computers (Hwang and Briggs, 1984). However, it was not until the late 1970s that research in parallel computations mushroomed in many engineering disciplines. In this section some previous works of parallel computations on structural mechanics applications are briefly reviewed.

Miranker (1971) surveyed concurrent numerical algorithms that can be used to solve engineering problems. The concurrent numerical algorithms of optimization, root finding, differential equation solving, and systems of linear equation solving were discussed.

In the earlier stage of computational research on multiprocessor computers, work on the development of numerical algorithms for vector computers was presented in some literature. The impact of the CDC Star-100 vector computer on the finite element method was addressed by Noor and Fulton (1975). In this paper major reorganization of a source program and alternate numerical algorithms were given to realize the full potential of these new vector computers. Later the CDC Star-100 vector computer was employed to evaluate element stiffness matrices (Noor and Hartley, 1978) and to analyze dynamic response by the finite element method (Noor and Lambiotte, 1979). Van Luchene et al. (1986) later proposed a linear and nonlinear finite element software for the Cyber 205 vector computer.

Recent advances in parallel computers have attracted great interest from researchers in structural analysis. Ortega and Voigt (1985) indicated that there is considerable activity in developing parallel algorithms for engineering computing problems, and structural analysis is one of the major application areas for parallel computing systems. Noor and Atluri (1987) surveyed the trends and impacts of parallel computation on structural mechanics. Some monographs also revealed the meaningfulness of parallel computation methods on various scientific and engineering mechanics problems. These representative studies were addressed in the ASME proceedings edited by Noor (1986, 1987) and the technical literature edited by Rodrigue (1982).

To date, applications of parallel computation to the finite element method have received great attention. Zave and Rheinboldt (1979) designed an adaptive parallel finite element system for a certain class of linear elliptic problems. A multi-microprocessor system was proposed by Jordan and Sawyer (1979) for finite element structural analysis to speed solutions. In the past decade a finite element machine was built at NASA's Langley Research Center (Storaasli et al., 1982) to support parallel solutions of structural analysis problems. Fulton (1986) also discussed some research in parallel computing associated with the NASA finite element machine (FEM) project. The focus in that work was eigenvalue and transient response analysis.

Law (1986) directly solved element displacements without forming the global stiffness matrix on an MIMD (multiple instruction–multiple data) multiprocessing system. Carey (1986) modeled physical processes and gave examples in which inherent parallelism is evident in finite element analysis. Zois (1988a) studied parallel processing techniques for the generation of element stiffnesses, element loads, the assembly process, and the evaluation of element stress. In a companion paper Zois (1988b) also presented a parallel system solution method called asynchronous systolic multiple Gauss-Jordan (ASMGJ) to solve a system of equations. Another equation solver using a triangular decomposition approach for finite element problems was given by Goehlich et al. (1989).

Recently Noor and Peters (1989) proposed a computational procedure for the nonlinear dynamic analysis of unsymmetric structures on multiprocessor computers. The idea of a linear combination of symmetric and antisymmetric response modes was presented in that paper. For hypercube multiprocessor computers the results of analysis of large plate-bending problems were reported by Farhat et al. (1987). About the same time a new computer program for solutions of the finite element system was also proposed by Farhat and Wilson (1987). Another experience with a hypercube multiprocessor computer was presented by Nour-Omid and Park (1987), in which some structural mechanics problems were solved by the finite element method on the 32-node Caltech hypercube multiprocessor computer. Carter et al. (1989) described a parallel implementation of the finite element method on a hypercube parallel system. The strategy, which does not require the formation of a global system of equations, was presented.

Element-by-element solution strategy also shows potential application on parallel computing systems. The preconditional conjugate gradient method associated with the element-by-element solution strategy was studied by Carey and Jiang (1986). Another application was given by King and Sonnad (1987). The implementation of an element-by-element solution algorithm arising from applying the finite element method on the loosely coupled array parallel computer located at IBM Kingston was illustrated. The results show that solution speedup efficiencies of approximate 95% can be readily achieved.

Sun and Mao (1988a,b) employed a global-local finite element analysis procedure to perform elastic and elastic-plastic stress analyses. This procedure involves a global analysis using a coarse mesh and the subsequent locally refined analyses of a number of subregions. These local analyses can be performed concurrently on a multiprocessor computer.

Many existing stress analysis methods are suitable for parallel computation. Chien (1989) and Chien and Sun (1990) demonstrated the efficiency of using orthogonal expansion and boundary element methods programmed in parallel algorithms. These methods are discussed in Sections 3 and 4.

Further research on developing numerical solution methods of nonlinear structural problems on parallel computing environments was undertaken extensively. Utku et al. (1982) studied Newton-Raphson types of iterations to solve nonlinear structural problems. The idea of assigning comparably sized substructures to parallel processors was exploited. An experimental software system design for an adaptive nonlinear finite element solver was described by Sledge and Rheinboldt (1985). Advantage was taken of the inherent parallelism in the system. White (1986) investigated the conditions of elliptic or parabolic partial differential equations, such that a parallel algorithm of a nonlinear Gauss-Seidel type converges to the unique solution. Moreover, a concurrent processing algorithm was developed for a material nonlinear analysis by Darbhamulla et al. (1987). Again, the NASA special-purpose computer (FEM) was employed in this study. Ou and Fulton (1988) examined several integration algorithms for nonlinear dynamic problems on parallel computers. The central difference method has shown high computational efficiency and parallel effectiveness in that work.

Some research publications involving structural dynamic analyses on parallel computing environments were discussed as well in the literature. Ortiz and Nour-Omid (1986) proposed a new class of algorithms for transient finite element analysis that is amenable to an efficient implementation in parallel computing systems. Another experience of solving transient response on the NASA experimental MIMD computer was reported by Storaasli et al. (1987), who presented computational speedups of up to 7.83 times for eight processors on test problems. Bostic and Fulton (1987) solved for natural frequencies of a large vibrating structure by implementing the Lanczos method on the FLEX/32 parallel computer. The parallel eigenvalue algorithm demonstrated significant speedups in calculation time over traditional sequential methods. Hajjar and Abel (1989) studied the implementation of the central difference algorithm for the fully nonlinear, transient dynamic analysis of three-dimensional framed structures.

The application of parallel computing to structural optimization has also gained attention from researchers. Svensson (1987) developed an active design strategy to improve the efficiency of sensitivity analysis in an iterative process for structural optimization. An application of the strategy exploiting multiprocessor architecture was outlined. Schnabel (1987) discussed some basic opportunities for the use of multiprocessing in the solution of optimization problems. Two fundamental optimization problems, including unconstrained optimization and global optimization, were considered there. Adeli and Kamal (1992a,b) developed parallel algorithms for the optimization of structures on shared-memory multiprocessor computers. The work load balance among multiprocessors was shown by introducing a substructuring algorithm. The applications on truss and frame problems by implementing the algorithm were also presented.

Several parallel numerical algorithms emphasized mathematical operations for structural mechanics analysis. Houstis et al. (1987) presented a method for coarse-grained partitioning of computations for parallel architectures and applied it to partial differential equation (PDE) applications. An approach to the construction of a posteriori error estimates for nonlinear finite element computations was shown by Rheinboldt (1985). In addition, Douglas and Miranker (1988) discussed parallel algorithms in the setting of iterative multilevel methods.

Another area of major application of parallel computing in structural analysis is the solving of systems of linear equations. Contributions in this area are discussed in the next section.

2 PARALLEL GAUSSIAN ELIMINATION AND STIFFNESS MATRIX-FORMING PROCEDURES IN FINITE ELEMENT ANALYSIS

2.1 Introduction

In this section attention is concentrated on the development of a parallel method to solve a large system of linear algebraic equations that usually results from the global stiffness equations of finite element solutions.

Much of the computational effort of the finite element method in a large structure involves the solution of a system of linear algebraic equations. The global stiffness matrix generated from this system of linear algebraic equations is symmetric, positive, definite, and banded. Also, the dimension of the global stiffness matrix is usually very large for a structure with a large number of degrees of freedom. The solution schemes that have been used to solve this large system of equations can generally be categorized as indirect (numerical) or direct (Gaussian elimination) method. Among these the Gaussian elimination method has been considered the most powerful solution scheme and has been widely used.

Traditional sequential computers are very expensive and time consuming in solving a large system of linear algebraic equations. According to Bostic and Fulton (1987), a processor that obtains natural frequencies spends most of its solution time in solving a system of equations. To optimize the solution time Noor et al. (1983) incorporated a special-purpose finite element machine and showed the potential of using highly parallel architectures to solve a large system of equations. The applications of different parallel computing systems in solving finite element problems were also discussed in their work.

A second way to optimize the solution time came from Pease (1967). A parallel operation of Gaussian elimination was suggested, the technique of replacing a row by a linear combination of two rows. Heller (1978) further discussed parallel methods for a system of linear equations and eigenvalue problems. Rodrigue (1986) proposed incorporating premultiplication with a system of equations associated with large-scale scientific computing. The numerical algorithm from this incorporation of a system of equations can be solved on parallel computers. Leuze and Saxon (1983) studied the parallel triangularization of a symmetric system of linear equations. In this paper a completely connected parallel machine with an arbitrarily large number of processors was assumed.

Utku and Salama (1986) presented the odd-even permutation and associated unitary transformation technique to solve closely coupled systems on parallel computers. Implementation of the Householder method for tridiagonalizing a real symmetric matrix into a parallel algorithm for a concurrent machine was discussed in the work of Change et al. (1988). Another equation solver was shown by Farhat and Wilson (1988) in which general-purpose parallel FORTRAN subroutines were developed for the solution of sparse and dense symmetric systems of linear equations.

Most of the methods just discussed are direct rather than iterative. Lawrie and Sameh (1983) studied the solution of a large system of equations on multiprocessor computers. The basic technique involves the subdivision of the problem into a number of independent systems. This subdivision allows each processor to proceed independently of the others. Bauget (1978) proposed a class of asynchronous iterative methods for solving a system of equations with a sufficient condition given to guarantee the convergence of any asynchronous iterations.

Many parallel solution schemes have been proposed, and more still are being undertaken. Some focus on purely mathematical manipulations; others study parallel solution methods from an engineering point of view. This section presents a parallel Gaussian elimination solution scheme accomplished by modifying the classic sequential Gaussian elimination procedures and integrating with the concept of SIMD (single instruction–multiple data) parallel computation.

The approach to concurrently solve large systems of linear algebraic equations, assuming a limited number of processors and constant bandwidth, is presented in this section. This method, described here, is implemented with SIMD parallel computing environments. Based on classic Gaussian elimination solution procedures, this method can be divided into two steps: (1) parallel forward reduction and (2) parallel backward substitution. The main idea of this method is that the rows of the global stiffness matrix of a finite element solution are divided among the multiprocessors and processed concurrently. The shared data are then transferred between rows through the global memory. Parallel forward reduction of the global stiffness matrix and the load vector is first carried out by the multiprocessors. Clearly, an upper triangular matrix and a new load vector are formed after the reduction. Parallel backward substitution is performed subsequently by the multiprocessors to obtain the solution. Thus, when the Gaussian elimination procedures are performed, the rows can process with the solution steps simultaneously. As a result the operating time decreases proportionally with the increasing number of processors.

2.2 Parallel Forward Elimination and Backward Substitution

Consider the simulation block diagram of the Sequent Balance 21000 parallel computer as shown in Figure 1. The shared data calculated among processors are

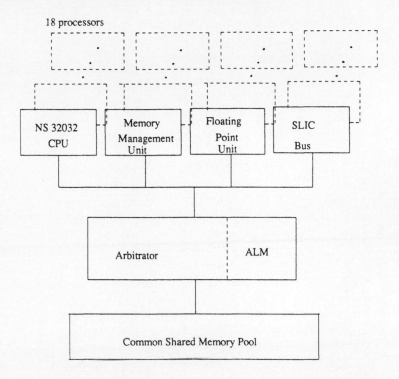

Figure 1 Simulation block diagram of the Sequent Balance 21000 tightly coupled parallel computing system.

stored in the global memory. These shared data, for example, the pivot row that contains the pivot diagonal element, are stored in the global memory for each processor to use. All the arithmetic operations in the parallel Gaussian elimination procedures are performed locally in each processor.

The processor i contains rows $i + np \cdot n$ of the global stiffness matrix at the beginning of the parallel Gaussian elimination steps as shown in Figure 2, where np is the number of total processors used and n is the integer running from zero to a number that can properly include the number of rows m of the global stiffness matrix in $i + np \cdot n$. For instance, with $m = 10$ and $np = 3$, the first processor ($i = 1$) contains rows 1, 4, 7, and 10, the second processor ($i = 2$) contains rows 2, 5, and 8, and the third processor ($i = 3$) contains rows 3, 6, and 9. Naturally n is observed as 0, 1, 2, and 3. Each individual row is stored with a logical stack corresponding to the row number. This logical stack is used to communicate with other processors and/or other rows.

At the beginning of each step of the parallel forward reduction, the pivot row, containing the pivot diagonal element needed for the rows below the current

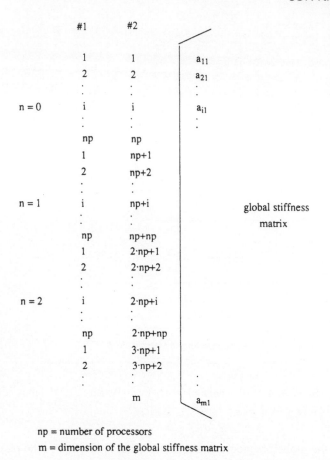

np = number of processors
m = dimension of the global stiffness matrix

Figure 2   Overall view of the storage of rows in each processor. Column 1 is the repeated processor's number. The numbers in column 2 corresponding to a particular number in column 1 are the rows stored in that particular processor.

pivot row to process parallel forward reduction, is calculated. The logical stack of the pivot row then toggles and sends the pivot row information to the global memory. Those processors that contain the rows below the pivot row can then access the pivot row in the global memory concurrently. Parallel forward reduction is then performed in these processors. Accordingly, the column elements below the pivot diagonal element become zeros simultaneously at the end of each step of parallel forward reduction. For instance, suppose the diagonal element of row j is used as the pivot element at the jth step of parallel forward reduction. The jth logical stack toggles and sends the jth row information to the global memory. The jth row is then stored in the global memory and is "ready" for other

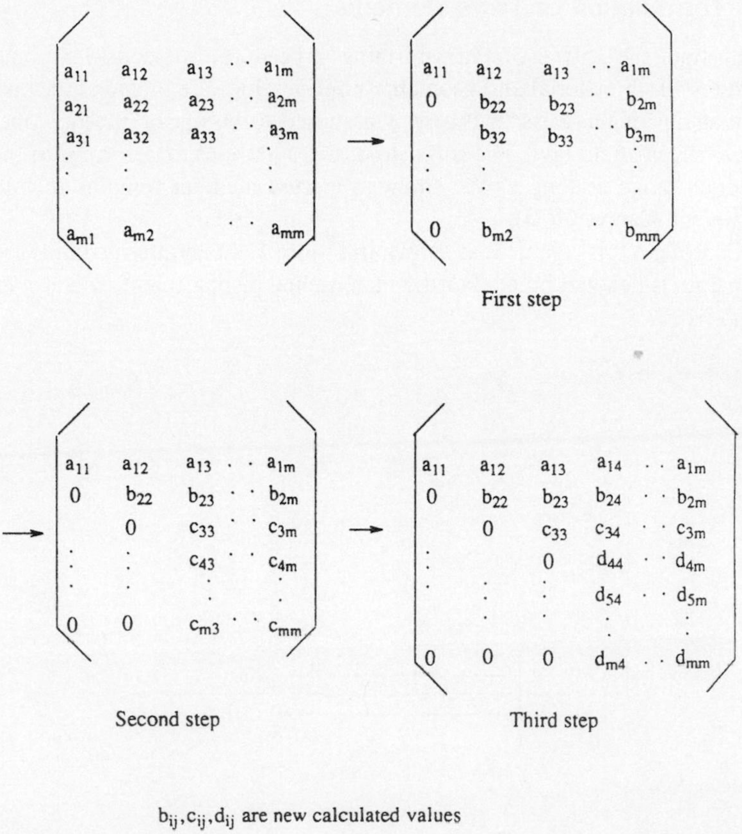

$$
\begin{pmatrix}
a_{11} & a_{12} & a_{13} & \cdots & a_{1m} \\
a_{21} & a_{22} & a_{23} & \cdots & a_{2m} \\
a_{31} & a_{32} & a_{33} & \cdots & a_{3m} \\
\vdots & \vdots & \vdots & & \vdots \\
a_{m1} & a_{m2} & & \cdots & a_{mm}
\end{pmatrix}
\longrightarrow
\begin{pmatrix}
a_{11} & a_{12} & a_{13} & \cdots & a_{1m} \\
0 & b_{22} & b_{23} & \cdots & b_{2m} \\
 & b_{32} & b_{33} & \cdots & b_{3m} \\
\vdots & \vdots & \vdots & & \vdots \\
0 & b_{m2} & & \cdots & b_{mm}
\end{pmatrix}
$$

First step

$$
\longrightarrow
\begin{pmatrix}
a_{11} & a_{12} & a_{13} & \cdots & a_{1m} \\
0 & b_{22} & b_{23} & \cdots & b_{2m} \\
 & 0 & c_{33} & \cdots & c_{3m} \\
 & & c_{43} & \cdots & c_{4m} \\
\vdots & \vdots & \vdots & & \vdots \\
0 & 0 & c_{m3} & \cdots & c_{mm}
\end{pmatrix}
\longrightarrow
\begin{pmatrix}
a_{11} & a_{12} & a_{13} & a_{14} & \cdots & a_{1m} \\
0 & b_{22} & b_{23} & b_{24} & \cdots & b_{2m} \\
 & 0 & c_{33} & c_{34} & \cdots & c_{3m} \\
\vdots & & 0 & d_{44} & \cdots & d_{4m} \\
 & \vdots & \vdots & d_{54} & \cdots & d_{5m} \\
 & & & \vdots & & \vdots \\
0 & 0 & 0 & d_{m4} & \cdots & d_{mm}
\end{pmatrix}
$$

Second step　　　　　　　　　　Third step

b_{ij}, c_{ij}, d_{ij} are new calculated values

Figure 3　The elimination wavefront of the parallel forward reduction of the first three steps.

processors to access it. All the processors receive this pivot row information concurrently and process the parallel forward reduction. As a result the lower off-diagonal elements in column j are then eliminated to zero simultaneously. Subsequently, similar parallel forward reduction processes are carried out on the $j + 1$ row. The "wavefront" of the parallel forward reduction moves from left to right in the global stiffness matrix at each step, which can be seen in Figure 3. When all the logical stacks of the global stiffness matrix have toggled, the global stiffness matrix is in upper triangular form. Obviously the new load vector is obtained.

Parallel backward substitution starts subsequently in a similar logical process but in the reverse row order. The solution is obtained after all the logical stacks have toggled again.

2.3   Formulation of Truss Elements

To illustrate the stiffness matrix-forming procedure, we consider plane truss structures. Both material and geometric nonlinearities are included in this study. The material of the truss members is assumed to be elastic-plastic, obeying a bilinear stress-strain law. The truss structure is allowed to deform into the large deflection range as long as the strain in a truss member remains infinitesimal (Martin and Carey, 1973).

Consider a truss element as shown in Figure 4. At an intermediate load P the axial strain is denoted by ϵ^0. For the incremental displacements u and v the total strain ϵ is

$$\epsilon = \epsilon^0 + \epsilon^a \tag{1}$$

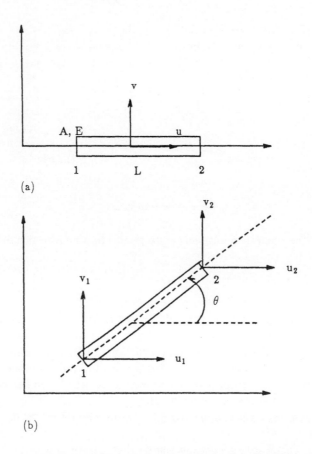

(a)

(b)

Figure 4   A two-dimensional truss element in (a) local and (b) global coordinate systems.

where the strain increment ϵ^a includes moderately large rotation and is given as

$$\epsilon^a = \frac{du}{dx} + \frac{1}{2}\left(\frac{dv}{dx}\right)^2 \tag{2}$$

For elastic deformation before yielding, the total elastic strain energy increment U is obtained from

$$U = \int_V \left(\int_{\epsilon^o}^{\epsilon^o + \epsilon^a} \sigma \, d\epsilon\right) dV \tag{3}$$

where V is the volume of the element.

In the elastic range the total strain energy increment in the truss element is obtained by integrating Eq. (3) over the element length L. Retaining only terms up to second order in Eq. (3) we obtain

$$U = \frac{1}{2}\int_0^L AE\left(\frac{du}{dx}\right)^2 dx + \int_0^L AE\epsilon^o\left[\frac{1}{2}\left(\frac{dv}{dx}\right)^2\right] dx$$
$$+ \int_0^L AE\epsilon^o \frac{du}{dx} dx \tag{4}$$

where A is the element cross-sectional area and E is the tangent modulus. Use the linear shape functions for the truss element

$$u = \left(1 - \frac{x}{L}\right) u_1 + \frac{x}{L} u_2$$
$$v = \left(1 - \frac{x}{L}\right) v_1 + \frac{x}{L} v_2 \tag{5}$$

where u_1, u_2, v_1 and v_2 are the incremental nodal displacement components as shown in Figure 4a. By substituting Eq. (5) into Eq. (4) and performing the necessary integrations and differentiations, the elastic stiffness matrix $[\mathbf{k}_E]$ and geometric stiffness matrix $[\mathbf{k}_G]$ for each individual element are obtained following the standard procedure:

$$[\mathbf{k}_E] = \frac{AE}{L}\begin{bmatrix} 1 & 0 & -1 & 0 \\ 0 & 0 & 0 & 0 \\ -1 & 0 & 1 & 0 \\ 0 & 0 & 0 & 0 \end{bmatrix} \tag{6}$$

$$[\mathbf{k}_G] = \frac{P}{L} \begin{bmatrix} 0 & 0 & 0 & 0 \\ 0 & 1 & 0 & -1 \\ 0 & 0 & 0 & 0 \\ 0 & -1 & 0 & 1 \end{bmatrix} \qquad (7)$$

where P is the intermediate axial load of the truss element at the current stage of loading.

For plastic deformation a good assumption is made by employing Eq. (5) as the shape functions. By doing so the stiffness matrices in Eqs. (6) and (7) can also be used when plastic deformation occurs except that E must be updated according to the bilinear law. Thus, the stiffness (6) and (7) are both valid in the deformation history of the material and geometrically nonlinear truss element.

Transforming the stiffness matrices (6) and (7) into the global coordinate by the appropriate transforming matrix yields

$$[\mathbf{k}_E] = \frac{AE}{L} \begin{bmatrix} \lambda^2 & \lambda\mu & -\lambda^2 & -\lambda\mu \\ \lambda\mu & \mu^2 & -\lambda\mu & -\mu^2 \\ -\lambda^2 & -\lambda\mu & \lambda^2 & \lambda\mu \\ -\lambda\mu & -\mu^2 & \lambda\mu & \mu^2 \end{bmatrix} \qquad (8)$$

and

$$[\mathbf{k}_G] = \frac{P}{L} \begin{bmatrix} \mu^2 & -\lambda\mu & -\mu^2 & \lambda\mu \\ -\lambda\mu & \lambda^2 & \lambda\mu & -\lambda^2 \\ -\mu^2 & \lambda\mu & \mu^2 & -\lambda\mu \\ \lambda\mu & -\lambda^2 & -\lambda\mu & \lambda^2 \end{bmatrix} \qquad (9)$$

where $\lambda = \cos\theta$ and $\mu = \sin\theta$. The truss element stiffness matrix in the global coordinate is $[\mathbf{k}] = [\mathbf{k}_E] + [\mathbf{k}_G]$.

The global equilibrium equations are

$$[\mathbf{K}]\{d\} = \{F\} \qquad (10)$$

where the global stiffness matrix $[\mathbf{K}]$ depends on P and θ, which are functions of the nodal displacements. Thus Eq. (10) represents a system of nonlinear equations.

The method of direct iteration (or successive approximations) is employed in the solution scheme. The total load $\{F\}$ is divided by the number of load increment steps. A constant load increment $\{\Delta F\}$ is then used in each load increment step. At the beginning of each increment step the global stiffness matrix is recalculated and remains constant through the step. The displacement increment $\{\Delta d\}$ is then solved from the global system of equations

$$[\mathbf{K}] \{\Delta d\} = \{\Delta F\} \tag{11}$$

The total displacements are thus made up of a sequence of incremental displacements.

2.4 Parallel Forming of the Nonlinear Global Stiffness Matrix

In generating the nonlinear global stiffness matrix at each incremental load step, element stiffnesses are calculated independently based on the deformed configuration. These stiffnesses are then placed at the appropriate locations of the global stiffness matrix. Generally the nonlinearities in the truss elements are different from each other. Thus computing all element stiffnesses becomes essential; this increases the time in forming the global stiffness matrix.

Because of the independence of the element stiffnesses, it would be helpful if the MIMD parallel computing systems could proceed to the assembly procedure concurrently. A potential problem occurs whenever an entry of the global stiffness matrix is contributed by different element stiffnesses. These different element stiffnesses are individually assembled locally by different processors of a parallel computing system. These processors may send their results to the location of the common global stiffness entry simultaneously and yield a misleading solution. We overcome this synchronization problem either by setting the local locks in the common shared-memory pool or by numbering the elements throughout the entire structure in a special manner.

Setting the Local Locks in Shared-Memory Pool. Local locks in the common shared-memory pool are very important in parallel numerical algorithms. A local lock at a particular location of the common shared-memory pool is needed whenever the result at this particular location comes partially from distinct processors of a parallel computing system. This local lock provides the function of organizing the order of contributions from distinct processors to the result. This local lock thus prevents the overlapping problem and ensures the correct solution. In the present study we set a local lock corresponding to the location of each entry of the global stiffness matrix. Because each local lock may delay a processor in making an entry to the global stiffness matrix, the speedup decreases as the number of processors increases.

Special Numbering for the Elements Throughout the Entire Structure.
The total elements of a large truss structure can be numbered in a special way for use on parallel computers. A large truss structure is divided nearly evenly by the number of processors to several substructures, so that the common

boundary nodes are minimized. The assemblage of the element stiffness matrices in each substructure is then carried out by a single processor.

Numbering the elements in each substructure is achieved in the following way. The element numbers start at the front end, sequentially running over the entire substructure. The far end of a substructure is the front end of the adjoining one. Thus when the global stiffness matrix-forming procedure starts, each processor may start concurrently from the front end of the respective substructure. By the time the forming procedure in each processor approaches the far end of the substructure, the entry made from the other processor for the adjacent substructure is completed. The overlapping problem is thereby avoided. However, the overlapping problem may be inevitable if a large number of processors is used and the substructure is small. In such an event the forming time for each substructure is not long enough to separate the forming of the common nodes between two adjacent substructures.

2.5   Example

To illustrate the parallel forming of a nonlinear global stiffness matrix and the parallel solution scheme, a two-dimensional cantilever truss is used. This truss is composed of 50 truss cells subjected to a single point load at the free end and analyzed by the finite element method. The truss dimensions and material properties are shown in Figure 5. Parallel global stiffness matrix forming is executed first. Then the parallel solution scheme of the global stiffness matrix is applied to obtain the solution at each load step. To approximate the loading history 200 load steps are used.

To evaluate the parallel algorithms the user time of arithmetic operations is recorded. The speedup is defined in most parallel algorithms as

$$\text{Speedup} = \frac{\text{user time for one processor}}{\text{user time for } N \text{ processors}}$$

Also, the efficiency for a parallel algorithm is defined by the ratio

$$\text{Efficiency} = \frac{\text{speedup}}{\text{number of processors used}}$$

The user time of the nonlinear global stiffness forming versus the number of processors is shown for two parallel forming cases. The speedups and efficiencies for these two cases are displayed for comparison in Figures 6 and 7, respectively. It is evident that parallel forming can achieve great time savings. It is also seen that the use of local locks may suffer some loss in efficiency as the number of processors increases.

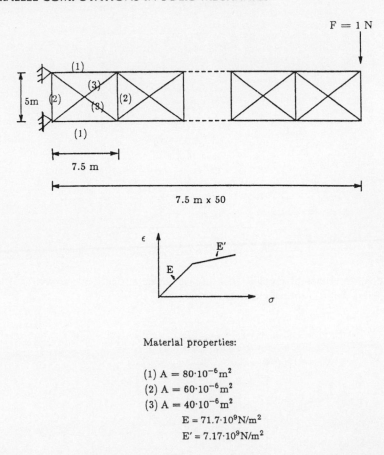

Figure 5 A two-dimensional nonlinear cantilever truss with 50 cells subjected to a concentrated load at the free end.

For this particular example the global stiffness matrix is very narrow banded (half-bandwidth = 7), and thus the speedup levels off beyond six processors. This occurs because relatively fewer computational operations are involved. The savings in user time in computational operations balances the system communication time when about six processors are used. However, for matrices with large bandwidths the speedup is more apparent as a larger number of processors is used. This is verified by the results shown in Figures 8 and 9 in which the speedups and efficiencies for solving the global system of equations with half-bandwidths of 7 and 50 are considered.

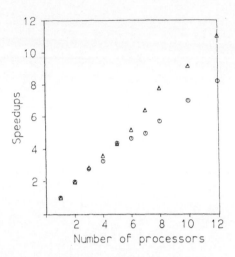

Figure 6   Speedups of forming the nonlinear global stiffness matrix with (○) local locks setting and (△) special elements numbering.

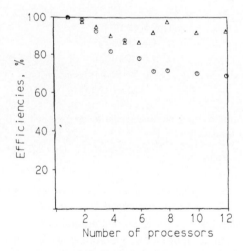

Figure 7   Efficiencies of forming the nonlinear global stiffness matrix with (○) local locks setting and (△) special elements numbering.

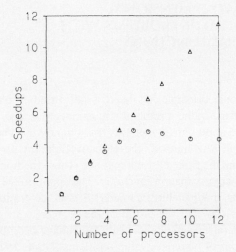

Figure 8 Speedups of parallel Gaussian elimination with 200 degrees of freedom and (○) half-bandwidth 7, (△) half-bandwidth 50.

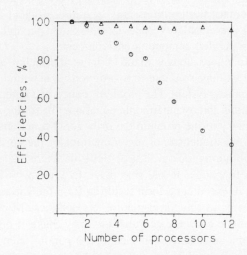

Figure 9 Efficiencies of parallel Gaussian elimination with 200 degrees of freedom and (○) half-bandwidth 7, (△) half-bandwidth 50.

3   SOLVING TWO-DIMENSIONAL BOUNDARY VALUE PROBLEMS WITH ORTHOGONAL FUNCTIONS

3.1   Introduction

The classic variational principle has been widely used to obtain approximate solutions of boundary value problems in engineering and applied sciences (Reddy, 1986). The approximate methods that employ the variational principle include the Reyleigh-Ritz method, the Bubnov-Galerkin method, the Petrov-Galerkin method, and the finite element method. These methods differ from each other in the form of approximation functions. The approximate solutions of these methods are in terms of adjustable parameters that are determined by minimizing the functional or solving the weak form of the problem.

A number of parallel computational methods have been proposed to obtain approximate solutions of boundary value problems (Noor, 1987). Most of these parallel processing techniques are constructed to parallelize the solution procedures in conjunction with the finite element method, which has long been recognized as a powerful numerical method for solving problems with arbitrary geometric domain. Nevertheless, for practical problems with regular geometric domains other numerical algorithms may be more efficient for multiprocessor computation. The present study focuses on invoking a parallel algorithm to solve these problems in parallel computing environments.

Orthogonal functions along with the variational principle are used in this section to obtain approximate solutions to two-dimensional boundary value problems on a parallel computing system. The problems concerned here are those governed by Poisson equations and by biharmonic equations. In addition, depending upon the domain geometry of the problem the orthogonal functions chosen are either one-dimensional eigenfunctions or eigenfunctions of the corresponding two-dimensional free vibration problem. It is obvious that there are some advantages of using these eigenfunctions. In the first place they are well known and square integrable. Second, the integral form resulting from the variational principle can be obtained in terms of the corresponding eigenvalues. More importantly, the orthogonalities of the chosen eigenfunctions may make it possible to independently calculate the adjustable coefficients on a parallel computing system.

3.2   Description of the Method

Consider the two-dimensional boundary value problem

$$Au = f \quad \text{in } \Omega \tag{12}$$

where A is a linear positive definite operator on the domain Ω in a separable Hilbert space H_A, and $f \in H_A$. The solution u is subjected to various boundary conditions. The generalized (weak) solution can be obtained by solving the associated variational (weak) form or by minimizing an associated quadratic functional. Thus the generalized solution is an element of H_A and satisfies the weak form

$$B\ (u,v)\ =\ (f,\ v) \tag{13}$$

where $B(u,\ v)\ =\ (Au,\ v)$, and $(\cdot,\cdot)$ is an inner product in H_A.

An approximate solution u_N to Eq. (12) in the variational method can be assumed in the form of a finite series,

$$u_N\ =\ \sum_{j=1}^{N}\ c_j\phi_j\ +\ \phi_0 \tag{14}$$

where ϕ_j are the linearly independent coordinate functions of the assumed solution and satisfy the homogeneous geometric boundary conditions, ϕ_0 is a function that satisfies the nonhomogeneous parts of the essential boundary conditions, and c_j are the adjustable parameters to be determined.

Since $\{\phi_1,\ \phi_2\ .\ .\ .\ \phi_N\}$ form a basis in H_A, the weak solution u_N can be approximated to arbitrary accuracy by increasing the number of terms. As a result the approximate solution u_N is expected to converge to the actual solution as N approaches infinity.

Substituting Eq. (14) into Eq. (13) and letting $v\ =\ v_N\ =\ \phi_i,\ i\ =\ 1,2,\ .\ .\ .$ N, yields

$$B(u_N,\ \phi_i)\ =\ (f,\ \phi_i) \tag{15}$$

where

$$B(u_N,\ \phi_i)\ =\ \int_\Omega\ (Au_N)(\phi_i)\ d\Omega$$

$$(f,\ \phi_i)\ =\ \int_\Omega f\phi_i\ d\Omega$$

Thus the adjustable parameters c_j can be determined by solving Eq. (15), and hence an approximate solution (14) is obtained.

3.3   Formulation of Parallel Algorithms

Second-Order Differential Equations.   Let operator A in Eq. (12) be the Laplace operator over Ω. The problem that occurs frequently in engineering applications is stated as follows.

$$\nabla^2 u + f = 0 \qquad \text{in } \Omega \tag{16}$$

$$u\ =\ g(s) \qquad \text{on } \Gamma_g \tag{17}$$

$$u_{,n} = h(s) \qquad \text{on } \Gamma_h \tag{18}$$

where $u_{,n}$ is the partial derivative of u with respect to outward normal coordinate n on Γ. It is also recognized that $g(s)$ and $h(s)$ are geometric and natural boundary conditions, respectively.

Substituting Eq. (14) into Eq. (16) and applying Eq. (15) gives

$$\sum_{j=1}^{N} c_j \int_{\Omega} \nabla^2 \phi_j(\mathbf{x}) \, \phi_i(\mathbf{x}) \, d\Omega + \int_{\Omega} \nabla^2 \phi_0(\mathbf{x}) \, \phi_i(\mathbf{x}) \, d\Omega$$

$$+ \int_{\Omega} f(\mathbf{x}) \, \phi_i(\mathbf{x}) \, d\Omega = 0 \qquad i = 1, 2, \ldots, N \tag{19}$$

where $\mathbf{x}$ denotes the position vector. It is also noted that the assumed function ϕ_0 equals $g(s)$ on Γ_g. Integrating by parts and using the divergence theorem, Eq. (19) becomes

$$\sum_{j=1}^{N} c_j \left(-\int_{\Omega} \nabla \phi_j(\mathbf{x}) \cdot \nabla \phi_i(\mathbf{x}) \, d\Omega + \int_{\Gamma} \nabla \phi_j(\mathbf{x}) \cdot \phi_i(\mathbf{x}) \mathbf{n} \, d\Gamma \right)$$

$$- \int_{\Omega} \nabla \phi_0(\mathbf{x}) \cdot \nabla \phi_i(\mathbf{x}) \, d\Omega + \int_{\Gamma} \nabla \phi_0(\mathbf{x}) \cdot \phi_i(\mathbf{x}) \mathbf{n} \, d\Gamma \tag{20}$$

$$+ \int_{\Omega} f(\mathbf{x}) \, \phi_i(\mathbf{x}) \, d\Omega = 0 \qquad i = 1, 2, \ldots, N$$

Since the assumed coordinate functions ϕ_j satisfy the homogeneous geometric boundary condition, that is, $\phi_j = 0$ on Γ_g, the boundary integral terms of Eq. (20) are valid only along $\Gamma_h = \Gamma - \Gamma_g$.

Multiplying the natural boundary condition (18) by ϕ_i and integrating along Γ_h gives

$$\int_{\Gamma_h} h(s) \, \phi_i(\mathbf{x}) \, d\Gamma = \sum_{j=1}^{N} c_j \int_{\Gamma_h} \nabla \phi_j(\mathbf{x}) \cdot \phi_i(\mathbf{x}) \mathbf{n} \, d\Gamma$$

$$+ \int_{\Gamma_h} \nabla \phi_0(\mathbf{x}) \cdot \phi_i(\mathbf{x}) \mathbf{n} \, d\Gamma \qquad i = 1, 2, \ldots, N \tag{21}$$

Substituting Eq. (21) into Eq. (20) yields a system of algebraic equations of c_j,

$$\sum_{j=1}^{N} c_j \int_{\Omega} \nabla \phi_j(\mathbf{x}) \cdot \nabla \phi_i(\mathbf{x}) \, d\Omega = \int_{\Omega} f(\mathbf{x}) \, \phi_i(\mathbf{x}) \, d\Omega$$

$$+ \int_{\Gamma_h} h(s) \, \phi_i(\mathbf{x}) \, d\Gamma - \int_{\Omega} \nabla \phi_0(\mathbf{x}) \cdot \nabla \phi_i(\mathbf{x}) \, d\Omega \qquad i = 1, 2, \ldots, N \tag{22}$$

It is seen that the coordinate functions ϕ_j can be chosen so that their gradients $\nabla\phi_j$ are orthogonal over the domain Ω. By doing so the system of Eqs. (22) is uncoupled and the adjustable parameters can be simultaneously calculated on a parallel computing system with the following result:

$$c_j = \frac{\displaystyle\int_\Omega f(\mathbf{x})\,\phi_j(\mathbf{x})\,d\Omega + \int_{\Gamma_h} h(s)\,\phi_j(\mathbf{x})\,d\Gamma - \int_\Omega \nabla\phi_0(\mathbf{x})\cdot\nabla\phi_j(\mathbf{x})\,d\Omega}{\displaystyle\int_\Omega (\nabla\phi_j)^2\,d\Omega} \tag{23}$$

Consider the Prandtl torsional problem of beams of rectangular cross section as an example of this parallel algorithm. The torsion of a beam of rectangular cross section $\Omega = \{(x,y): -a < x < a, -b < y < b\}$ can be determined by

$$\begin{aligned}\nabla^2_u + 2 &= 0 \quad \text{in } \Omega\\ u &= 0 \quad \text{on } \Gamma_g\end{aligned} \tag{24}$$

where u is the Prandtl stress function and $\Gamma_g = \Gamma$.

For the purpose of parallel computation an approximate solution to Eq. (24) can be assumed as

$$u_N = \sum_{j=1}^{N} c_j\phi_j \tag{25}$$

where

$$\phi_j(x, y) = \cos\frac{m\pi x}{2a}\cos\frac{n\pi y}{2b} \qquad m, n = 1, 3, 5, \ldots \tag{26}$$

Clearly, Eq. (26) satisfies the homogeneous geometric boundary conditions and the gradients $\nabla\phi_j$ are orthogonal over the domain Ω. Thus Eq. (23) can be applied to solve the adjustable parameters in Eq. (25). These assumed coordinate functions coincide with the double Fourier series in the Galerkin method (Sokolnikoff, 1956).

Substituting Eq. (26) into Eq. (23), the adjustable parameters

$$c_j = \frac{128(-1)^{(m+n)/2-1}}{\pi^4 mn\,(m^2/a^2 + n^2/b^2)} \tag{27}$$

can be calculated simultaneously on a parallel computing system.

Fourth-Order Differential Equations. In plate-bending problems the operator A in Eq. (12) is the biharmonic operator $\nabla^2\nabla^2$. Following procedures similar to those described in the previous section, a system of algebraic equations for c_j can be derived. For simplicity the geometric boundary conditions are assumed to be homogeneous. Integrating by parts twice and applying the divergence theorem, the system of algebraic equations of c_j is obtained as

$$\sum_{j=1}^{N} c_j \int_{\Omega} (\nabla^2 \phi_i(\mathbf{x})) \, (\nabla^2 \phi_j(\mathbf{x})) \, d\Omega = \int_{\Omega} f \, \phi_i(\mathbf{x}) \, d\Omega \qquad i = 1, 2,, \ldots, N \qquad (28)$$

It is evident that if the coordinate functions ϕ_j are chosen so that their Laplacian $\nabla^2 \phi_j$ are orthogonal over the domain Ω, then the system of equations (28) can be uncoupled. Hence,

$$c_j = \frac{\displaystyle\int_{\Omega} f(\mathbf{x}) \, \phi_j(\mathbf{x}) \, d\Omega}{\displaystyle\int_{\Omega} (\nabla^2 \phi_j(\mathbf{x}))^2 \, d\Omega} \qquad (29)$$

Without this orthogonality property of $\nabla^2 \phi_j$ the expression given by Eq. (29) is not valid, and solving a system of coupled algebraic equations becomes inevitable.

One of the functions that can be used for parallel computation in Eq. (29) is the set of eigenfunctions of the corresponding free vibration problem. The nature of these eigenfunctions enables one to calculate the stiffness, the denominator of Eq. (29), with relatively little effort.

3.4   Examples

Bending problems of circular and rectangular plates are chosen as numerical examples. For the circular plate the free vibration eigenfunctions, which involve Bessel functions, are used as coordinate functions. On the other hand, for the rectangular plate the coordinate functions are composed of 2 one-dimensional beam eigenfunctions.

Example 1

A clamped-clamped circulate plate with radius R subjected to a linearly varying load is shown in Figure 10. The corresponding free vibration eigenfunctions are (Kirchhoff, 1850)

$$\left[J_m(k_{mn}r) - \frac{J_m(k_{mn}R)}{I_m(k_{mn}R)} I_m(k_{mn}r) \right] \cos \left[m(\theta - \gamma) \right] \qquad \begin{array}{l} m = 0, 1, 2, \ldots, \infty \\ n = 1, 2, 3, \ldots, \infty \end{array} \qquad (30)$$

where J_m and I_m are Bessel functions of the first kind of order m and a modified Bessel function of the first kind of order m, respectively, k_{mn} is the eigenvalue, and γ is a constant.

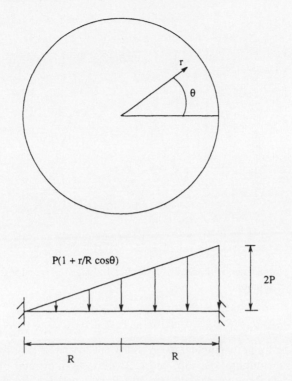

Figure 10 Clamped-clamped circular plate subjected to a linearly varying load $f = P(1 + r/R \cos \theta)$.

The proposed eigenfunctions satisfy the required homogeneous geometric boundary conditions and meet the orthogonality requirement in Eq. (29). Since the corresponding free vibration problem is self-adjoint, the integrated form of the denominator of Eq. (29) can be expressed as

$$\int_\Omega (\nabla^2 \phi_j(r, \theta))^2 \, d\Omega = k_{mn}^4 \int_\Omega (\phi_j(r, \theta))^2 \, d\Omega \qquad \begin{array}{l} m = 0, 1, 2, \ldots, \infty \\ n = 1, 2, 3, \ldots, \infty \end{array} \qquad (31)$$

Applying the recurrence relations of Bessel functions and of modified Bessel functions, Eq. (31) can also be written in terms of J_m and I_m explicitly.

The linearly varying load is described as

$$f = P\left(1 + \frac{r}{R} \cos \theta\right) \tag{32}$$

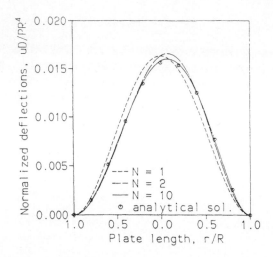

Figure 11 Normalized deflections of Example 1 along $\theta = 0$ and π for $N = 1, 2, 10$, and the analytic solution.

where P is the force magnitude and r and θ are the polar coordinates shown in Figure 10. Substituting Eqs. (30) and (32) into the numerator of Eq. (29), the analytic form in terms of eigenvalues and Bessel functions is obtained. In this problem the forcing function (32) includes only constant and $\cos \theta$ terms; thus all m vanish except $m = 0$ and 1 in Eq. (30). Also, $\gamma = 0$ because of the symmetry of loading with respect to $\theta = 0$; it follows that c_j in Eq. (29) are simultaneously calculated on a parallel computer.

The approximate solution of deflections along $\theta = 0$ and π from Eq. (29) is compared with the analytic solution (Ugural, 1981) in Figure 11. The deflections are normalized with respect to P, R, and D (flexural rigidity of the plate). The 1-term and 2-term as well as 10-term approximations are displayed. As can be seen, when $N = 10$ the result agrees with the analytic solution very well. The normalized bending moment M_{rr} with respect to P and R for $N = 1, 2, 10$ and the analytic solution along $\theta = 0$ and π are shown in Figure 12. The maximum error of M_{rr} for the case $N = 10$ occurring at the edge $r = R$ and $\theta = 0$ is about 4%.

Example 2

The same circular plate in Example 1 subjected to an eccentric concentrated load P applied at $r = 0.5R$ and $\theta = 0$ is shown in Figure 13.

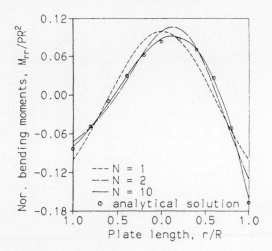

Figure 12 Normalized bending moments M_{rr}/PR^2 of Example 1 along $\theta = 0$ and π for $N = 1, 2, 10$, and the analytic solution.

As a result of the same geometric domain and boundary conditions, Eq. (30) is also used as the coordinate functions. Likewise, Eq. (31) is valid in the present problem. The numerator of Eq. (29), involving a Dirac δ function, is calculated as

$$\int_\Omega P\, \phi_j(r,\, \theta)\, \delta(b,\, \theta) r\, dr\, d\theta = P\Bigg(J_m(k_{mn}\, b)$$

$$-\, \frac{J_m(k_{mn}\, R)}{I_m(k_{mn}\, R)}\, I_m\, (k_{mn}\, b)\Bigg) \qquad \begin{array}{l} m = 0, 1, 2, \ldots, \infty \\ n = 1, 2, 3, \ldots, \infty \end{array} \tag{33}$$

where $b = 0.5R$ and $\theta = 0$.

The approximate solutions of deflections along $\theta = 0$ and π for $N = 1, 2$, and 30 are shown in Figure 14. The analytic solution (Michell, 1902) is also shown for comparison. Again the deflections are normalized with respect to D (flexural rigidity), P, and R. It is clear that when $N = 30$ the proposed approximate solution approaches the analytic solution. Figure 15 shows the normalized bending moment M_{rr} with respect to P for $N = 1, 30, 60$ and the analytic solution along $\theta = 0$ and π. The agreement between $N = 60$ and the analytic solution is also seen except at the singular point $r = 0.5R$, where the eccentric concentrated load is applied.

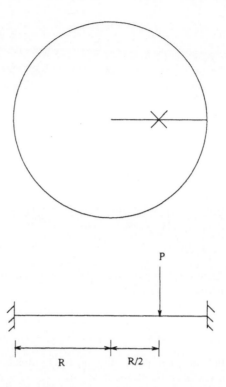

Figure 13 Clamped-clamped circular plate subjected to an eccentric concentrated load at $r = 0.5R$ and $\theta = 0$.

It follows that the speedups and efficiencies of Example 2 with $N = 60$ versus number of processors are shown in Figures 16 and 17, respectively.

Example 3

The last example illustrates the use of Eq. (28) and the parallel Gaussian elimination solution scheme developed in Section 2 to solve the problem of the rectangular plate subjected to a central point load.

Each coordinate function is composed of the product of 2 one-dimensional beam eigenfunctions. It appears that the beam eigenfunctions may provide the flexibility for problems with different boundary conditions. In addition, by applying the orthogonalities of beam eigenfunctions, each element in the coefficient matrix of Eq. (28) is shown in terms of the eigenvalues of the two cross-beam eigenfunctions (Kantorovich and Krylov, 1964). As a result the elements in the coefficient matrix can be simultaneously calculated on a parallel computing system without incurring a synchronization problem.

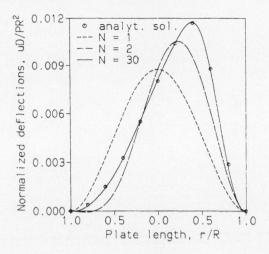

Figure 14 Normalized deflections of Example 2 along θ = 0 and π for N = 1, 2, 30, and the analytic solution.

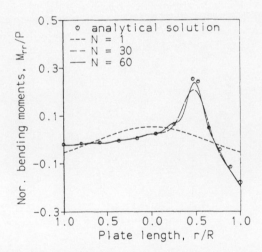

Figure 15 Normalized bending moments M_{rr}/P of Example 2 along θ = 0 and π for N = 1, 30, 60, and the analytic solution.

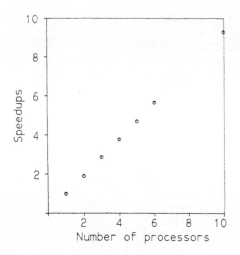

Figure 16   Speedups of the proposed parallel method in Example 2 for $N = 60$.

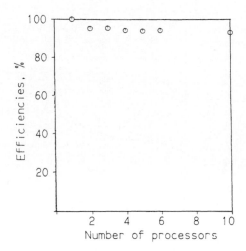

Figure 17   Efficiencies of the proposed parallel method in Example 2 for $N = 60$.

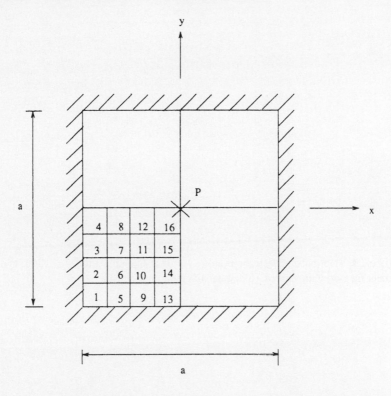

Figure 18 Clamped-clamped square plate subjected to a concentrated load at the center with a finite element model.

Consider a clamped-clamped square plate with length a subjected to a central point load as shown in Figure 18. Each ϕ_j is the product of 2 one-dimensional normalized eigenfunctions for a fixed-fixed beam,

$$X_m(x)Y_n(y) \qquad m, n = 1, 2, \ldots, \infty$$

Each of these two orthogonal functions $X_m(x)$ and $Y_n(y)$ satisfies the corresponding homogeneous geometric boundary conditions in its own direction. Hence the coordinate function $\phi_j(x,y)$ meets the requirements of the boundary conditions as mentioned in Eq. (14).

The point load P of this problem is also expressed as a Dirac δ function. Substituting the Dirac δ function into Eq. (28), a system of algebraic equations is obtained. Then the parallel Gaussian elimination solution scheme in Section 2 is applied to solve the adjustable coefficients.

The normalized deflections of the proposed parallel method along $y = 0$ are shown in Figure 19 for two cases, $N = 1$ and 25. The deflections of the finite

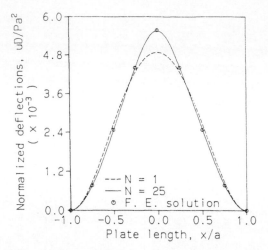

Figure 19   Normalized deflections along $y = 0$ of Example 3 for $N = 1, 25$, and finite element method with sixteen 16 degrees of freedom elements.

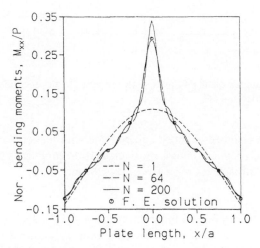

Figure 20   Normalized bending moments M_{xx}/P along $y = 0$ of Example 3 for $N = 1, 64, 200$, and finite element method with sixteen 16 degrees of freedom elements.

element analysis with sixteen 16 degrees of freedom elements are also shown for comparison. The result with $N = 25$ agrees very well with the finite element solution. Normalized bending moments M_{xx} with respect to P along $y = 0$ for $N = 1, 64, 200$ and the finite element solution are shown in Figure 20. It shows that at the singular point $x = 0$ M_{xx} approaches infinity as N increases.

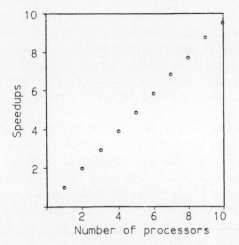

Figure 21 Speedups of the proposed parallel method in Example 3 for $N = 64$.

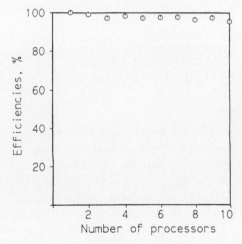

Figure 22 Efficiencies of the proposed parallel method in Example 3 for $N = 64$.

Finally, the speedups and efficiencies in calculating the coefficient matrix and solving the system of equations for $N = 64$ on the Sequent parallel computer are shown in Figures 21 and 22, respectively.

3.5 Discussion

In this section the use of eigenfunctions with the variational principle shows the linear speedups of execution time versus the number of processors for the example problems.

Regular geometric domains, that is, circular and rectangular, are employed to demonstrate the potential of using eigenfunctions for parallel computation. For simplicity, the boundary conditions assumed here are all clamped-clamped. Nevertheless, the proposed parallel techniques are also applicable to problems with other types of boundary conditions. This can be accomplished simply by using the eigenfunctions associated with the appropriate boundary conditions as the coordinate functions.

It is shown in the circular plate problem that the corresponding eigenfunctions make it possible to calculate the adjustable parameters concurrently on a parallel computing system. The same situation occurs in the rectangular domain if the corresponding eigenfunctions are used. The Prandtl torsional problem is an example. However, one-dimensional eigenfunctions are employed as coordinate functions for biharmonic problems with rectangular domain. These one-dimensional eigenfunctions uncouple the elements of the coefficient matrix. Thus the elements can be simultaneously calculated on a parallel computing system. Although a system of equations is to be solved, speedup is achieved by invoking the parallel Gaussian elimination solution scheme, which is presented in Section 2. It is also found that high efficiency can be achieved even if 10 processors are employed.

4 PARALLEL COMPUTATION USING BOUNDARY ELEMENT METHOD

4.1 Introduction

The boundary element method offers many features that are natural for parallel computations. First of all, this method needs to model only the boundary geometry and hence only the boundary must be divided into elements. The number of discrete elements along the boundary is always much smaller than the number of elements necessary to describe the entire problem domain. Thus the system of equations to be solved is much smaller than the system of equations required by the finite element method. After the boundary information is concurrently solved by the parallel Gaussian elimination solution scheme, the solution of the entire domain can then be calculated concurrently from the solved boundary information. In other words, instead of solving the large system of equations as in a domain method (e.g., finite element method), the boundary element method needs to solve only a much smaller system of equations resulting from the boundary conditions. The domain solutions at all the points of interest are then calculated concurrently based on the solved boundary information. Although the system of equations is no longer sparse, the efficiency of applying a parallel Gaussian elimination solution scheme is still high.

In addition each element of the global coefficient matrix of the system of equations in the boundary element method is a solution of the fundamental problem. Clearly, all the elements of the global coefficient matrix are uncoupled and can be independently calculated on a parallel computing system. In other words, the synchronization problem in the forming process of the finite element method does not exist in the boundary element method. The parallelism of calculating and hence forming the global coefficient matrix in the boundary element method is obvious.

In this section, the direct boundary element method with linear element formulation is employed on the Sequent Summary S81 parallel computing system to solve two-dimensional linear elastic boundary value problems. The displacements and tractions are assumed to vary linearly between the nodal points of straight-line segments along the boundary. Three computational components are parallelized in this method to show the speedup and efficiency in computation The global coefficient matrix is first formed by the inherent parallelism. Then the parallel Gaussian elimination solution scheme is applied to solve the resulting system of equations. Finally, and more importantly, the domain solutions of a given boundary value problem are calculated concurrently. An illustrated boundary value problem consisting of two different geometric cases is solved in this study. This boundary value problem is an infinite plate with an elliptic hole at the center under uniaxial tension at infinity. Two different ratios of short axis to long axis of the elliptic hole are considered here. The speedups and efficiencies of the proposed parallel algorithms for solving the illustrated boundary value problems are also presented.

4.2 The Direct Linear Boundary Element Formulation

Consider a two-dimensional domain Ω encompassed by its boundary contour C and subjected to two different linear elastic fields $\{1\}$ and $\{2\}$ without body forces. Betti's reciprocal theorem (Betti, 1872), which linked these two two-dimensional linear elastic boundary value problems, states

$$\oint_C \mathbf{t}^{\{1\}} \cdot \mathbf{u}^{\{2\}} \, ds = \oint_C \mathbf{t}^{\{2\}} \cdot \mathbf{u}^{\{1\}} \, ds \qquad (34)$$

where $\mathbf{t}$ and $\mathbf{u}$ are the traction vector and the displacement vector, respectively, along the boundary contour C. In physical terms Betti's reciprocal theorem indicates that the work done by the tractions of field $\{1\}$ acting through the displacements of field $\{2\}$ over the entire boundary contour C equals the work done over C with $\{1\}$ and $\{2\}$ interchanged.

To begin deriving the direct boundary element method, consider the two-dimensional plane strain Kelvin's singular solutions in rectangular Cartesian coordinate system x—y. Let P and Q be any arbitrary two distinct points on the boundary contour C of a domain Ω. The displacement component u_i^Q at a field point Q in the i direction due to a concentrated force F_j^P applied at a load point P in the j direction from outside the domain Ω is

$$u_i^Q = G_{ij}^{QP} F_J^P \tag{35}$$

where G_{ij}^{QP} is the displacement influence coefficient, which specifies the displacement at a field point Q in the i direction due to a unit concentrated force applied at a load point P in the j direction from outside the domain Ω, and is expressed as

$$G_{ij}^{QP} = \frac{1}{2G} \left[(3 - 4v)g\,\delta_{ij} - (x_j - c_j)\frac{\partial g}{\partial x_i} \right] \tag{36}$$

where

$$g(x, y) = \frac{-1}{4\pi(1 - v)} \ln \left[(x - c_x)^2 + (y - c_y)^2 \right]^{1/2} \tag{37}$$

G is the shear modulus, v is the Poisson's ratio, δ_{ij} is the Kronecker δ, c_x and c_y are the distances along x and y axes, respectively, between points P and Q, and $i, j = x, y$.

Likewise, the traction component t_i^Q at a field point Q due to a concentrated force F_j^P applied at a load point P in the j direction from outside the domain Ω is

$$t_i^Q = H_{ij}^{QP} F_j^P \tag{38}$$

where H_{ij}^{QP} is the traction influence coefficient, which specifies the traction at a field point Q in the i direction due to a unit concentrated force applied at a load point P in the j direction from outside the domain Ω, and is expressed as

$$H_{ij}^{QP} = S_{xij}n_x + S_{yij}n_y \tag{39}$$

where

$$S_{ijk} = (1 - 2v)\left(\frac{\partial g}{\partial x_i}\delta_{jk} + \frac{\partial g}{\partial x_j}\delta_{ki} \right) + 2v\frac{\partial g}{\partial x_k}\delta_{ij} - (x_k - c_k)\frac{\partial^2 g}{\partial x_i\,\partial x_j} \tag{40}$$

n_x and n_y are the x and y components, respectively, of the unit normal vector $\mathbf{n}$ at point Q and $i, j, k = x, y$.

These two uncoupled influence coefficients G_{ij}^{QP} and H_{ij}^{QP} are the key to constructing the resulting global coefficient matrix of the direct boundary element method. It appears later that these two uncoupled influence coefficients also characterize the parallel computing and hence form the resulting global coefficient matrix.

Substituting Eqs. (35) and (38) into Eq. (34) as the second linear elastic field, Betti's reciprocal theorem becomes

$$\oint_C t_i^{\{1\}} G_{ij}^{QP} F_j^P\, ds(Q) = \oint_C u_i^{\{1\}} H_{ij}^{QP} F_j^P\, ds(Q) \qquad i, j = x, y \tag{41}$$

Dropping the superscript $\{1\}$ in Eq. (41) for simplicity and canceling the common term F_j^P on both sides of Eq. (41), the direct boundary integral equation of a two-dimensional linear elastic boundary value problem yields

$$\oint_C t_i\, G_{ij}^{QP}\, ds(Q) = \oint_C u_i\, H_{ij}^{QP}\, ds(Q) \qquad i, j = x, y \tag{42}$$

Discretizing the boundary contour C in Figures 23, Eq. (42) becomes

$$\sum_{L=1}^{N} \left(\int_{-a_L}^{a_L} t_x^L(\bar{x})\, G_{xx}^{Lm}(\bar{x} - c_{\bar{x}}^L, c_{\bar{y}}^L)\, d\bar{x} + \int_{-a_L}^{a_L} t_y^L(\bar{x})\, G_{yx}^{Lm}(\bar{x} - c_{\bar{x}}^L, c_{\bar{y}}^L)\, d\bar{x} \right)$$

$$= \sum_{L=1}^{N} \left(\int_{-a_L}^{a_L} u_x^L(\bar{x})\, H_{xx}^{Lm}(\bar{x} - c_{\bar{x}}^L, c_{\bar{y}}^L)\, d\bar{x} + \int_{-a_L}^{a_L} u_y^L(\bar{x})\, H_{yx}^{Lm}(\bar{x} - c_{\bar{x}}^L, c_{\bar{y}}^L)\, d\bar{x} \right)$$

$$\sum_{L=1}^{N} \left(\int_{-a_L}^{a_L} t_x^L(\bar{x})\, G_{xy}^{Lm}(\bar{x} - c_{\bar{x}}^L, c_{\bar{y}}^L)\, d\bar{x} + \int_{-a_L}^{a_L} t_y^L(\bar{x})\, G_{yy}^{Lm}(\bar{x} - c_{\bar{x}}^L, c_{\bar{y}}^L)\, d\bar{x} \right)$$

$$= \sum_{L=1}^{N} \left(\int_{-a_L}^{a_L} u_x^L(\bar{x})\, H_{xy}^{Lm}(\bar{x} - c_{\bar{x}}^L, c_{\bar{y}}^L)\, d\bar{x} + \int_{-a_L}^{a_L} u_y^L(\bar{x})\, H_{yy}^{Lm}(\bar{x} - c_{\bar{x}}^L, c_{\bar{y}}^L)\, d\bar{x} \right) \tag{43}$$

$$m = 1, 2, \ldots, N$$

where $-a_L \le \bar{x} \le a_L$ and N is the number of total straight-line elements assumed along the boundary contour C. Equation (43) is a system of $2N$ simultaneous algebraic equations in terms of $2N$ boundary unknowns. These $2N$ boundary unknowns are the boundary displacements or tractions that are not prescribed as boundary conditions on a straight-line element $[L]$ of the two-dimensional linear elastic boundary value problem of interest. In addition, Eq. (43) provides the starting point to construct a numerical algorithm for the direct boundary element method.

To build the linear element for the direct boundary element method, the displacements and tractions in a straight-line element $[L]$ are then assumed to vary linearly between the two end nodes. In such a case the displacements $u_x^L(\bar{x})$ and $u_y^L(\bar{x})$ and the tractions $t_x^L(\bar{x})$ and $t_y^L(\bar{x})$ in Eq. (43) can be written in the form of y

$$u_x^L(\bar{x}) = \frac{1}{2}\left(1 - \frac{\bar{x}}{a_L}\right) u_x^L(l) + \frac{1}{2}\left(1 + \frac{\bar{x}}{a_L}\right) u_x^L(l + 1) \tag{44}$$

where the notations l and $(l + 1)$ indicate the values at nodes l and $l + 1$ of element $[L]$. Also, a node l is shared by two adjacent straight-line elements $[L - 1]$ and $[L]$ as shown in Figure 23.

Since the values of displacements and tractions in two adjacent elements $[L - 1]$ and $[L]$ depend on their values at the common node l, the continuity requirements of displacements and tractions at node l must be specified. Clearly, at node l the displacements are single valued; that is,

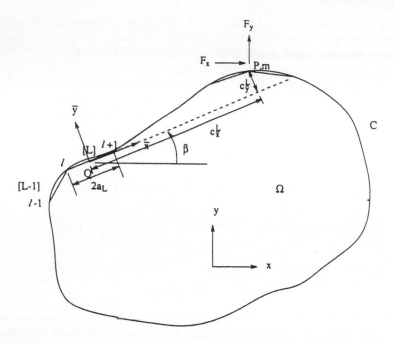

Figure 23 Discretized boundary contour C of the domain Ω with straight-line boundary elements.

$$u_x^{L-1}(l) = u_x^L(l) = u_x(l)$$
$$u_y^{L-1}(l) = u_y^L(l) = u_y(l)$$

$$(45)$$

However, in general, because of the different element inclinations, the tractions at node l of element $[L-1]$ are not the same as those at node l of element $[L]$; that is,

$$t_x^{L-1}(l) \neq t_x^L(l)$$
$$t_y^{L-1}(l) \neq t_y^L(l)$$

$$(46)$$

As a result, Eqs. (45) and (46) provide six boundary parameters, two displacement and four traction components, associated with each nodal point. To obtain $2N$ unknowns in the system of $2N$ simultaneous algebraic equations (43), four boundary parameters in Eqs. (45) and (46) at each node must be prescribed.

Some approaches to determine the $t_x^{L-1}(l)$, $t_x^L(l)$, $t_y^{L-1}(l)$, and $t_y^L(l)$ at a nodal point l are discussed by Crouch and Starfield (1983). Nevertheless, selection of four suitable traction components at each nodal point in linear boundary element formulation remains a topic of much current research.

Substitute Eq. (44) into Eq. (43), and rearrange the resulting equations so that the summations are performed over the nodal points rather than the elements. Equation (43) then becomes a system of $2N$ algebraic equations in terms of six boundary parameters, $u_x(l)$, $u_y(l)$, $t_x^{L-1}(l)$, $t_x^L(l)$, $t_y^{L-1}(l)$, and $t_y^L(l)$.

$$
\sum_{l=1}^{N} \left(\begin{bmatrix} U_{xx}^{L-1,m}(l) & U_{xx}^{Lm}(l) & U_{yx}^{L-1,m}(l) & U_{yx}^{Lm}(l) \\ U_{xy}^{L-1,m}(l) & U_{xy}^{Lm}(l) & U_{yy}^{L-1,m}(l) & U_{yy}^{Lm}(l) \end{bmatrix} \begin{Bmatrix} t_x^{L-1}(l) \\ t_x^L(l) \\ t_y^{L-1}(l) \\ t_y^L(l) \end{Bmatrix} \right)
$$

$$
= \sum_{l=1}^{N} \left(\begin{bmatrix} T_{xx}^{L-1,m}(l) + T_{xx}^{Lm}(l) & T_{yx}^{L-1,m}(l) + T_{yx}^{Lm}(l) \\ T_{xy}^{L-1,m}(l) + T_{xy}^{Lm}(l) & T_{yy}^{L-1,m}(l) + T_{yy}^{Lm}(l) \end{bmatrix} \begin{Bmatrix} u_x(l) \\ u_y(l) \end{Bmatrix} \right) \tag{47}
$$

$$
m = 1, 2, \ldots, N
$$

where the elements $U_{xx}^{L-1,m}(l), \ldots,$ and $T_{xx}^{L-1}(l), \ldots$ on both sides of the coefficient matrix are functions of the displacement influence coefficient (36) and traction influence coefficient (39), respectively, and are computed in detail (Chien, 1989). Equation (47) is the centerpiece of the two-dimensional direct linear boundary element algorithm.

A typical two-dimensional direct linear boundary element algorithm can be characterized as the governing system of algebraic equations.

$$
\begin{Bmatrix} Y_x^m(l) \\ Y_y^m(l) \end{Bmatrix} = \sum_{l=1}^{N} \left(\begin{bmatrix} C_{xx}^{lm} & C_{yx}^{lm} \\ C_{xy}^{lm} & C_{yy}^{lm} \end{bmatrix} \begin{Bmatrix} X_x^l \\ X_y^l \end{Bmatrix} \right) \qquad m = 1, 2, \ldots, N \tag{48}
$$

where Y_x^m and Y_y^m are certain linear combinations of the prescribed boundary conditions, and $C_{xx}^{lm}, \ldots$ are the appropriate elements in both coefficient matrices in Eq. (47) associated with the unknown boundary information X_x^l and X_y^l.

Solving Eq. (48), the complete boundary information along the boundary contour C of a two-dimensional linear elastic boundary value problem is then obtained.

Besides the development of boundary information as mentioned, the calculation of tangential stress σ_t along boundary contour C is also an important matter. For a plane strain boundary value problem the tangential stress σ_t^L on a boundary element $[L]$ can be derived by using Hook's law:

$$
\sigma_t^L = \frac{2G}{1-v} \frac{\partial u_{\bar{x}}^L}{\partial \bar{x}} + \frac{v}{1-v} \sigma_{\bar{y}}^L \tag{49}
$$

where $\delta u_{\bar{x}}^L / \delta \bar{x}$ is the rate of change of tangential displacement $u_{\bar{x}}^L$ of a boundary element $[L]$ along the traversal direction $\bar{x}$, and $\sigma_{\bar{y}}^L$ is the normal component of tractions on $[L]$. The computations of derivative $\partial u_{\bar{x}}^L / \partial \bar{x}$ and $\sigma_{\bar{y}}^L$ on a straight-line element $[L]$ are accomplished by using the standard finite difference approximations given by

$$\frac{\partial u_{\bar{x}}^L}{\partial \bar{x}} = \frac{u_{\bar{x}}^L(l + 1) - u_{\bar{x}}^L(l)}{2a_L} \qquad (50)$$

$$\sigma_{\bar{y}}^L = \frac{\sigma_{\bar{y}}^L(l + 1) + \sigma_{\bar{y}}^L(l)}{2} \qquad (51)$$

where

$$u_{\bar{x}}^L(l + 1) = u_x(l + 1) \cos \beta_L + u_y(l + 1) \sin \beta_L$$

$$u_{\bar{x}}^L(l) = u_x(l) \cos \beta_L + u_y(l) \sin \beta_L$$

and

$$\sigma_{\bar{y}}^L(l + 1) = -t_{\bar{x}}^L(l + 1) \sin \beta_L + t_{y_L}(l + 1) \cos \beta_L$$

$$\sigma_{\bar{y}}^L(l) = -t_{x_L}(l) \sin \beta_L + t_{y_L}(l) \cos \beta_L$$

It can be seen that Eq. (49) represents a constant value of tangential stress on each straight-line element $[L]$. This constant tangential stress has the characteristic that all the tangential stress components in the boundary elements along boundary contour C are discontinuous.

Finally, Somigliana's formulas (Somigliana, 1885–1886) are applied to find the solutions in domain Ω. For a two-dimensional linear elastic boundary value problem Somigliana's formulas give the displacements u_x^p and u_y^p at an interior point p in Ω as

$$u_x^p = -\oint_C \mathbf{u}(s) \cdot \mathbf{t}(F_x^p) \, ds + \oint_C \mathbf{t}(s) \cdot \mathbf{u}(F_x^p) \, ds$$

$$u_y^p = -\oint_C \mathbf{u}(s) \cdot \mathbf{t}(F_y^p) \, ds + \oint_C \mathbf{t}(s) \cdot \mathbf{u}(F_y^p) \, ds \qquad (52)$$

where $\mathbf{u}(s)$ and $\mathbf{t}(s)$ are the displacements and tractions along the boundary contour C, which are found from the prescribed boundary conditions and by solving Eq. (48). Also, $\mathbf{u}(F_i^p)$ and $\mathbf{t}(F_i^p)$ denote the displacements and tractions developed on boundary contour C by a concentrated force F_i^p $(i = x, y)$ applied at an interior point p in the i direction as shown in Figure 24.

Equations (52) can be calculated numerically as before by dividing the entire boundary contour C into N straight-line elements. By doing so, Eq. (25) becomes

$$u_x^p = -\sum_{L=1}^{N} \left(\int_{-a_L}^{a_L} u_x^L(s) \cdot t_x^L(F_x^p) \, ds + \int_{-a_L}^{a_L} u_y^L(s) \cdot t_y^L(F_x^p) \, ds \right)$$

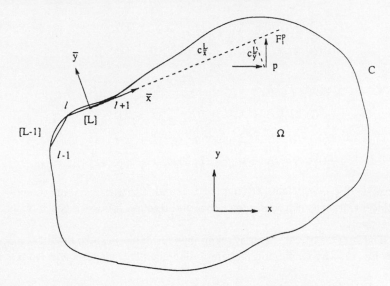

Figure 24 Relation between an interior point p and a boundary element $[L]$.

$$+ \sum_{L=1}^{N} \left(\int_{-a_L}^{a_L} t_x^L(s) \cdot u_x^L(F_x^p) \, ds + \int_{-a_L}^{a_L} t_y^L(s) \cdot u_y^L(F_x^p) \, ds \right)$$

$$u_y^p = - \sum_{L=1}^{N} \left(\int_{-a_L}^{a_L} u_x^L(s) \cdot t_x^L(F_y^p) \, ds + \int_{-a_L}^{a_L} u_y^L(s) \cdot t_y^L(F_y^p) \, ds \right)$$

$$+ \sum_{L=1}^{N} \left(\int_{-a_L}^{a_L} t_x^L(s) \cdot u_x^L(F_y^p) \, ds + \int_{-a_L}^{a_L} t_y^L(s) \cdot u_y^L(F_y^p) \, ds \right) \tag{53}$$

Furthermore, the displacements $u_x^L(s)$ and $u_y^L(s)$ and tractions $t_x^L(s)$ and $t_y^L(s)$ in Eq. (53) are assumed to change linearly between the two end nodes of a straight-line element $[L]$.

The displacements u_x^p and u_y^p at an interior point p are thus expressed as

$$u_x^p = \sum_{L=1}^{N} \Phi^L$$

$$u_y^p = \sum_{L=1}^{N} \Psi^L \tag{54}$$

where

$$\Phi^L = -(u_x^L(l) \cdot T_{xx}^{Lp}(l) + u_x^L(l+1) \cdot T_{xx}^{Lp}(l+1))$$

$$- (u_y^L(l) \cdot T_{yx}^{Lp}(l) + u_y^L(l+1) \cdot T_{yx}^{Lp}(l+1))$$

$$+ (t_x^L(l) \cdot U_{xx}^{Lp}(l) + t_x^L(l + 1) \cdot U_{xx}^{Lp}(l + 1))$$

$$+ (t_y^L(l) \cdot U_{yx}^{Lp}(l) + t_y^L(l + 1) \cdot U_{yx}^{Lp}(l + 1))$$

$$\Psi^L = - (u_x^L(l) \cdot T_{xy}^{Lp}(l) + u_x^L(l + 1) \cdot T_{xy}^{Lp}(l + 1))$$

$$- (u_y^L(l) \cdot T_{yy}^{Lp}(l) + u_y^L(l + 1) \cdot T_{yy}^{Lp}(l + 1))$$

$$+ (t_x^L(l) \cdot U_{xy}^{Lp}(l) + t_x^L(l + 1) \cdot U_{xy}^{Lp}(l + 1))$$

$$+ (t_y^L(l) \cdot U_{yy}^{Lp}(l) + t_y^L(l + 1) \cdot U_{yy}^{Lp}(l + 1))$$

Likewise, the coefficients $U_{xx}^{Lp}(l), \ldots,$ and $T_{xx}^{Lp}(l), \ldots,$ are functions of displacement influence coefficients (36) and traction influence coefficients (39), respectively.

The stresses σ_{xx}^p, σ_{yy}^p, and σ_{xy}^p at an interior point p can be directly computed from Eq. (54) by employing the strain-displacement relations and Hook's law.

4.3   Parallel Algorithms of the Direct Linear Boundary Element Method

The direct linear boundary element formulas already discussed are partitioned into three components to be computed on a parallel computing system. First, since the elements in the coefficient matrices on both sides of Eq. (47) are uncoupled, all the elements in both coefficient matrices can be calculated concurrently and formed with a system of algebraic equations. The parallel forming strategy of Eq. (48) is described as follows.

The total number of elements in the coefficient matrices on both sides of Eq. (47) are mapped into the multiprocessors such that each processor stores approximately the same number of elements of the coefficient matrices in its local memory locations. These elements are then computed locally within each processor. Since these elements in the two coefficient matrices in Eq. (47) are not coupled, each individual processor can simultaneously compute the share of elements stored in its local memory. Subsequently, all the processors put their computed results into the global memory pool concurrently to form Eq. (48) without incurring synchronization between any two processors. Furthermore, the work load balance among the processors is also ensured to obtain a high speedup in the forming procedure.

The boundary element analysis normally generates a nonsymmetric, nonpositive definite, fully populated coefficient matrix for a single-region boundary value problem as well as a nonsymmetric, nonpositive definite, block-banded coefficient matrix for a multiple-region boundary value problem. This matrix is more densely populated than those that are normally suitable for iterative methods. Moreover, although diagonally strong it is not necessarily diagonally dominant, and iterative convergence cannot be ensured. A direct reduction method is therefore suggested (Banerjee and Butterfield, 1981) to best solve the resulting system of algebraic equations in the boundary element analysis. Thus the parallel Gaussian elimination solution scheme developed in Section 2 is most suitable as a solving procedure in boundary element-formulated problems. The second phase of parallelization of boundary element methods is thus accomplished by invoking the parallel Gaussian elimination technique described in Section 2.2 to solve Eq. (48).

Finally, the desired interior solutions given in Eq. (54) at distinct interior points in Ω are also obtained concurrently in the multiprocessors. The total number of desired solution points in Ω are divided into the number of processors. As a result each individual processor contains approximately the same number of interior points where the solutions are to be calculated by using Eq. (54). Computation of the solution at a point is independent of that at other points as a result of the nature of Somigliana's formulas and is performed individually. Each processor, at this stage, calculates the solutions of those interior points assigned to its local memory address. Thus the solution phase in the boundary element method by employing Somigliana's formulas can be performed on a parallel computing system to achieve a linear speedup.

The boundary element method has thus shown the features of parallel computation, which include forming Eq. (48) and computing Eq. (54) and stresses. These features are the bases for the linear speedups and high efficiencies achieved in solving boundary value problems using the boundary element method.

4.4 Examples

A two-dimensional infinite panel contains a central elliptic hole with short axis a and long axis b. The panel is subjected to uniaxial tension q along the direction of the short axis at infinity, as shown in Figure 25. This elastostatic problem is solved by using the direct linear boundary element algorithm along with the proposed parallel processing techniques on the Sequent S81 parallel computing system. Since the problem is symmetric with respect to both the x and y axes, only the first quadrant of the boundary of the elliptic hole is discretized to obtain the solution. Two cases with different ratios a/b are considered. Besides solving the boundary information, for example the stress concentration factor at the tip of long axis b, stresses at 40 interior points are computed concurrently for Case (1), as well as 60 interior points for the other case. The analytic solution (Timoshenko and Goodier, 1951) for each case is also shown for comparison. In addition, speedup and efficiency of the parallel computation are presented for Case 2.

Case 1   a/b = 1

This special case coincides with a circular hole with radius $R = a = b$ at the center of the infinte panel. Five linear elements are used to describe a quarter (θ

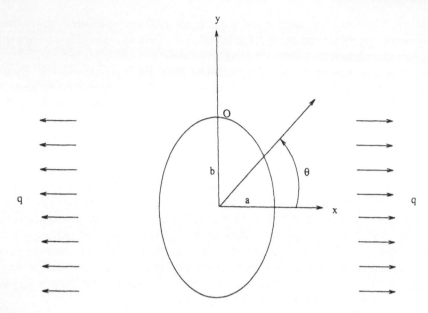

Figure 25   Infinite plate containing an elliptic hole at the center subjected to uniaxial tension q at infinity.

= 0–90°) of the circular boundary. The dimensionless circumferential (hoop) stress σ_t/q and the analytic solution are shown in Figure 26. To obtain the domain solutions 20 interior points from the boundary of the circular hole along the x and y axes are used. Figure 27 gives nondimensional stress σ_{xx}/q and σ_{yy}/q along the x axis running from boundary $x/R = 1$ to $x/R = 4$. The nondimensional stresses σ_{xx}/q and σ_{yy}/q along the y axis running from boundary $y/R = 1$ to $y/R = 4$ are presented in Figure 28. Excellent agreement between the linear boundary element solution and the analytic solution is evident.

In this case the five linear elements result in a 12×24 coefficient matrix and a 12×12 coefficient matrix on the left and the right sides of Eq. (47), respectively. The total number of elements in these two coefficient matrices is 432 and is mapped into the multiprocessor system to be computed. Each processor thus contains 72 elements when six processors are used. The computations of the values of all elements in the two coefficient matrices are then performed simultaneously among six processors to form Eq. (48). Linear speedup is noticeably observed in this phase. Solutions at 40 desired interior points are calculated here. Of six processors four compute the solutions at 7 points and the other two processors at 6 points. Each processor thus repeatedly applies Eq. (54)

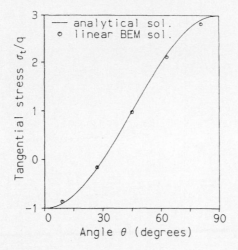

Figure 26 Nondimensional tangential stress σ_t/q and the analytic solution along the circular boundary of Case 1.

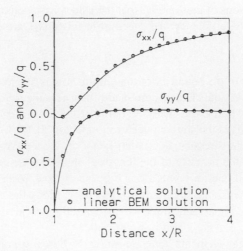

Figure 27 Nondimensional stresses σ_{xx}/q and σ_{yy}/q and the analytic solutions along the x axis of Case 1.

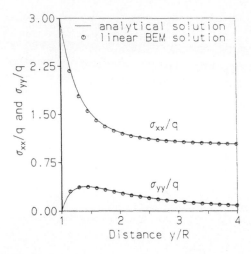

Figure 28 Nondimensional stresses σ_{xx}/q and σ_{yy}/q and the analytic solutions along the y axis of Case 1.

to obtain the solutions of those points that have been assigned to its local memory address. Again, the efficiency and speedup in this phase are expected.

Case 2   a/b = 0.1

The first quadrant of the boundary of the elliptic hole is discretized into 40 linear elements. Moreover, the elements are made finer near the tip of long axis b than in the rest of the boundary. The tangential stress in the end element, which includes the tip, gives the stress concentration factor 20.13. Comparing with the analytic value of the stress concentration factor of 21, the error is about 4%. As in Case 1, within the distance of the length of short axis a, stresses σ_{xx} and σ_{yy} are computed at 30 interior points along the x and y axes, respectively. Figure 29 shows the nondimensional stresses σ_{xx}/q and σ_{yy}/q along the x axis in the region $1 \leq x/a \leq 2$. The logarithmic values of σ_{xx}/q and σ_{yy}/q along the y axis in the region $1 \leq y/b \leq 1.1$ are given in Figure 30.

These 40 linear elements generate an 82×82 fully populated coefficient matrix. The performance of parallel computation, including forming the coefficient matrix, solving the resulting system of equations, and computing the interior solution at 60 points, is indicated as speedups in Figure 31 and as efficiencies in Figure 32.

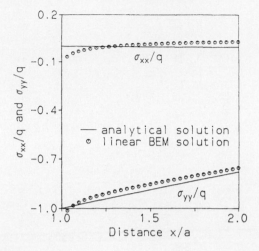

Figure 29   Nondimensional stresses σ_{xx}/q and σ_{yy}/q and the analytic solutions along the x axis of Case 2.

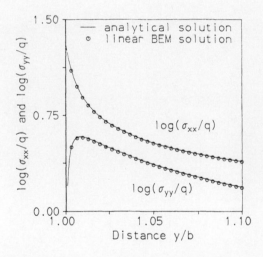

Figure 30   Logarithmic values of nondimensional stresses σ_{xx}/q and σ_{yy}/q and the analytic solutions along the y axis of Case 2.

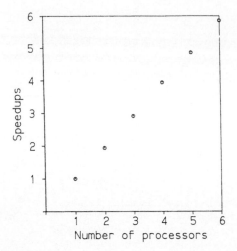

Figure 31   Speedups of proposed parallel computational strategy using linear boundary element algorithms in Case 2.

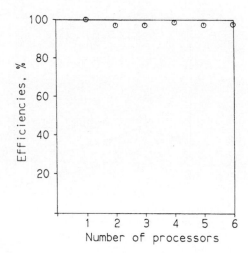

Figure 32   Efficiencies of proposed parallel computational strategy using linear boundary element algorithms in Case 2.

4.5 Discussion

The boundary element method has beem employed in a parallel computing environment for solving solid mechanics problems. Linear variation of displacements and tractions within a straight-line boundary element is assumed in the present study. Furthermore, in lieu of using numerical integration methods, the analytic results of the elements in coefficient matrices (47) and Somigliana's formulas for interior solutions are presented (Chien, 1989) by resorting to the MACSYMA symbolic program. The significance of potential parallelism of the boundary element method on multiprocessor computation to expedite the solution of engineering mechanics problems is demonstrated.

For Case 1, five linear boundary elements give a fairly accurate stress distribution around the circular hole. For this problem the linear boundary element algorithm generates a small system of equations.

For Case 2, the stresses along the y axis obtained by discretizing the boundary into 40 elements seem to be satisfactory compared with the analytic solution. However, the results of stress distribution along the x axis near the boundary can be further improved by employing additional elements on the boundary.

Attention of this study is focused on demonstrating the parallelism of the boundary element method rather than on further study of the method itself. Three components in the computation flowchart of the boundary element method have been parallelized. Besides the parallel solution scheme, the other two computational components, that is, forming the coefficient matrices and computing the interior solutions, also present the characteristics of concurrent processing. Moreover, these two computational components have shown that the synchronization problems in parallel computing are completely avoided on SIMD parallel computing environments. Figures 31 and 32 exhibit the significance of the boundary element method in parallel computation.

The two-dimensional Kelvin's singular solution furnished the Green's function used in this study. The use of this Green's function makes possible parallelization of the boundary element method. Speedups of parallel computing should also be expected with the use of other types of Green's functions for different problems, such as dislocation functions for crack problems.

REFERENCES

Adeli, H., and Kamal, O. (1992a). Concurrent optimization of large structures. Part 1. Algorithms. ASCE J. of Aerospace Engineering. To be published January 1992.

Adeli, H., and Kamal, O. (1992b). Concurrent optimization of large structures. Part 2. Applications. ASCE J. of Aerospace Engineering. To be published January 1992.

Banerjee, P. K., and Butterfield, R. (1981). *Boundary Element Methods in Engineering Science*. McGraw-Hill, London.

Bauget, G. M. (1978). Asynchronous iterative methods for multiprocessors. J. Association for Computing Machinery, 25(2):226.

Betti, E. (1872). Il Nuovo Cimento, Ser. 2, Vols. 7 and 8.

Bostic, S. W., and Fulton, R. E. (1987). Implementation of the Lanczos method for structural vibration analysis on a parallel computer. Computers and Structures, 25(3):395.

Carey, G. F. (1986). Parallelism in finite element modelling. Communications in Applied Numerical Methods, 2:281.

Carey, G. F., and Jiang, B. N. (1986). Element-by-element linear and nonlinear solution schemes. Communications in Applied Numerical Methods, 2:145.

Carter, W. T., Jr., Sham, T. L., and Law, K. H. (1989). A parallel finite element method and its prototype implementation on a Hypercube. Computers and Structures, 31(6): 921.

Chang, H. Y., Utku, S., Salama, M., and Rapp, D. (1988). A parallel Householder tridiagonalization stratagem using scattered row decomposition. Int. J. for Numerical Methods in Engineering, 26:857.

Chien, L. S. (1989). Parallel computational methods on structural mechanics analysis. Ph.D. thesis, Purdue University.

Chien, L. S., and Sun, C. T. (1990). Parallel computation using boundary elements in solid mechanics. Proceedings of the 31st AIAA Structures, Structural Dynamics and Materials Conference, Long Beach, California, April 2–4, 1990, 644.

Crouch, S. L., and Starfield, A. M. (1983). *Boundary Element Methods in Solid Mechanics*. George Allen and Unwin, London.

Darbhamulla, S. P., Razzaq, Z., and Storaasli, O. O. (1987). Concurrent processing for nonlinear analysis of hollow rectangular structural sections. Engineering with Computers, 2:209.

Douglas, C. C., and Miranker, W. L. (1988). Constructive interference in parallel algorithms. SIAM J. Numer. Anal. 25(2):376.

Farhat, C., and Wilson, E. (1987). A new finite element concurrent computer program architecture. Int. J. for Numerical Methods in Engineering, 24(9):1771.

Farhat, C., and Wilson, E. (1988). A parallel active column equation solver. Computers and Structures, 28(2):289.

Farhat, C., Wilson, E., and Powell, G. (1987). Solution of finite element systems on concurrent processing computers. Engineering with Computers, 2:157.

Fulton, R. E. (1986). The finite element machine: An assessment of the impact of parallel computing on future finite element computations. Finite Elements in Analysis and Design, 2:83.

Goehlich, D., Komzsik, L., and Fulton, R. E. (1989). Application of a parallel equation solver to static FEM problems. Computers and Structures, 31(2):121.

Hajjar, J. F., and Abel, J. F. (1989). Parallel processing of central difference transient analysis for three-dimensional nonlinear framed structures. Communications in Applied Numerical Methods, 5:39.

Heller, D. (1978). A survey of parallel algorithms in numerical linear algebra. SIAM Review, 20(4):740.

Houstis, C. E., Houstis, E. N., and Rice, J. R. (1987). Partitioning PDE computations: Methods and performance evaluation. Parallel Computing, 5(1–2):141.

Hwang, K., and Briggs, F. A. (1984). *Computer Architecture and Parallel Processing*. McGraw-Hill, New York.

Jordan, H. F., and Sawyer, P. L. (1979). A multi-microprocessor system for finite element structural analysis. Computers and Structures, 10(1/2):21.

Kantorovich, L. V., and Krylov, V. I. (1964). *Approximate Methods of High Analysis* (translated by C. D. Benster). Interscience (Noordhoff, The Netherlands), New York.

King, R. B., and Sonnad, V. (1987). Implementation of an element-by-element solution algorithm for the finite element method on a coarse-grained parallel computer. Computer Methods in Applied Mechanics and Engineering, 65:47.

Kirchhoff, G. R. (1850). Uber das gleichgewicht und die bewegung einer elastischen scheibe. J. Mathematik (Crelle), Vol. 40.

Law, K. H. (1986). A parallel finite element solution method. Computers and Structures, 23(6):845.

Lawrie, D. H., and Sameh, A. H. (1983). *Applications of Structural Mechanics on Large-Scale Microprocessor Computers*. ASME, New York, pp. 55–64.

Leuze, M. R., and Saxon, L. V. (1983). On minimum parallel computing time for Gaussian elimination. Congressus Numerantium, 40:169.

Martin, H. C., and Carey, G. F. (1973). *Introduction to Finite Element Analysis, Theory and Application*. McGraw-Hill, New York.

Michell, J. H. (1902). The flexure of a circular plate. Proc. London Math. Soc., 34:223.

Miranker, W. L. (1971). A survey of parallelism in numerical analysis. SIAM Review, 13(4):524.

Noor, A .K., Ed. (1986). *Computational Mechanics, Advances and Trends*. ASME, New York, AMD-Vol. 75.

Noor, A. K., Ed. (1987). Parallel Computations and Their Impact on Mechanics. ASME, New York, AMD-Vol. 86.

Noor, A. K., and Atluri, S. N. (1987). Advances and trends in computational structure mechanics. AIAA, 25(7):977.

Noor, A. K., and Fulton, R. E. (1975). Impact of CDC STAR-100 computer on finite element systems. J. of The Structural Division, ASCE, 101(ST4):731.

Noor, A. K., and Hartley, S. J. (1978). Evaluation of element stiffness matrices on CDC STAR-100 computer. Computers and Structures, 9:151.

Noor, A. K., and Lambiotte, J. J. (1979). Finite element dynamic analysis on CDC STAR-100 computer. Computers and Structures, 10:7.

Noor, A. K., and Peters, J. M. (1989). A partitioning strategy for efficient nonlinear finite element dynamics analysis on multiprocessor computers. Computers and Structures, 31(5):795.

Noor, A. K., Storaasli, O. O., and Fulton, R. E. (1983). Impact of new computing systems on finite element computations. In: *Impact of New Computing Systems on Computational Mechanics*. Noor, A. K., Ed. ASME Publication H00275, New York, pp. 1–32.

Nour-Omid, B., and Park, K. C. (1987). Solving structural mechanics problems on the Caltech hypercube machine. Computer Methods in Applied Mechanics and Engineering, 61:161.

Ortega, J. M., and Voigt, R. G. (1985). Solution of parallel differential equations on vector and parallel computers. NASA Report CR 172500 or ICASE Report No. 85-1.

Ortiz, M., and Nour-Omid, B. (1986). Unconditionally stable concurrent procedures for transient finite element analysis. Computer Methods in Applied Mechanics and Engineering, 58:151.

Ou, R., and Fulton, R. E. (1988). An investigation of parallel numerical integration methods for nonlinear dynamics. Computers and Structures, 30(1/2):403.

Pease, M. C. (1967). Matrix inversion using parallel processing. J. Assoc. Comput. Mach., 14:757.

Reddy, J. N. (1986). Applied Functional Analysis and Variational Methods in Engineering. McGraw-Hill, New York.

Rheinboldt, W. C. (1985). Trends in numerical analysis and parallel algorithms: Error estimates for nonlinear finite element computations. Computers and Structures, 20(1–3):91.

Rodrigue, G., Ed. (1982). Parallel Computations. Computation Department, Lawrence Livermore National Laboratory, California.

Rodrigue, G. (1986). Advances and Trends in Parallel Algorithms. Lawrence Livermore National Laboratory, California.

Schnabel, R. B. (1987). Concurrent function evaluations in local and global optimization. Computer Methods in Applied Mechanics and Engineering, 64:37.

Sledge, F. R., and Rheinboldt, W. C. (1985). Trends in numerical analysis and parallel algorithms: A program design for an adaptive, nonlinear finite element solver. Computers and Structures, 20(1–3):85.

Sokolnikoff, L. S. (1956). Mathematical Theory of Elasticity. McGraw-Hill, New York.

Somigliana, C. (1885–1886). Il Nuovo Cimento, Ser. 3, tt. 17–20.

Storaasli, O. O., Peebles, S. W., Crockett, T. W., Knott, J. D., and Adams, L. (1982). The finite element machine: An experiment in parallel processing. NASA TM 84514, or NASA CP 2245, pp. 201–217.

Storaasli, O. O., Ransom, J., and Fulton, R. (1987). Structural dynamics analysis on a parallel computer: The finite element machine. Computers and Structures, 26(4): 551.

Sun, C. T., and Mao, K. M. (1988a). A global-local finite element method suitable for parallel computations. Computers and Structures, 29(2):309.

Sun, C. T., and Mao, K. M. (1988b). Elastic-plastic crack analysis using a global-local approach on a parallel computer. Computers and Structures, 30(1/2):395.

Svensson, B. (1987). A substructuring approach to optimum structural design. Computers and Structures, 25(2):251.

Timoshenko, S., and Goodier, J. N. (1951). Theory of Elasticity. McGraw-Hill, New York.

Ugural, A. C. (1981). Stresses in Plates and Shells. McGraw-Hill, New York.

Utku, S., and Salama, M. (1986). Parallel solution of closely coupled systems. Int. J. for Numerical Methods in Engineering, 23:2177.

Utku, S., Melosh, R., Islam, M., and Salama, M. (1982). On nonlinear finite element analysis in single-, multi-, and parallel-processors. Computers and Structures, 15(1): 39.

Van Luchene, R. D., Lee, R. H., and Meyers, V. J. (1986). Large scale finite element analysis on a vector processor. Computers and Structures, 24(4):625.

White, R. E. (1986). A nonlinear parallel algorithm with application to the Stefan problem. SIAM J. Numer. Anal. 23(3):639.

Zave, P., and Rheinboldt, W. C. (1979). Design of an adaptive, parallel finite-element system. ACM Transactions on Mathematical Software, 5(1):1.

Zois, D. (1988a). Parallel processing techniques for FE analysis: Stiffnesses, loads and stresses evaluation. Computers and Structures, 28(2):247.

Zois, D. (1988b). Parallel processing techniques for FE analysis: System solution. Computers and Structures, 28(2):261.

10
Domain Decomposition
A Unified Approach for Solving Fluid Mechanics Problems on Parallel Computers

Garry Rodrigue   *University of California—Davis, Livermore, California*

1   COMPUTERS

The traditional computer consists of three basic units: the control unit, the processing unit, and the main memory (Hwang and Briggs, 1984; Rodrigue et al., 1982). Generally speaking, the control unit causes data and instructions to flow through the machine, the processor carries out the arithmetic and logical operations of a program, and the main memory stores the program and the data to be processed. Traditionally the term "central processing unit" (CPU) has been used to refer loosely to the control unit and processor to distinguish them from the main memory and other peripheral memory devices.

An important feature associated with any computer for measuring the efficiency of a numerical model is the clock: an independent operating device that produces control pulses at regular intervals. These intervals are often called the cycle time or clock periods. These pulses are distributed throughout the computer system at control points to regulate the flow of information.

Demands from the scientific community for faster processing on larger amounts of data have driven the computing field from the preceding rudimentary description to a number of different directions. However, over the past two decades the dominant direction has been toward computer configurations that can execute operations simultaneously, that is, parallel processing. Machines vary widely on how this simultaneity is carried out, but over the last few years two forms of parallel processing have become popular on the commercial market: the pipelined vector processor (SIMD, single instruction–multiple data stream) and the multiprocessor (MIMD, multiple instruction–multiple data stream). In this chapter we concentrate only on the multiprocessor.

2 MULTIPROCESSORS

It is very difficult to characterize multiprocessors in any simple manner. However, for our purposes we can assume they are individual processor and memory modules that are interconnected in some way. This interconnection can occur in a number of ways, but generally processor-memory modules communicate with one another directly or through a common shared memory. The processing unit in the model can be a simple bit processor, a scalar processor, or a vector processor. The memory unit in the module can be a few registers or a cache memory. Because of the nonlinearities of fluid mechanics it is important that the interaction between the computer modules in a multiprocessing system be controlled by a single operating system.

Multiprocessors take on two forms: the loosely coupled or distributed memory multiprocessors and the tightly coupled or shared-memory multiprocessors. In a loosely coupled system each computer module has a relatively large local memory where it accesses most of the instructions and data. In a loosely coupled system processes executing on different computer modules communicate by exchanging messages through an interconnection network since there is no shared memory. In fact, the communication topology of this interconnection network is the crucial factor of these systems. Because of this, loosely coupled systems are usually efficient when the interaction between computational tasks is minimal.

Tightly coupled multiprocessor systems communicate through a globally shared memory. Hence the rate at which data can communicate from one computer module to the other is of the order of the bandwidth of the memory. Complete connectivity exists between the computer modules and memory so that one of the limiting factors of tightly coupled systems is the performance degradation due to memory contentions.

3 NUMERICAL MODELS AND MULTIPROCESSORS

Ideal numerical models for multiprocessors are those that can be broken down into algorithmic tasks, each of which can be executed independently on a computer module without ever having to obtain or pass data between the modules during the course of the execution. An example is the inversion of an independent sequence of matrices (as often required in any alternating-direction-implicit scheme for solving partial differential equations). However, for most fluid dynamics problems it is difficult (and even impossible) to define such totally independent subtasks. The reason, of course, is that the nonlinear properties of the problem that the numerical model is attempting to simulate are so intercoupled that it is not known how to break them apart. However, over the past 5 years a framework has been developed that enables the problem to be decomposed among different processors.

This framework also allows a mechanism for analyzing the movement of data within a multiprocessing system. The basic idea is to regard the computational tasks being performed by the individual computer modules as numerical solutions of individual boundary value problems. In this way numerical data being obtained or transmitted between computer modules are the initial and boundary data of the differential equations. The solution of the overall mathematical model is then provided by ''piecing'' together each of the subproblems.

To illustrate this idea, let us consider the numerical solution of the general partial differential equation

$$\frac{\partial u}{\partial t} = F (x, u, t, D_x u, D_x^2 u, \ldots)$$ (1)

where $x \in \Omega(t) \subset R^k$, $0 \le \tau \le \tau$, and $u(x, t) \in R^m$ satisfies given boundary and initial conditions. The classic approach is to solve Eq. (1) at a given time t_n on $\Omega(t_n)$, increment the time variable t by Δt_n, and then solve Eq. (1) again on $\Omega(t_{n+1}) = \Omega(t_n + \Delta t_n)$.

Clearly, if several problems of the type Eq. (1) were defined, then parallelism could be achieved by solving each one of them on a processor. The goal, then, is to define several subproblems from the given global problem Eq. (1).

One method of defining the equations to be solved in the individual computer module is to first decompose the spatial domain at a given time t^*. That is, the domain $\Omega(t^*)$ is expressed as a union of subdomains (not necessarily disjoint):

$$\Omega(t^*) = \bigcup_{j=1}^{k(t^*)} \Omega_j(t^*)$$ (2)

and on each subdomain a related initial boundary value equation

$$\frac{\partial u}{\partial t} = F_j(x, u) \qquad j = 1, 2, \ldots, k(t^*)$$ (3)

is defined (Fig. 1). Each processor then assumes the task of solving one or more of these partial differential equations over a prespecified time interval Δt^*. At the end of this time interval a new substructuring of the domain is performed; that is

$$\Omega(t^* + \Delta t^*) = \bigcup_{j=1}^{k(t^* + \Delta t^*)} \Omega_j(t^* \, \Delta t^*)$$ (4)

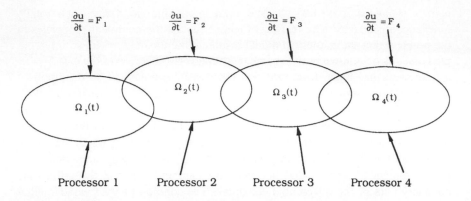

Figure 1   Processor subproblem allocation.

and the process is repeated. The numerical and mathematical relationship between the computed subdomain solutions in Eq. (3) and the solution of the global problem Eq. (1) is delicate and is a function of the partial differential equation being solved. However, it is precisely this relationship that determines the efficiency of the computation on a multiprocessing system.

4   STATIC DOMAIN DECOMPOSITION AND ELLIPTIC PDEs

For elliptic partial differential equations there exist a mathematical precedent based around the ideas given earlier (Schwartz, 1890). In this situation the global partial differential equation is

$$F(u, D_x u, D_x^2 u, \ldots) = 0 \tag{5}$$

with u satisfying a given piecewise continuous function f on the contour Γ of Ω. The domain Ω is partitioned $\Omega = U^m_{j} = 1 \ \Omega_j$, as in Figure 2, with this partitioning remaining fixed throughout the course of the analysis. The boundary Γ_i of Ω_i is given by

$$\Gamma_i = \alpha_i \cup \beta_i \cup \gamma_i$$

$$\alpha_i = \Gamma_i \cap \Omega_{i-1}$$

$$\beta_i = \Omega_i \cup \Gamma$$

$$\gamma_i = \Gamma_{i \cup \Omega_{i+1}}$$

(see Fig. 2). On Ω_1 boundary values are known only on β_1. So we assign arbitrarily on γ_1 a function ϕ_1 and together with the function f we have a

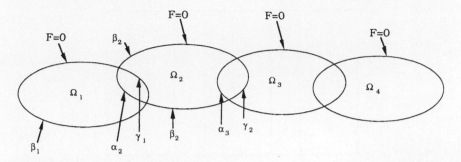

Figure 2 Internal boundary conditions.

piecewise continuous function on L_1. We then construct the function $u_1^{(0)}$ (x, y) by solving Eq. (5) on Ω_1 under the boundary condition

$$u_1^{(0)} (x, y) = \begin{cases} f & \text{on } \beta_1 \\ \phi_1 & \text{on } \gamma_1 \end{cases} \tag{6}$$

In the same manner we assign arbitrarily on γ_2 a function ϕ_2 so that together with the function f we have a piecewise continuous function on L_2. We then construct the function $u_2^{(0)}$ (x, y) by solving Eq. (5) on Ω_2 under the boundary conditions

$$u_1^{(0)} = \begin{cases} u_{i-1}^{(0)} & \text{on } \alpha_2 \\ f & \text{on } \beta_2 \\ \phi_2 & \text{on } \gamma_2 \end{cases} \tag{7}$$

We proceed in the same manner to obtain functions $u_3^{(0)}, u_4^{(0)}, \ldots, u_p^{(0)}$ on $\Omega_3^{(0)}$, $\Omega_4^{(0)}, \ldots, \Omega_p^{(0)}$, respectively. The process is then repeated to obtain $u_i^{(k)}, i = 1, \ldots, p, k = 1, \ldots,$. That is, we construct the function $u^{(k)}{}_i$ by solving the problem on Ω_i under the boundary conditions

$$u_i^{(k)} = \begin{cases} u_{i-1}^{(k)} & \text{on } \alpha_i \\ f & \text{on } \beta_i \\ u_{i+1}^{(k-1)} & \text{on } \gamma_i \end{cases} \tag{8}$$

In Kantorovich and Krylov (1958) it was proved that if Eq. (5) satisfied certain maximum principle conditions then the solution $u_i^{(k)}$ converged to the restriction of the solution u on Ω_i. For second-order elliptic partial differential equations an asychronous version of this above process was derived in Kang (1981) whereby the iterates $u_i^{(k)}$ are not defined in a fixed pattern as here but in a somewhat random fashion.

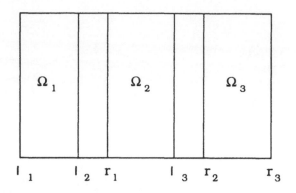

Figure 3   Simple domain decomposition.

5   NUMERICAL SCHWARZ PROCESSES

Numerical analogs of these processes are obvious. In this case partitioning of the domain is carried out from a superimposed quadrilateral grid on the global domain Ω. This decomposition could be determined from the geometry of the problem as in finite element structures calculations, or they can be determined purely on the mathematical properties of the problem. To illustrate the latter approach, let $\Omega = [0,1] \times [0,1]$. Then for a given positive integer k, define the grid $G = \{(x_i, y_j): x_i = i/k + 1, y_j = j/k + 1, 0 \leq i, j \leq k + 1\}$. Let $0 < p < n + 1$ be a given integer (the integer p is typically the number of computer modules in the multiprocessing system). Define a sequence of integers r_i, l_i, $1 \leq i \leq p$, such that

$$0 = l_1 < l_2 < r_1 < l_3 < r_2 < \cdots < l_p < r_{p-1} < r_p = k + 1 \qquad (9)$$

These integers delimit the right and left boundaries on the abscissa of the subdomains (see Fig. 3):

$$\Omega_i = [x_{r_i}, x_{l_i}] \times [0, 1] \qquad 1 \leq i \leq p$$

Let G_i be the restriction of the grid G on Ω_i. A sequence of solutions $\bar{u}_i^{(k)}$ is then defined on G_i using a discrete Schwarz process.

For example, consider the solution of the second-order linear elliptic problem

$$F(u, x, y) = a\frac{\partial^2 u}{\partial x^2} + 2b\frac{\partial^2 u}{\partial x\,\partial y} + c\frac{\partial^2 u}{\partial y^2} - g(x, y) = 0 \quad (x, y) \in \Omega = [0,1] \times [0,1] \qquad (10)$$

with $u = \phi$ on the boundary of Ω. Discrete approximations to Eq. (10) are made on the subgrids G_i to generate a set of p matrix problems. For example, central difference approximations to Eq. (10) yield matrix equations

$$A_i^{(k)} \, \bar{u}_i^{(k)} = b_i^{(k)} \qquad \text{in } \Omega_i \tag{11}$$

where each $A_i^{(k)}$ is an $\eta_i^{(k)} \times \eta_i^{(k)}$ matrix and $\bar{b}_i^{(k)}$ is a vector constructed from the values of f_i on the grid, the boundary conditions of Eq. (10), and "pseudoboundary" conditions furnished by the neighboring solutions. In this case the vectors $\bar{u}_i^{(k)}$ are simply the solution of the matrix (11). Convergence of the Schwarz iterates (8) for this method can be found in Kang (1981), Rodrigue (1985), and Rodrigue et al. (1989). The numerical task for each of the processing units is the numerical inversion of the systems in Eq. (11). Consequently, since the Schwarz process is inherently parallel the computational time to obtain the numerical solution of Eq. (10) is the product of the number of Schwarz iterations with the sum of the solution time for the linear systems solvers and the communication time for the passage of pseudoboundary data between processors. Hence a means for accelerating this process would be to reduce the number of Schwarz iterations needed for convergence and to reduce the computational time needed to provide the approximations $\bar{u}_i^{(k)}$. One such means is to use the coarse grid acceleration techniques that have been so successful on the classic Gauss-Siedel iterative methods (Briggs, 1987).

6 COARSE GRID ACCELERATION

Coarse grid accelerations owe their origin to observations made by Brandt (1977) that with certain iterative methods the magnitude of the high Fourier frequencies of the error is diminished considerably after only a few iterations. Since the number of frequencies present in the error is determined by the mesh size $1/\eta$ of the grid G_η, low frequencies in one grid are high frequencies in a coarser grid. Hence by successively changing the grid from fine to coarse, one is able to achieve an acceleration of the original iteration. In this section we present an algorithm that uses such a coarse grid technique to accelerate the Schwarz method. The parallel performance of this idea was studied on an Alliant FX/8 multiprocessor.

To demonstrate the idea we apply the algorithm to obtain a numerical solution of Eq. (10) in which $a = c = 1$, $b = 0$, and $f(x, y) = 0$ (i.e., Poisson's equation). Let $\eta_1 = 64$, and define the grid G_{η_1} with mesh size $1/\eta$. Similarly, let $\eta_2 = 32$ and again define the grid G_{η_2} with mesh size $1/\eta_2$ That is, G_{η_1} and G_{η_2} represent the fine and coarse grids, respectively. Since our computations are carried on an Alliant FX/8 multiprocessor, we take $p = 16$ and we apply the following algorithm to solve the Poisson problem. Let Ω_i by given by (9), where $\alpha = r_i - l_{i+1}$, $i = 1, 2, \ldots, 15$, is fixed.

Algorithm 1

For $k = 1, 2, \ldots$, do until $\| u_i^{(k)} - u \|_\infty \equiv \| e_i^{(k)} \|_\infty < 10^{-6}$,

For $j = 1, L$ do

If $j = 1$ and $L = 2$, then $M = 3$ else $M = 30$;

For $m = 1, 2, \ldots, M$ do

If $j = 1$ then

For $i = 1, 3, \ldots, 15$ do in parallel

Solve numerically:

$$\begin{cases} \Delta u_i^{(k)} = g_i & \text{on } \Omega_i \cap G_{\eta_j} \\ u_i^{(k)} = f & \text{on } \Gamma \cap \Gamma_i \\ u_i^{(k)} = u_{i-1}^{(k-1)} & \text{on } \Gamma_i \cap \Omega_{i-1} \\ u_i^{(k)} = u_{i+1}^{(k-1)} & \text{on } \Gamma_i \cap \Omega_{i+1} \end{cases} \tag{12}$$

For $i = 2, 4, \ldots, 16$ do in parallel

Solve numerically:

$$\begin{cases} \Delta u_i^{(k)} = g_i & \text{on } \Omega_i \cap G_{\eta_j} \\ u_i^{(k)} = f & \text{on } \Gamma \cap \Gamma_i \\ u_i^{(k)} = u_{i-1}^{(k)} & \text{on } \Gamma_i \cap \Omega_{i-1} \\ u_i^{(k)} = u_{i+1}^{(k)} & \text{on } \Gamma_i \cap \Omega_{i+1} \end{cases} \tag{13}$$

Else if $j = 2$ then

For $i = 1, 3, \ldots, 15$ do in parallel

Solve numerically:

$$\begin{cases} \Delta e_i^{(k)} = 0 & \text{on } \Omega_i \cap G_{\eta_j} \\ e_i^{(k)} = 0 & \text{on } \Gamma \cap \Gamma_i \\ e_i^{(k)} = e_{i-1}^{(k-1)} & \text{on } \Gamma_i \cap \Omega_{i-1} \\ e_i^{(k)} = e_{i+1}^{(k-1)} & \text{on } \Gamma_i \cap \Omega_{i+1} \end{cases} \tag{14}$$

For $i = 2, 4, \ldots, 16$ do in parallel

Solve numerically:

$$\begin{cases} \Delta e_i^{(k)} = g_i & \text{on } \Omega_i \cap G_{\eta_j} \\ e_i^{(k)} = f & \text{on } \Gamma \cap \Gamma_i \\ e_i^{(k)} = e_{i-1}^{(k)} & \text{on } \Gamma_i \cap \Omega_{i-1} \\ e_i^{(k)} = e_{i+1}^{(k)} & \text{on } \Gamma_i \cap \Omega_{i+1} \end{cases} \tag{15}$$

If $L = 1$ Algorithm 1 is called the parallel multigrid algorithm (PMG), and if $L = 2$ it is called the parallel odd-even Schwarz method (POES).

As we mentioned in the previous section, carrying out Eqs. (11) through (15) involves the solution of linear systems of the form (11), and in the case of Poisson's equation

$$A_i^{(k)} = \begin{bmatrix} B & C & & \\ C & B & \ddots & \\ & \ddots & \ddots & C \\ & & C & B \end{bmatrix} \qquad i = 1, 2, \ldots, 16 \tag{16}$$

$$B = \begin{bmatrix} -4 & -1 & & \\ -1 & \ddots & & \ddots \\ & \ddots & \ddots & -1 \\ & & -1 & -4 \end{bmatrix} \tag{17}$$

$$C = \begin{bmatrix} -1 & & \\ & \ddots & \\ & & -1 \end{bmatrix} \tag{18}$$

In the results to follow the numerical solution process for each of these linear systems is the classic point-Gauss-Siedel iterative method (Fletcher, 1988). The initial vector for the iterative method on Ω_i at the kth step of the Schwarz process is the vector $v_i^{(0)}$ given by

Table 1 Comparison of Parallel Multigrid Method (PGM) and Parallel Odd-Even Schwarz Method (POES)

	$s = 1$		$s = 3$		$s = 10$		$s = 20$	
	$\bar{k}$	t	$\bar{k}$	t	$\bar{k}$	t	$\bar{k}$	t
$\alpha = 0$								
PMG	237	12.5	102	14.6	68	31.6	64	58.9
POES	6250	24.2	2250	25.6	1000	37.7	850	64
$\alpha = 2$								
PMG	189	12.8	82	15.6	50	30.9	41	50.4
POES	4800	25.6	1700	26.8	550	28.8	350	36.6
$\alpha = 4$								
PMG	112	9.29	54	12.7	34	26.3	26	40.1
POES	3250	22.2	1100	22.2	350	23.5	200	26.8

$$v_i^{(0)} = \begin{cases} u_{i-1}^{(k-1)} & \text{on } \Omega_i \cap \Omega_{i-1} \\ u_i^{(k-1)} & \text{on } \Omega_i - (\Omega_{i-1} \cap \Omega_{i+1}) \\ u_{i+1}^{(k-1)} & \text{on } \Omega_i \cap \Omega_{i+1} \end{cases} \tag{19}$$

for $i = 1, 3, 5, \ldots$, and

$$v_i^{(0)} = \begin{cases} u_{i-1}^{(k-1)} & \text{on } \Omega_i \cap \Omega_{i-1} \\ u_i^{(k-1)} & \text{on } \Omega_i - (\Omega_{i-1} \cup \Omega_{i+1}) \\ u_{i+1}^{(k)} & \text{on } \Omega_i \cap \Omega_{i+1} \end{cases} \tag{20}$$

for $i = 2, 4, 6, \ldots$. In addition, only a finite number s of Gauss-Siedel iterations is taken on each subregion.

Computational comparisons of PMG and POES are given in Table 1. All timings are in seconds. $\bar{k}$ is the value of k when termination of the algorithm occurs.

As a reference point we solved the problem on G_{η_1} by a sequential point-Gauss-Siedel method (SGS) and a sequential multigrid method (SMG). The sequential multigrid method used was a three-level method with 3 Gauss-Siedel iterations on G_{η_1} and G_{η_2} and 30 Gauss-Siedel iterations on G_{η_3} (a 16 × 16 grid) with injection and bilinear interpolation of the approximations between the grids (Briggs, 1987). Table 2 lists the results. The algorithms were terminated when the absolute difference between the computed and the exact solutions was less than 10^{-6}.

Table 2 Comparison of Sequential Point-Gauss-Seidel Method (SGS) and Sequential Multigrip Method (SMG)

	$\bar{k}$ •	t
SGS	6250	164
SMG	239	41

7 ALTERNATING DIRICHLET-NEUMANN PSEUDOBOUNDARIES

In this section we reexamine the basic formulation put forth by Schwarz with the goal of redefining it to achieve a faster iteration. We see that this redefinition does not destroy the inherent parallelism of the Schwarz methods, nor does it preclude the possibility of applying any of the conjugate gradient or coarse grid acceleration techniques that have already been successful on the classic Schwarz methods. Looking at the basic iteration of Eq. (8), we see that along the "pseudoboundaries" $\Gamma_i - \Gamma$ Dirichlet boundary data are used. This leads to the question of whether it is possible to use different boundary conditions with the hope of achieving faster convergence. We consider the two-domain situation first. Then if α_i and β_i, $i = 1, 2$, are real constants, we consider the following variation of iteration (8):

$$-\Delta u_1^{(k)} = f \qquad \text{on } \Omega_1$$

$$u_1^{(k)} \mid \Gamma = g$$

$$\alpha_1 u_1^{(k)} + \beta_1 \frac{\partial u_1^{(k)}}{\partial n} \mid \Gamma_1 - \Gamma = \alpha_1 u_2^{(k-1)} + \beta_1 \frac{\partial u_2^{(k-1)}}{\partial n} \mid \Gamma_1 - \Gamma \qquad (21)$$

$$-\Delta u_2^{(k)} = f \qquad \text{on } \Omega_2$$

$$u_2^{(k)} \mid \Gamma = g$$

$$\alpha_2 u_2^{(k)} + \beta_2 \frac{\partial u_2^{(k)}}{\partial n} \mid (\Gamma_2 - \Gamma) \cap \Omega_1 = \alpha_2 u_1^{(k-1)} + \beta_2 \frac{\partial u_1^{(k-1)}}{\partial n} \mid \Gamma_2 - \Gamma$$

where n refers to the normal outward direction. Since iteration (21) constitutes a new subdomain iterative process, questions of convergence and rates of convergence for both the two-domain and the multidomain situations must be addressed. In addition, comparisons with the standard iteration (8) must be made. In the following section we address these issues for the solution of Laplace's equation.

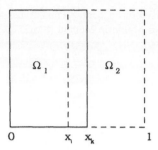

Figure 4   Simple alternating domain decomposition.

As a test problem consider

$$- \Delta u = 0 \qquad (x, y) \in \Omega = [0, 1] \times [0, 1] \tag{22}$$

with

$$u(x, 0) = 0 \qquad u(x, 1) = x \qquad \text{for } 0 \leq x \leq 1$$

$$u(0, y) = 0 \qquad u(1, y) = y \qquad \text{for } 0 \leq x \leq 1 \tag{23}$$

In this case the exact solution is $u = xy$.

A uniform grid is placed on Ω with $\Delta x = \Delta y = \frac{1}{16}$. Algorithm (21) is used to solve Eq. (22), where the Δ operators are replaced by central difference operators. The domain decomposition of Ω is given by Figure 4, where $x_k = i\Delta x$, $x_i = j\Delta x$ for some integers $j < i$. The individual subproblems in algorithm (21) are solved to an error tolerance of 10^{-11} using the preconditioned conjugate gradient method with a point-Jacobi preconditioner. In Table 3 we compare the number of iterations for algorithms DD ($\beta_1 = \beta_2 = 0$, $\alpha_1 = \alpha_2 = 1$) and DN

Table 3   Convergence of DD and DN Schemes

	Iterations					
	$d = 0.125$		$d = 0.1875$		$d = 0.25$	
j	DD	DN	DD	DN	DD	DN
2	43	23	32	20	22	18
4	52	35	36	29	27	24
6	51	44	37	34	29	28
8	53	52	37	39	28	32
10	50	63	33	45	25	37
12	41	70	25	63		

$(\beta_1 = \alpha_2 = 0, \alpha_1 = \beta_2 = 1)$ to achieve $\| u_{\text{computed}} - u_{\text{exact}} \|_\infty < 10^{-11}$. In all cases $d = x_k - x\|$.

Note that on a parallel computer the best efficiency is obtained when the size of each subdomain is the same and the overlap of the subdomain is minimal so that efficiency is achieved when $j = 6$ and $d = 0.125$ in the present case. Table 3 reveals that algorithm DN is then preferred.

8   THREE SUBDOMAINS

The three-subdomain case in Figure 5 is now investigated. Equation (22) is to be solved via an extension of iteration (21) on the three domains $\Omega_1 = [0, x_{k_1}] \times [0, 1]$, $\Omega_2 = [x_{1_2}, x_{k_2}] \times [0, 1]$, and $\Omega_3 = [x_{1_3}, 1] \times [0,1]$, where $0 < x_{1_2} < x_{k_1} < x_{k_2} < x_{k_2} < 1$. On the partitioning we consider the following iteration:

$$
\left\{
\begin{aligned}
-\Delta u_1^{(k)} &= f \quad \text{on } \Omega_1 \\
u_1^{(k)} |_\Gamma &= g \\
\alpha_1 u_1^{(k)} + \beta_1 \frac{\partial u_1^{(k)}}{\partial x} \Big|_{\Gamma_1 - \Gamma} &= \alpha_1 u_2^{(k-1)} + \beta_1 \frac{\partial u_2^{(k-1)}}{\partial x} \Big|_{\Gamma_1 - \Gamma}
\end{aligned}
\right. \tag{24a}
$$

$$
\left\{
\begin{aligned}
-\Delta u_2^{(k)} &= f \quad \text{on } \Omega_1 \\
u_1^{(k)} |_\Gamma &= g \\
\alpha_2 u_2^{(k)} + \beta_2 \frac{\partial u_2^{(k)}}{\partial x} \Big|_{\Gamma_2 - \Gamma} &= \alpha_2 u_1^{(k-1)} + \beta_2 \frac{\partial u_1^{(k-1)}}{\partial x} \Big|_{\Gamma_2 - \Gamma \cap \Omega_1} \\
\alpha_3 u_2^{(k)} + \beta_3 \frac{\partial u_2^{(k)}}{\partial x} \Big|_{(\Gamma_2 - \Gamma) \cap \Omega_3} &= \alpha_3 u_3^{k-1} + \beta_3 \frac{\partial u_3^{k-1}}{\partial x} \Big|_{(\Gamma_2 - \Gamma) \cap \Omega_3} \quad (24b)
\end{aligned}
\right.
$$

$$
\left\{
\begin{aligned}
-\Delta u_3^{(k)} &= f \quad \text{on } \Omega_3 \\
u_3^{(k)} |_\Gamma &= g \\
\alpha_4 u_3^{(k)} + \beta_4 \frac{\partial u_3^{(k)}}{\partial x} \Big|_{\Gamma_3 - \Gamma} &= \alpha_4 u_2^{(k-1)} + \beta_4 \frac{\partial u_2^{(k-1)}}{\partial x} \Big|_{\Gamma_3 - \Gamma}
\end{aligned}
\right. \tag{24c}
$$

Specifically, the classic situation with Dirichlet conditions on the pseudoboundaries, defined by

$$\alpha_i = 1 \quad \beta_i = 0, \quad i = 1, 2, 3, 4 \quad \text{DD}$$

is compared with the mixed Neumann-Dirichlet situation

$$\alpha_2 = \beta_1 = \beta_3 = \alpha_4 = 0 \quad \beta_2 = \alpha_1 = \alpha_3 = \beta_4 = 1 \quad \text{DN}$$

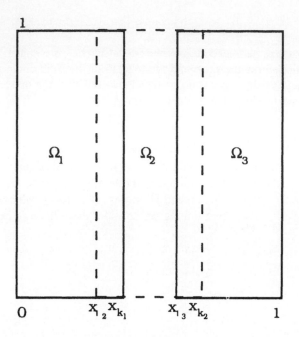

Figure 5   Three subdomains.

The computational parameters are the same as in Section 7. In the numerical experiment we let $x_{k(1)} = 7\Delta x$ and $x_{l(3)} = 9\Delta x$ be fixed and

$$| x_k(2) = 8\Delta x | = | x_{l(2)} - 8\Delta x | = d$$

be allowed to vary. The number of iterations necessary for convergence of algorithims DD and DN are recorded in Table 4. Again we see that algorithm DN is faster for the optimal parallel situation of equally sized subdomains and minimal overlap (i.e., $d = 0.1875$).

Table 4   Iterative Performance of Derechlet-Derichlet (DD) and Derichlet-Neumann (DN) Pseudoboundary Conditions

d	DD	DN
0.1875	69	49
0.25	43	40
0.3125	33	34
0.375	25	29
0.4375	19	26

9 DOMAIN AND OPERATOR DECOMPOSITION

In contrast to the decomposition methods described in the previous section, the partitioning of the domain Ω can be performed according to the heuristics of the analytic solution of the partial differential equation. Then different forms of the problem are defined in different subregions. One example of such an idea is furnished by the solution of the convection-diffusion equation

$$F(u, x, y) = \frac{\partial u}{\partial x} - \in \left(\frac{\partial^2 u}{\partial x^2} + \frac{\partial^2 u}{\partial y_2} \right) = 0$$

$$(x, y) \in \Omega = [0, 1] \times [0, 1]$$

$$u(0, y) = u(x, 1) = 1 \tag{25}$$

$$u(x, 0) = 0$$

$$\frac{\partial u}{\partial x}(1, y) = 0$$

This problem is of interest primarily because it is a linearization of the Navier-Stokes equations, where ϵ is a positive parameter representing the Reynolds number of fluid mechanics. If the flow velocity is strong or the viscous effects are small but nonnegligible, ϵ is small.

10 ASYMPTOTIC BEHAVIOR

The solution of Eq. (25) is known to possess boundary layers, where, for example, there can be regions of Ω such that

$$u \simeq \sum_{n=0}^{m} \epsilon^n \phi_n$$

$$\frac{\partial \phi_0}{\partial x} = 0$$

$$\frac{\partial \phi_n}{\partial x} = \epsilon \left(\frac{\partial^2 \phi_{n-1}}{\partial x^2} + \frac{\partial^2 \phi_{n-1}}{\partial y^2} \right) \quad n > 0$$

(called ordinary boundary layers) or there can be regions such that

$$u \simeq \sum_{n=0}^{m} \epsilon^n \Psi_n$$

$$\frac{\partial \Psi_0}{\partial x} - \frac{\partial^2 \Psi_0}{\partial \xi^2} = 0$$

$$\frac{\partial \Psi_n}{\partial x} - \frac{\partial^2 \Psi_n}{\partial \xi^2} = \epsilon \frac{\partial^2 \Psi_{n-1}}{\partial x^2} \qquad n > 0$$

(called parabolic boundary layers) and ξ is a local "stretched" coordinate (Eckhaus, 1979).

In the following section we define a domain decomposition on Ω based on the boundary layer behavior of the solution and then apply a variant of the Schwarz alternating procedure to obtain a numerical approximation of the solution of Eq. (25).

11 PARABOLIC LAYERS

In Rodrigue and Reiter (1990) it was established that the sequence of functions $\{u^{(n)}\}$ defined by the iteration

$$\begin{cases} \dfrac{\partial u^{(0)}}{\partial x} - \epsilon \dfrac{\partial^2 u^{(0)}}{\partial^2 y} = 0 \\[3mm] \dfrac{\partial u^{(n)}}{\partial x} - \epsilon \dfrac{\partial^2 u^{(n)}}{\partial y^2} = \dfrac{\partial u^{(n-1)}}{\partial x_2} \qquad n > 0 \end{cases} \tag{26}$$

$$u^{(n)}(x, 1) = 1, \qquad 0 \le x \le 1 \tag{27a}$$

$$u^{(n)}(x, 0) = 0, \qquad 0 \le x \le 1 \tag{27b}$$

$$u^{(n)}(a, y) = g(y), \qquad 0 \le y \le 1 \tag{27c}$$

satisfies

$$\| u - u^{(n)} \|_\infty = O(\epsilon^{n+1})$$

when

$$\frac{d^k g}{\partial y^k}(0) = \frac{d^k g}{\partial y^k}(1) \qquad k = 0, 1, \ldots, n \tag{28}$$

Since the boundary conditions in Eq. (25) do not satisfy Eq. (28), a domain decomposition strategy would be to split Ω into the subregions $\Omega = \Omega_1 \cup \Omega_2$ where

$$\Omega_1 = [0, l_1] \times [0, 1] \qquad l_1 > 0$$
$$\Omega_2 = [l_2, 1] \times [0, 1] \qquad l_2 < l_1$$

(see Fig. 6) and numerically solve Eq. (25) on Ω_1. This is followed by numerically carrying out the iteration (26) with boundary conditions (27a–c) and using

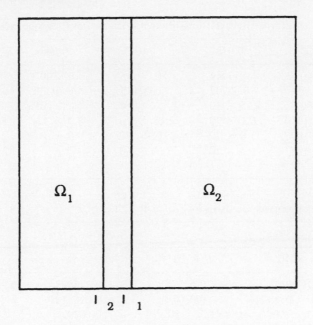

Figure 6 Domain decomposition for parabolic boundary layers.

$$g(y) = u(l_2, y) \tag{29}$$

where u is the computed solution in Ω_1.

To test the feasibility of the approach we take several values of $l_1 > 0$ and for each of these values we solve Eq. (25) with $\epsilon = 0.0005$, the boundary conditions (27a–c), and

$$\frac{\partial u}{\partial x}(l_1, y) = 0 \qquad 0 < y < 1$$

We then carry out the iteration (26) on $\Omega_2 = [l_2, 1] \times [0, 1]$, $l_2 = l_1 - k\Delta x$, and

$$u^{(n)}(x, 1) = 1 \qquad l_2 \leq x \leq 1$$

$$u^{(n)}(x, 0) = 0 \qquad l_2 \leq x \leq 1$$

$$u^{(n)}(l_2, y) = u(l_2, y) \qquad 0 \leq y \leq 1$$

On Ω_1, Eq. (25) is approximated by second-order finite differences on a grid with $\Delta x = 0.0015$ and $\Delta y = 0.001$ and the resultant linear system is solved with a direct matrix solver. Equation (27a–c), on the other hand, requires the solution

Table 5 Convergence Behavior for Different Locations of Pseudoboundary

n	0.003	0.018	0.033	0.048	0.063
1	0.90401	0.77714	0.71635	0.67007	0.63575
2	0.01379	0.00287	0.00143	0.00100	7.8×10^{-4}
3	0.00601	1.15×10^{-4}	4.3×10^{-5}	2×10^{-5}	1.3×10^{-5}
4	3.9×10^{-5}	6.1×10^{-6}	2.08×10^{-6}	8.9×10^{-7}	5×10^{-7}
5	3.1×10^{-6}	3.8×10^{-7}	1.2×10^{-7}	5×10^{-8}	2×10^{-8}
6	2.8×10^{-7}	3×10^{-8}	1×10^{-8}		
7	2×10^{-8}				

Table 6 Convergence Behavior for Different Values of ϵ

n	ϵ			
	5×10^{-5}	5×10^{-4}	5×10^{-3}	5×10^{-2}
1	0.73911	0.63575	0.277827	0.037807
2	0.000178	0.00078	0.010518	0.8499
3	4.1×10^{-7}	1.3×10^{-5}	0.00195	2.6304
4	3×10^{-9}	5×10^{-7}	0.00085	21.944
5		2×10^{-8}	4.6×10^{-4}	184.6
6			2.9×10^{-4}	∞

of inhomogeneous heat equations with the x variable interpreted as the time variable. In this situation the Crank-Nicholson method is used on a grid with $\Delta x = 0.01$ and $\Delta y = 0.001$. Table 5 lists the results. Each entry of the table is $\| U^{(n)} - U^{(n-1)} \|_\infty$, where $U^{(n)}$ is the computed approximation to $u^{(n)}$. Note that convergence in all cases is quite rapid.

In Rodrigue and Reiter (1990) it was established that for fixed Δx and Δy there exists $\epsilon > 0$ so that divergence occurs. To determine such values we use the same grid structure as before, take $l_2 = 0.041$, vary the ϵ, and list the errors in Table 6. As can be seen, larger values of ϵ result in divergence.

12 ORDINARY LAYERS

We again consider Eq. (25). In Eckhaus (1979) it was established that the sequence of functions $u^{(n)}$ defined by the iteration

$$u^{(0)} (x, y) = 1 \tag{30a}$$

$$\frac{\partial u^{(n)}}{\partial x} = \epsilon \left(\frac{\partial^2 u^{(n-1)}}{\partial x^2} + \frac{\partial^2 u^{(n-1)}}{\partial y^2} \right) \qquad n > 0 \tag{30b}$$

$$u^{(n)}(x, 0) = f(x) \qquad 0 \le x \le 1 \tag{30c}$$

such that

$$\frac{d^k f}{dx^k}(0) = 1 \qquad k = 0, 1, \ldots, n \tag{31}$$

satisfies

$$\|u - u^{(n)}\|_\infty = O(\epsilon^{n-1})$$

[u is the solution to Eq. (25)].

As before, since the boundary conditions of Eq. (25) do not satisfy Eq. (31), a domain decomposition strategy would be to split Ω into the subregions $\Omega = \Omega_1 \cup \Omega_2$, where

$$\Omega_1 = [0, 1] \times [0, l_1] \qquad l_1 > 0$$
$$\Omega_2 = [0, 1] \times [l_2, 1] \qquad l_2 < l_1 \tag{32}$$

(see Fig. 7) and numerically solve Eq. (25) on Ω_1, followed by numerically carrying out the iteration (30) with

$$f(x) = u(x, l_2) \tag{33}$$

where u is the computed solution in Ω_1.

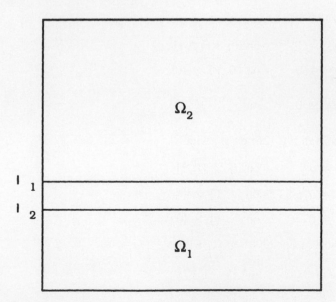

Figure 7 Domain decomposition for ordinary boundary layers.

To test the feasibility of this approach we take several values of $l_2 > 0$ and carry out Eqs. (30) through (33). As a test case we take $\epsilon = 2 \times 10^{-4}$, use the exact solution of (25) on Ω_1 (see Rodrigue and Reiter, 1990), and then carry out Eqs. (30) through (33) on Ω_2 using a backward Euler method on a grid with $\Delta x = \Delta y = 10^{-2}$. Convergence to an error of $\| U^n - U^{(n-1)} \|_\infty < 10^{-5}$ occurred at 30 iterations for each value of $l_2 = 10^{-2}, 10^{-1}, 4 \times 10^{-1}$. Also, for $l_2 = 10^{-2}$ convergence occurred for $\epsilon = 10$ whereas divergence occurred for $\epsilon = 5 \times 10^{-4}$.

13   SCHWARZ METHOD

In this section we develop a Schwarz alternating procedure for solving Eq. (25) based on the results of the previous two sections. That is, we split $\Omega = \Omega_1 \cup \Omega_2 \cup \Omega_3$, where

$$\Omega_1 = [0, 1] \times [b, 1]$$
$$\Omega_2 = [0, r] \times [0, t] \qquad b < t$$
$$\Omega_3 = [l, 1] \times [0, t] \qquad l < r$$

(see Fig. 8). Let $u_1^{(0)} = u_2^{(0)} = u_3^{(0)} = 1$, on Ω_1, Ω_2, and Ω_3, respectively. We then define the sequences $\{u_1^{(i)}\}$, $\{u_2^{(i)}\}$, and $[u_3^{(i)}]$ as follows. For $i = 1, 2, \ldots$,

1. $u_2^{(i)}$ solves Eq. (25) on Ω_2 with

$$u_2^{(i)} = u_1^{(i)} \qquad \text{on } [0, r] \times \{t\}$$
$$u_2^{(i)} = 1 \qquad \text{on } \{0\} \times [0, t]$$
$$u_2^{(i)} = 0 \qquad \text{on } [0, r] \times \{0\}$$
$$\frac{\partial}{\partial x} u_2^{(i)} = 0 \qquad \text{on } \{r\} \times [0, t]$$

2. $u_3^{(i)}$ solves Eq. (26) on Ω_3 with

$$u_3^{(i)} = u_2^{(i)} \qquad \text{on } \{l\} \times [0, t]$$
$$u_3^{(i)} = u_1^{(1-l)} \qquad \text{on } [l, 1] \times \{t\}$$
$$u_3^{(i)} = 0 \qquad \text{on } [l, 1] \times \{0, t\}$$

3. $u_1^{(i)}$ solves Eq. (30b) on Ω_1 with

$$u_1^{(i)} = u_2^{(i)} \qquad \text{on } [0, r] \times \{b\}$$
$$u_1^{(i)} = u_3^{(i)} \qquad \text{on } [r, l] \times \{b\}$$

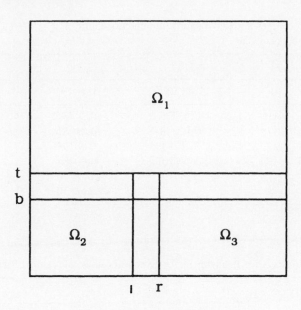

Figure 8   Domain decomposition for Schwarz method.

We carried out these iterations under different scenarios to examine their convergence behavior to the solution of Eq. (25). In all cases the following mesh sizes were used:

$$\Delta x = \Delta y = 10^{-2} \qquad \text{on} \qquad \Omega_1$$
$$\Delta x = \Delta y = 10^{-3} \qquad \text{on} \qquad \Omega_2$$
$$\Delta x = 10^{-2} \qquad \Delta y = 10^{-3} \quad \text{on} \quad \Omega_3$$

The numerical method for computing the solution was the same as that in Sections 11 and 12.

In the first experiment the lower boundary of Ω_1 is fixed and the boundary between Ω_2 and Ω_3 is varied. In this case,

$$\epsilon = 10^{-4}$$
$$b = 0.03$$
$$t = 0.031$$
$$r = 0.41$$
$$l = r - k \times 10^{-3}$$

Table 7 records the results.

Table 7 Convergence Behavior for Different Locations of
Horizontal Pseudoboundary

	ϵ			
n	5×10^{-5}	5×10^{-4}	5×10^{-3}	5×10^{-2}
1	0.73911	0.63575	0.277827	0.037807
2	0.000178	0.00078	0.010518	0.8499
3	4.1×10^{-7}	1.3×10^{-5}	0.00195	2.6304
4	3×10^{-9}	5×10^{-7}	0.00085	21.944
5		2×10^{-8}	4.6×10^{-4}	184.6
6			2.9×10^{-4}	∞

Table 8 Convergence Behavior for Different Locations of Vertical Pseudoboundary

	ϵ		
n	1	10	20
1	0.2580	0.3363	0.5799
3	0.6844	0.2859	0.0682
5	0.9212	0.4923	0.0846
7	0.7625	0.4753	0.0623
9	0.4705	0.3134	0.0325
11	0.2133	0.1752	0.0132
13	0.0734		
15	0.02		

In the second experiment the boundary between Ω_2 and Ω_3 is fixed and the boundary of Ω_1 is varied. In this case,

$$\epsilon = 2 \times 10^{-4}$$

$$t = 0.31$$

$$b = t - k \times 10^{-2}$$

$$r = 0.41$$

$$l = 0.04$$

Table 8 records the results.

Table 9 Convergence Behavior for Different Values of ϵ

	ϵ				
n	5×10^{-5}	2×10^{-4}	2.5×10^{-4}	3×10^{-4}	3.5×10^{-4}
1	0.6684	0.6718	0.6659	0.6561	0.6432
2	10^{-4}	0.0184	0.0304	0.0502	0.0802
3	7×10^{-5}	0.0146	0.0428	0.0978	0.1907
4	8×10^{-7}	0.0148	0.0571	0.1634	0.3838
5	10^{-7}	0.0129	0.0658	0.2344	0.6615
6	10^{-8}	0.0101	0.0676	0.2986	1.0091
7		0.0072	0.0629	0.3437	∞
8		0.0047	0.0536	0.3621	
9		0.003	0.0438	0.3647	
10		0.0018	0.0337	0.3459	
11		0.001	0.0241	0.3044	

In the final experiment the boundaries of Ω_1, Ω_2, and Ω_3 are held fixed and the value of ϵ is varied. In this case,

$$b = 0.3$$
$$t = 0.31$$
$$r = 0.41$$
$$l = 0.30$$

Table 9 records the results.

14 DYNAMIC DOMAIN DECOMPOSITION

For time-dependent partial differential equations the decomposition of the spatial domain can be coupled into the different evolutionary processes of the solution. Then the decomposition of the domain becomes time dynamic. Specifically, at a given time t_1 the properties of the solution may dictate a spatial decomposition

$$\Omega = \bigcup_{j=1}^{n(t_1)} \Omega_j(t_1)$$

whereas at a later time t_2 an altogether different spatial decomposition might be used. For example, consider the numerical solution of the one-dimensional equation of gas dynamics

$$\frac{\partial u}{\partial t} + \frac{\partial}{\partial x} F(u) = 0 \tag{34}$$

$$u = \begin{bmatrix} \rho \\ m \\ e \end{bmatrix}, \qquad F(u) = \begin{bmatrix} m \\ p + m^2/\rho \\ m(e+p)/\rho \end{bmatrix}$$

where ρ is the density, u is the velocity, $m = \rho u$ is the momentum, p is the pressure, and e is the total energy. To make the system of equations consistent with the number of unknowns, we add the equation of state

$$p = (\gamma - 1)\left(e - \frac{1}{2}u^2\right)$$

where γ is a given "gas" constant. Solutions of such problems are known to have regions where the components of the solution are either discontinuous or have large gradient changes. These regions propagate with time, and the dynamics of the solution need to be computationally monitored so that an appropriate domain decomposition can be performed at a later time.

To give an example of how this procedure might be accomplished, suppose a uniform grid G_1 ($t = 0$) with spatial mesh size Δx_1 is specified on $\Omega(t)$. This grid remains fixed throughout the duration of the calculation. As the computation proceeds, local subdomains are adaptively created in response to an estimation of the error. More specifically, if an approximate solution $u(t)$ is given at time t on grid $G_1(t)$, then an estimation of the truncation error at $u(t + \Delta t_1) = u(_2)$ is made and regions $\Omega_j(t_2)$, $j = 1, \ldots, n(t_2)$ of low accuracy are isolated. A new uniform grid $G_2(t)$ of spatial mesh size $\Delta x_1 \leq \Delta x_1$ can be placed on some of the subdomains $\Omega_j(t_2)$. A new computational problem is then defined on each of these subgrids with initial and boundary data specific to each of these grids (see Fig. 9). A possible scenario is given in Figure 9. Initial and boundary data are provided on $\Omega_1(t_3)$ and $\Omega_3(t_3)$ from the previous time step; we use the data (segment BC) in Figure 9. On $\Omega_2(t_3)$ we create an initial data finer grid by interpolating data from $\Omega_1(t_3)$ and $\Omega_3(t_3)$, as on segment DC. Similarly, the boundary data, such as DC, are obtained by interpolation. Unused data from $\Omega_1(t_3)$ are purged (segment AB). An iteration is then performed on each subdo-

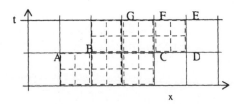

Figure 9 Dynamic domain decomposition.

main. An exception to the previous algorithm is made if $t = 0$ or if any of the domains abuts the actual physical boundary in which case the initial or boundary data are provided by the actual problem. The time step size on each of the subdomains is constant and is determined by satisfying a stability criterion. Typically the time step size is specified so that a local disturbance does not propagate more than one spatial mesh width per time step. Consequently the number of spatial zones in $\Omega_2(t_3)$ must be large enough that a disturbance has moved out of $\Omega_2(t_3)$ before the next domain decomposition is made.

This creation and initialization of subdomains is a recursive process so that, in turn, finer subdomains of $\Omega_2(t_3)$ can be created, initialized, and so on. In this way a data structure tree is formed with each of the subdomains as its nodes.

15 COMPUTATIONAL EXAMPLE

We now discuss some of the numerical details of the mesh refinement scheme used to obtain the results in the following section. On each level i a mesh spacing h_i is prespecified with $h_i/h_{i+1} = M_{i+1}$, a positive integer, along with a numerical scheme. Here the time integration scheme on each level is the same, the first-order accurate Godunov two-step method:

$$U_{i+1/2}^{n+1/2} = 1/2 \, (U_{i+1}^n + U_i^n) - \frac{k}{h} \, (F_{i+1/2}^n - F_i^n)$$

$$U_i^{n+1} = U_i^n - \frac{k}{h} \, (F_{i+1/2}^{n+1/2} - F_{i-1/2}^{n+1/2})$$

(35)

which has a stability criterion of $(\,|\,u\,| + c)k/h < 1$, where c is the speed of sound and k is the time step. An artificial viscosity term

$$\frac{\partial u}{\partial t} = V\frac{k}{h} \, h^2 \, \frac{\partial^2 u}{\partial x^2}$$

was added and numerically approximated by a second-order accurate odd-even hopscotch scheme (Berger et al., 1983):
for $n + j =$ even,

$$U_j^{n+1} = U_j^n + V\frac{k}{h} \, (U_{j-1}^{n+1} - 2U_j^n + U_{j+1}^n)$$

(36)

for $n + j =$ odd,

$$U_j^{n+1} = U_j^n + V\frac{k}{h} \, (U_{j-1}^{n+1} - 2U_j^n + U_{j+1}^{n+1})$$

Only the local truncation error of the density variable was estimated, and for this a Richardson extrapolation was used. Regridding is performed when the local error is $> \epsilon_{tol}$. Fine-grid initial and boundary data were obtained by linear interpolation.

16   NUMERICAL RESULTS

The parameters $\epsilon_{tol} = 10^{-3}$, $M_2 = 4$, $V = 0.5$, and $\gamma = 1.4$ were used in the following test runs. In this situation an error estimation is performed at every time step. The computations were performed on a Cray-1 with the program being written to promote easy vectorization by the compiler.

Numerical Experiment: One-Dimensional Shock Tube

Initial data: $m(x, 0) = 0$.

$$\rho(x, 0) = \begin{cases} 1.0 & x < 0.5 \\ 0.1 & x \le 0.5 \end{cases}$$

$$\rho(x, 0) = \begin{cases} 1.0 & x < 0.5 \\ 0.125 & x \ge 0.5 \end{cases}$$

Boundary conditions were reflecting. Three experiments were run with $0 \le x \le 1$ and $0 \le t \le 0.3$:

1.   Course-grid run:

$$k_1^{(c)} = 0.01 \qquad h_1^{(c)} = 0.01$$

2.   Fine-grid run:

$$k_1^{(f)} = 4/k_1^{(c)} \qquad h_1^{(f)} = 4/h_1^{(c)}$$

3.   Adaptive mesh run with two levels:

$$k_1 = k_1^{(c)} \qquad h_1 = h_1^{(c)}$$
$$k_2 = k_1^{(f)} \qquad h_2 = h_1^{(f)}$$

The density profiles at $t = 0.2$ are given in Figure 10 for each of the respective runs. The portions of the curve marked by 1s and 2s in Figure 10c correspond to the grid levels from which the computed solution originates.

Computer timings (Cray-1):

Run 1. 0.48
Run 2. 7.57
Run 3. 2.9

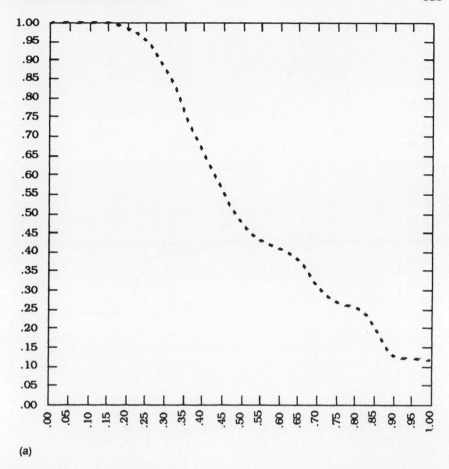

(a)

Figure 10   (a) Coarse-grid computation. (b) Fine-grid computation. (c) Dynamic domain decomposition computation.

Hence, run (2) is roughly 14 times more expensive than run (1), but more accurate answers are obtained. Mesh refinement, on the other hand, is only 6 times more expensive with apparently the same answers as run (2).

17   ADVECTION-DIFFUSION EQUATIONS

We develop a dynamic domain decomposition algorithm for the solution of advection-diffusion equations; specifically we solve the two-dimensional Burger's equation.

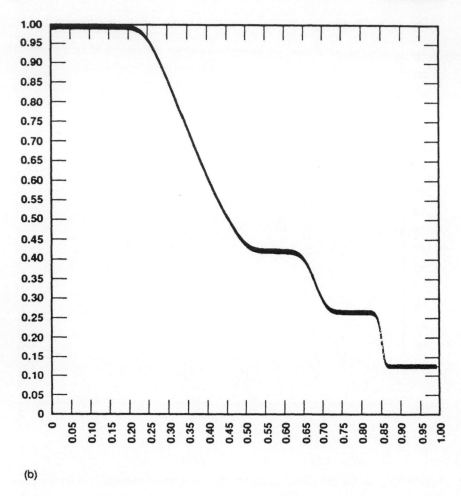

(b)

Figure 10 (continued)

$$\vec{v}_t + \vec{v} \cdot \nabla \vec{v} = \varepsilon \, \Delta \vec{v} \tag{37}$$

The advection-diffusion equations are first advanced on a coarse mesh subject to the hyperbolic stability condition. Significant parabolic regions are identified with a parabolic threshold criterion, and the coarse mesh is dynamically decomposed and refined. An ''Euler-Lagrange'' method advances these refined subdomains independently, allowing coarse-grained parallel scheduling.

The solution method is heterogeneous in the sense that the problem formulation and solution method vary from mesh to mesh. It uses an Eulerian reference

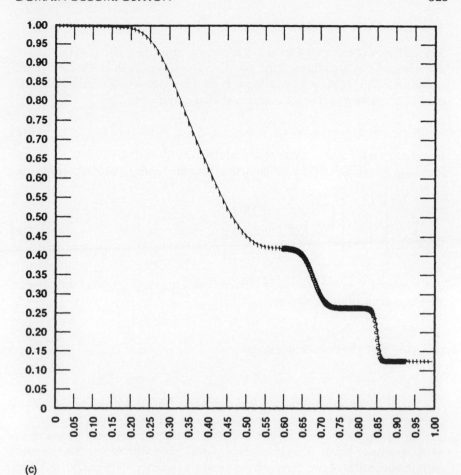

(c)

frame to formulate the equations on the coarse grid and a Lagrangian reference
frame to formulate the equations on the refined grids. The explicit coarse mesh
advancement uses upwind advection differences, centered diffusion differences,
and forward time differences. The implicit Lagrange scheme is an extension of
the method given in Rodrigue et al. (1982). The Lagrange grid is interpolated on
the underlying Eulerian mesh when the grid becomes excessively distorted or the
Lagrangian grid moves past its underlying Eulerian coarse mesh refinement area.
The refined values are conservatively averaged onto the coarse mesh at every
coarse time step (Berger and Colella, 1989).

18 DOMAIN DECOMPOSITION

The grid decomposition is a local uniform mesh refinement (Berger et al., 1983) about regions of significant diffusion. Local uniform mesh refinement is frequently used for hyperbolic equations. Here we apply it to a mixed hyperbolic parabolic equation. For $\vec{x}$ the Eulerian spatial coordinates, let

$$F_E(\vec{x}, \vec{v}, \varepsilon, t) \equiv \vec{v}_t + \vec{v} \cdot \nabla_{\vec{x}} \vec{v} - \varepsilon \Delta_{\vec{x}} \vec{v} = 0 \tag{38}$$

be defined on the coarse rectangular discretized Eulerian mesh $\Omega_d \subset R_d^2$, where the subscript reminds us that the domain is discrete, with given initial conditions

$$\vec{v} = \begin{bmatrix} u \\ v \end{bmatrix} = \begin{bmatrix} \begin{cases} 1 & \text{if } x \le 1.0 \\ 0 & \text{if } x > 1.0 \end{cases} \\ 0 \end{bmatrix} \tag{39}$$

and Neumann boundary conditions. For $\vec{x}'$ the Langrangian spatial coordinates, with coordinate transformation

$$\nabla_{\vec{x}'} = J^{-1} \nabla_{\vec{x}} \tag{40}$$

the associated Lagrangian operator is

$$F_L(\vec{x}', \vec{v}, \varepsilon, t) \equiv \vec{v}_t - \varepsilon(J^{-1} \nabla_{\vec{x}'} \cdot J^{-1} \nabla_{\vec{x}'}) = 0 \tag{41}$$

The initial conditions for the refined Lagrangian grids are given by Eq. (39) at time zero and are interpolated from the coarse mesh at advanced time steps.

Grid points with curvature greater than the parabolic threshold are marked as above-threshold points. We cluster these points into subdomains. The clustering algorithm consists of two parts, cluster analysis and overlap detection. We use a threshold that generates refined grids aligned with the coarse mesh. For each fixed above-threshold point (i, j) we cluster it with another above-threshold point (k, l) when $s_{ij}^{kl} = 1$ for

$$s_{ij}^{kl} = \begin{cases} 1 & |i-k| \le T_x \quad \text{and} \quad |j-1| \le T_y \\ 0 & \text{otherwise} \end{cases} \tag{42}$$

This allows us to construct rectangular clusters on one pass through the above-threshold points. Thresholds T_x and T_y are used to create rectangular zones customed to the application and may vary throughout the domain as well. The clustering algorithm is used to create nonoverlapping meshes; this is accomplished by resetting all above-threshold points once they are clustered. The overlap detection algorithm decides whether the clustered points are a new grid or an extension of an old grid, whether an existing grid is separating, or whether two existing grids are merging. Once the new grids are created we append domain of dependence buffer zones. These buffer zones are stripped off before merging back onto the coarse mesh.

After the grid points are clustered into subdomains a check is made to compare the number of available processors with the number of refined subdomains. At this point primitive load balancing is used to distribute the work load equitably among the available processors.

19 NUMERICAL RESULTS

For diffusion coefficient $\epsilon = 5 \times 10^{-3}$ we use a coarse grid spatial step of 0.1 and a temporal step of 0.04 and advance the coarse mesh to time 0.16. The spatial and temporal refinement ratio is 4. The solution on the coarse grid is shown at time 0.0 and 0.16 in Figure 11a and b. The domain decomposition is shown in Figure 12, and a refined mesh solution at time 0.16 is shown in Figure 11c. The speedup is 3.55 for the four processors available on the Cray-2. The method yields good-quality results with significant improvement over serial throughput.

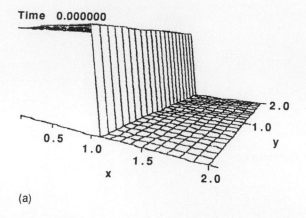

(a)

Figure 11   (a) Initial solution on coarse grid. (b) Solution on coarse grid. (c) Solution on fine grid.

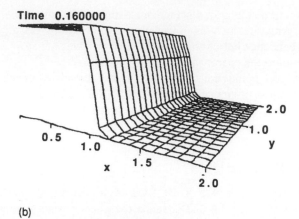

(b)

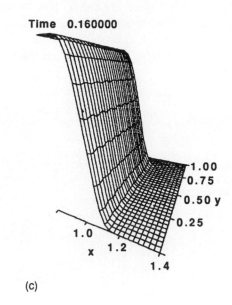

(c)

Figure 11 (continued)

Figure 12 Domain decomposition at final time level.

ACKNOWLEDGMENT

This work was performed under the auspices of the U.S. Department of Energy by the Lawrence Livermore National Laboratory under Contract No. W-7405-ENG-48.

REFERENCES

Berger, M., and Colella, P. (1989). Local adaptive mesh refinement for shock hydrodynamics. J. Comput. Phys. 82:64–84.

Berger, M., Hedstrom, G., Oliger, J., and Rodrigue, G. (1983). Adaptive mesh refinement for 1-dimensional gas dynamics. In: Viehnevetsky and Stepleman, Eds. "Proceedings of the IMACS Conference on Scientific Computing, North-Holland, Amsterdam.

Brandt, A. (1977). Multi-level adaptive solutions to boundary value problems. Mathematics of Computation, 31:333–390.

Briggs, W. (1987). *A Multigrid Tutorial*. SIAM, Philadelphia.

Eckhaus, W. (1979). *Asymptotic Analysis of Singular Perturbations*. North-Holland, New York.

Ferretta, T. A. (1989). Parallel multigrid method for solving elliptic partial differential equations. Lawrence Livermore National Laboratory Report UCRL-53918.

Fletcher, C. A. J. (1988). *Computational Techniques for Fluid Dynamics*, Vol. 1. Springer-Verlag, New York.

Hwang, K., and Briggs, F. (1984). *Computer Architecture and Parallel Processing.* McGraw-Hill, New York.

Kang, L. S. (1981). The Schwarz algorithm. Wuhan University Journal, National Science Edition, Special Issue of Mathematics, China, pp. 77–88.

Kantorovich, L. V., and Krylov, V. I. (1958). *Approximate Methods of Higher Analysis.* P. Noorfodd, Groningen, The Netherlands.

Kuck, D. (1978). *The Structure of Computers and Computation.* John Wiley, New York.

McCormick, S. (1989). *Multilevel Adaptive Methods for Partial Differential Equations.* SIAM, Philadelphia.

Perkins, A. (1989). Parallel heterogeneous mesh refinement for multidimensional convection-diffusion equations using an Euler-language method. Lawrence Livermore National Laboratory Report UCRL-53950.

Rodrigue, G. (1985). Inner/outer iterative methods and numerical Schwarz algorithms. Parallel Computing 2:205–208.

Rodrigue G., and Reiter, E. (1988). A domain decomposition method for parabolic boundary layer problems. Lawrence Livermore National Laboratory Report UCRL-98467.

Rodrigue, G., and Reiter, E. (1990). An asymptotically induced domain decomposition method for parabolic boundary layer problems. To be published in Proceedings of Workshop on Asympotic Analysis and Numerical Methods for Partial Differential Equations, Argonne National Laboratory.

Rodrigue, G., and Shah, S. (1989). Pseudo-boundary conditions to accelerate parallel Schwarz methods. In: Carey, G., Ed. *Parallel Supercomputing Methods, Algorithms and Applications,* John Wiley & Sons, New York.

Rodrigue, G., Hendrickson, C., and Pratt, M. (1982). An implicit numerical solution of the two-dimensional diffusion equation and vectorization experiments. In: Rodrigue, G., Ed. *Parallel Computations.* Academic Press, New York, pp. 101–128.

Rodrigue, G., Kang, L., and Liu, Y. (1989). Convergence and comparison analysis of some numerical Schwarz methods. Numerische Mathematik, 56:123–138.

Schwarz, H. A. (1890). *Gesammelte Mathematische Abhandlungen,* Vol. 2. pp. 133–1134. Springer-Verlag, Berlin.

Varga, R. (1962). *Matrix Iterative Analysis.* Prentice-Hall, Englewood Cliffs, New Jersey.

11
Parallel Methods for the Solution of Partial Differential Equations with Applications in Groundwater Hydrology

Alexander Peters   *IBM Deutschland GmbH, Heidelberg, Germany*

1   INTRODUCTION

Considerable activity has recently been observed in the development of parallel numerical methods. Although many new and very efficient algorithms, expecially for linear algebraic problems, are now available, many existing computer programs for the solution of partial differential equations for engineering applications perform poorly on vector processors and multiprocessors.

A group of applications that requires supercomputer power is the numerical modeling of groundwater contamination in large regional basins. The governing equation for the majority of contamination models is the advection-diffusion equation:

$$\frac{\partial c}{\partial t} + \frac{\partial}{\partial x_i}(v_i c) - \frac{\partial}{\partial x_i}\left(D_{ij}\frac{\partial c}{\partial x_j}\right) = f \tag{1}$$

defined over the domain V with Dirichlet and Neuman conditions on the boundary. In Eq. (1) the unknown function c represents the contaminant concentration, v_i is the flow velocity, D_{ij} the dispersion coefficient, and f is a sink or source term.

Discretization methods (e.g., finite differences, collocation, and finite elements) based on the approximation of the unknown function by a set of numbers or piecewise continuous functions on subdomains are commonly applied to the

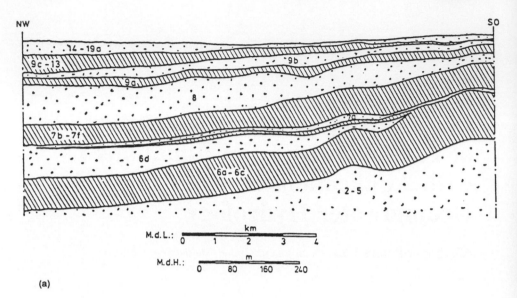

(a)

Figure 1   Example of a multilayered hydrogeologic basin: (a) cross section; (b) FE discretization.

solution of the advection-diffusion equation. Groundwater applications have several unique characteristics that must be taken into account during the implementation of any method of solution on parallel computers.

The problem domain V has usually a complicated structure, consisting of subregions with quasi-random shape, heterogeneous materials, and a large number of variables. Figure 1a illustrates a simplified cross section through a multilayered basin consisting of strata of high permeability (aquifers) and low permeability (aquitards). Figure 1b visualizes a possible finite element discretization in which the aquifers are represented by horizontally layered triangles and the aquitards by vertically oriented prisms. Regional models of this kind involve sereral thousand variables. The data are organized in blocks with irregular structures matched by complicated index maps.

The location of subregions containing singularities and steep gradient variations changes relatively fast and requires the dynamic and local refinement of time and spatial discretization. Local refinement techniques are based on hierarchic trees, conditional statements, variably dimensioned arrays, and indirect indexing.

The nonlinear and transient nature of the physical problem requires the application of solution methods based on iteration. Iteration is a recurrent process.

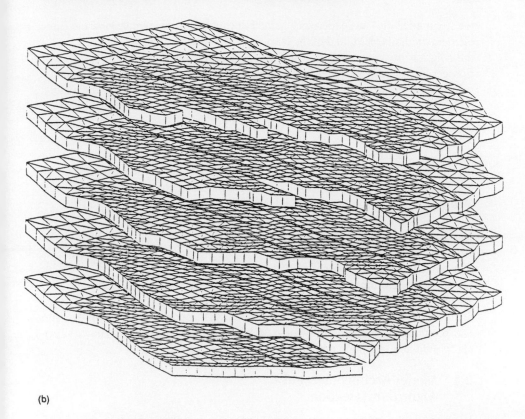

(b)

Figure 1b (continued)

As is well known, neither conditional processes nor relationships based on re-current computations and indirect indexing are suitable for exploiting parallelism. We claim that despite these inherent difficulties numerical techniques based on discretization applied to the advection-diffusion equation have an almost unlimited parallelization potential (for a general discussion see Rice, 1986).

This chapter is an attempt to survey neither discretization methods for partial differential equations (PDEs) nor parallel algorithms of linear algebra. Readers interested in detailed presentations of these topics are directed to the other chapters of this book and to several excellent review materials (Hockney and Jesshope, 1981; Sameh, 1983; Ortega and Voight, 1985; Van der Vorst, 1990). Rather we summarize our own experience gained during the implementation of the finite element method (FEM) and of the least-squares collocation method (LESCO) for the solution of the advection-diffusion equation on vector processors and multiprocessors. Further implementation aspects, included detailed algorithm description and benchmark results, are given in the references (Pelka and Peters, 1986; Peters, 1988a, b, 1990; Peters et al., 1988; Peters and Pinder, 1990).

In Section 2 a general approach for algorithm selection is outlined and several implementation issues are addressed. A pattern of data references valid for FEM and LESCO is introduced in Section 3. Based on this pattern we identify computational tasks that should be supported by parallel algorithms. In Sections 4 and 5 we analyze some issues related to the selection of parallel algorithms for the solution of the advection-diffusion equation [Eq. (1)].

2   GENERAL CONSIDERATIONS

Our aim is to design a system of program modules that are combined to provide the accurate solution of the advection-diffusion equation [Eq. (1)] defined over heterogeneous and randomly shaped domains. Each module should be able to achieve a reasonable fraction of the optimal performance on vector processors and multiprocessors without major modifications. The majority of the commercially available parallel computers are of this variety.

2.1   Model Architecture and Algorithm Selection

We address the "task-to-architecture" mapping problem according to the "algorithm development life cycle graph" introduced by Jamieson (1986). First we define the "target" or "model" architecture as having the following features:

The system consists of several identical processors.
The system has a global shared memory, and each processor accesses an associated local memory.
The processors can execute different algorithms concurrently.
Each processor is capable of performing vector operations.

A solution strategy based on the following steps is then applied to our problem:

Selection of a tentative algorithm: At this point a solution approach has been selected and the main computational tasks of the solution approach are identified. A set of candidate algorithms for each task is chosen.

Selection of the ideal algorithm: The potential for parallelization of the candidate algorithms is analyzed with respect to the model architecture and without taking into account any constraints. The algorithms with the highest theoretical potential for parallelization are identified. Some of these algorithms are not implemented on the available computer, but they provide a frame of reference during the selection process.

Binding of the algorithms to the available architecture: Algorithms are assembled into a global solution approach taking into account the limitations of the available computer.

2.2 Implementation Issues

We use FORTRAN for the implementation of the algorithms mentioned in Sections 4 and 5 as much as possible and try to isolate machine-specific code within a few modules. Many scientists consider FORTRAN a programming language with limited possibilities of expression for parallel algorithms. FORTRAN has the benefit of being well known, however, and used in the majority of numerical applications. The availability of FORTRAN compilers warrants the portability of FORTRAN codes on a wide variety of architectures.

The current generation of FORTRAN parallel compilers is able to transform subroutine structures and loop level constructs into concurrent code. Subroutine-level parallelism is obtained by manually allocating blocks of computations, structured in subroutines, to different processors. This kind of parallelism requires that the subroutines do not interfere. In addition, to balance the processor activity roughly the same number of computations should be performed on every processor.

Loop-level parallelism is achieved by dividing the loop iterations into groups of equal size and running each group on a separate processor. Processors capable of performing vector instructions divide each loop operation into segments, and further parallelism is achieved by maintaining the operation flow in different stages of completion. Loop-level parallelism is more restrictive than subroutine-level parallelism because it leaves the final responsability of concurrency detection to the compiler. To exploit loop-level parallelism algorithms must be coded in simple loops that perform long strings of arthmetic operations free of recurrent or indirect indexing.

The majority of the algorithms discussed in this chapter make possible the design of program modules based on both independent routines and/or parallel-vector loops. The latter implementation requires the careful analysis of every loop of the program, but the computational savings warrant the efficient implementation on both vector and multiprocessors. In addition, the programs are not exposed to load-balance problems introduced through inappropriate data partitions. For this reason we design our code such that the parallelism at the loop level is fully exploited. Subroutine-level parallelism is employed only when loop vectorization fails.

3   A PATTERN OF DATA REFERENCES

We begin the search for parallel algorithms for the solution of the advection-diffusion equation [Eq. (1)] with the identification of a general pattern of data references, valid for many discretization methods. FEM and LESCO belong to the general class of CAB methods. CAB methods are composed of three basic steps: compute, aggregate, and broadcast (Nelson and Snyder, 1986). With respect to discretization methods we identify these three steps with the following computational phases:

Discretization or compute phase: The problem domain is partitioned in subdomains, and some basic linear algebra computations are performed independently on each subdomain. A subdomain may be an area corresponding to one element or to several elements.

Assemblage or aggregate phase: The independent processes on subdomains are connected. The result is a large system of equations, which solution displays a global piece of information for the entire domain.

The broadcast phase: The result of the assemblage phase is interpreted and returned to the processes. This may be a message for the iteration to continue or an initial value for the next compute phase.

The potential for parallelization of the compute phase is unlimited. Each block of computations associated with a subdomain can be viewed as an independent process. Theoretically, the number of independent processes may be increased to infinity through discretization. In practice, however, an acceptable discretization pattern can be found for any given application, beyond which further partitions do not improve significantly either the efficiency or the accuracy of the discretization phase with respect to the entire solution.

The assemblage phase is the critical step of FEM and LESCO. A large number of index operations is required to connect the subdomains. In many cases the cost of assembling pieces of the partitioned domain grows faster than the benefit of partitioning the computations into independent blocks during the discretization phase. Moreover, solving the global system of linear equations is the most expensive computational step of the entire solution. We note that the efficiency of this step can be increased by exploiting several numerical properties of the coefficient matrix, such as grid-related sparsity, symmetry, and diagonal dominance.

Based on these considerations we select algorithms for the FEM and LESCO solutions of the advection-diffusion equation such that:

The most efficient available methods of linear algebra are applied to the solution of the global system of linear equations.

The potential for parallelization of the discretization phase is fully exploited. The assemblage process is based on a reduced set of index operations.

4 ALGORITHMS FOR FEM

4.1 The System of Linear Equations

The advection-diffusion equation is often solved over irregular domains using the finite element method based on the Galerkin approach. One of the most significant advantages of the Galerkin-FEM with respect to groundwater contamination problems is its versatility in handling domains with random geometries and material discontinuities. The application of the standard Galerkin-FEM to the solution of Eq. (1) yields a system of linear equations of which the matrix is nonsymmetric.

A first group of candidate algorithms for the solution of the system of linear equations is based on Gauss elimination or Crout decomposition (see, e.g., Press et al., 1986). Although direct methods are quite robust, they can hardly be applied to the solution of large FE applications because of their often unacceptable huge storage requirements. In addition to resubstitution procedures associated with them are recursive and very difficult to parallelize.

The storage requirements of iterative methods are quite small compared to those of direct methods. Moreover, they are easy to code and parallelize. The main difficulty related to iterative methods is the selection of a suitable algorithm that provides a satisfactory solution within an acceptable number of iterations for any application. We note that methods of solution that find widespread application in the context of finite difference schemes, like the alternating direction method (Varga, 1962) or multigrid methods (Hackbusch, 1985), are difficult to apply to the solution of Eq. (1) defined over randomly shaped regions since they require some regularity in the grid geometry.

A very efficient iterative method that gained widespread attention in recent years is the preconditioned conjugate gradients (PCG) method. Several authors (Obeyscare et al., 1987; Pini et al., 1988) have used some generalized variants of the PCG method to solve the nonsymmetric system of linear equations resulting from FE discretizations of the advection-diffusion equation [Eq. (1)].

We do not present the PCG methods in this chapter. The literature on this topic is huge and continues to grow rapidly. An excellent review was published by Van der Vorst (1990) recently. We note that the PCG variants developed for nonsymmetric systems are less attractive than those developed for symmetric systems since they require the storage of a certain number of previously iterated search directions. Moreover, the smaller the number of stored search directions, the larger is the number of iterations needed to reach convergence. The disadvantages of the nonsymmetric schemes can be eliminated by discretizing Eq. (1) such that a symmetric positive definite system is obtained.

Leismann and Frind (1989) recently proposed a time integration scheme that places the nonsymmetric advective component at the old time level and introduces an artificial dispersion term to compensate for the errors. Equation (1) is replaced at each step t through an equivalent PDE identical to the classic diffusion equation

$$\frac{\partial c}{\partial t} - \frac{\partial}{\partial x_i}\left(K_{ij}\frac{\partial c}{\partial x_j}\right) = F \tag{2}$$

where

$$F = f + \frac{\partial}{\partial x_i}(v_i c^{(0)}) \tag{2a}$$

$$K_{ij} = D_{ij} + D_{ij}^* \tag{2b}$$

In Eq. (2) $c^{(0)}$ denotes the concentration at the old time level $t - \Delta t$. The artificial diffusion term $D_{ij}^* = v_i v_j \Delta t/2$ is integrated spatially over the elements like the physical dispersion. Leismann and Frind (1989) show that their scheme is comparable in terms of accuracy and stability to the Crank-Nicolson scheme.

The application of the Galerkin-FEM to Eq. (2) yields a symmetric, positive definite system of linear equations that can be solved by the most efficient variants of the PCG method. Results of numerical experiments performed with different vectorized PCG solvers applied to the diffusion equation over randomly shaped domains are presented by Peters et al. (1988) and by Pini and Gambolati (1990). We assert that the time integration scheme of Leismann and Frind (1989) coupled with the PCG method for symmetric positive matrices is one of the most efficient approaches available for the solution of large groundwater contamination problems.

4.2   The Discretization Phase

The parallelization potential of the discretization phase is illustrated through an example. For simplicity we consider the solution of Eq. (2) defined over one of the aquifers of the system shown in Figure 1. The discretization of the first aquifer of the multilayered basin is illustrated in Figure 2. It consists of $n = 948$ nodes and $m = 1833$ triangular elements with linear approximation functions.

A preprocessing step is required to arrange the data in a convenient form for parallelization. Three index vectors $\vec{K}^1$, $\vec{K}^2$, and $\vec{K}^3$ of length m giving the relations between nodes and elements are defined. The ordering rule for the elements of $\vec{K}^1$, $\vec{K}^2$, and $\vec{K}^3$ is illustrated for a small example ($m = 4$) in Figure 3. These index vectors are used to prepare sets of elementwise stored data sets for the discretization phase. The expressions

$$X^i(e) = X(K(e)) \qquad e = 1, m \qquad i = 1, 3 \tag{3a}$$

$$Y^i(e) = Y(K(e)) \qquad e = 1, m \qquad i = 1, 3 \tag{3b}$$

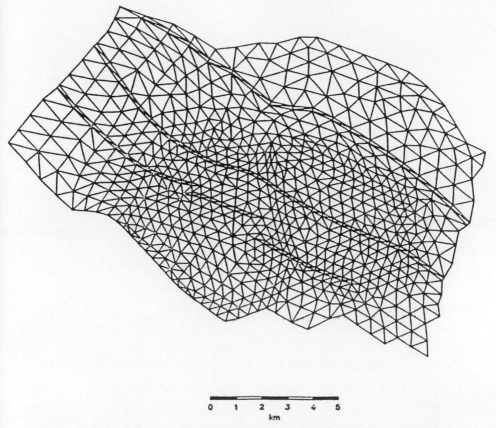

Figure 2 FE discretization of the first aquifer of the basin illustrated in Figure 1.

show how the n components the vectors of nodal coordinates $\vec{X}$ and $\vec{Y}$ are rearranged into six vectors of length m: $\vec{X^1}$, $\vec{X^2}$, $\vec{X^3}$, $\vec{Y^1}$, $\vec{Y^2}$, and $\vec{Y^3}$. Similar relations are employed to rearrange the dispersion coefficients defined in Eq. (2) (for details see Pelka and Peters, 1986). We note that the so-called Gather-type indirect index operation used in expression (3) does not involve recurrences and is amenable to parallelization.

The areas of the m elements are computed and stored into a vector $\vec{F}$ according to the expression

$$F(e) = 0.5 \times (X^1(e) \times (Y^2(e) - Y^3(e)) + X^2(e) \times (Y^3(e) - Y^1(e)) + X^3(e) \times (Y^1(e) - Y^2(e))) \qquad e = 1, m \qquad (4)$$

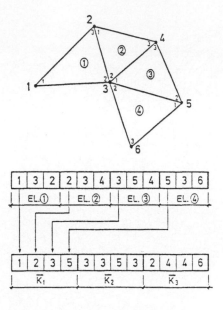

Figure 3 Defining the index vectors $\vec{K}_1$, $\vec{K}_2$, $\vec{K}_3$ (example).

The matrices of element derivatives are computed and stored in six vectors:

$$B^{(11)}(e) = (Y^2(e) - Y^3(e)) \times F(e)\,0.5 \qquad e = 1, m \qquad (5a)$$

$$B^{(12)}(e) = (Y^3(e) - Y^1(e)) \times F(e)\,0.5 \qquad e = 1, m \qquad (5b)$$

$$B^{(13)}(e) = (Y^1(e) - Y^2(e)) \times F(e)\,0.5 \qquad e = 1, m \qquad (5c)$$

$$B^{(21)}(e) = (X^3(e) - X^2(e)) \times F(e)\,0.5 \qquad e = 1, m \qquad (5d)$$

$$B^{(22)}(e) = (X^1(e) - X^3(e)) \times F(e)\,0.5 \qquad e = 1, m \qquad (5e)$$

$$B^{(23)}(e) = (X^2(e) - X^1(e)) \times F(e)\,0.5 \qquad e = 1, m \qquad (5f)$$

Each of the expressions (4) and (5) can be either split into as many as m independent tasks or vectorized. Moreover, the computation of the expressions (5) can be executed concurrently.

Figure 4 shows two basic possibilities of computing the element matrices. In Figure 4a the multiplications are performed element by element, but in Figure 4b they are grouped into long vectors of length m. Both partitions make possible a maximal speedup of m on a machine with m processors. The former ordering (Fig. 4a) requires that a subroutine construct is employed to represent each matrix multiplication. The subroutine is allocated to all processors simultaneously. In the latter ordering (Fig. 4b) the coefficients are grouped into long

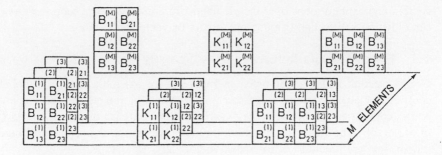

$$\underline{D}^{(k)} = \underline{B}^{(k)^T} * \underline{K}^{(k)} * \underline{B}^{(k)} \quad, \quad k = 1, M$$

(a)

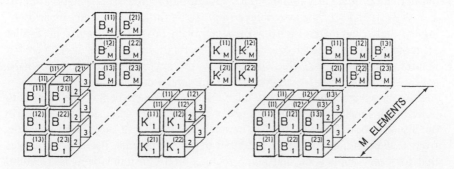

$$\vec{D}^{(ij)} = \vec{B}^{(1j)} * (\vec{B}^{(1i)} * \vec{K}^{(11)} + \vec{B}^{(2i)} * \vec{K}^{(21)}) + \vec{B}^{(2j)} * (\vec{B}^{(1i)} * \vec{K}^{(12)} + \vec{B}^{(2i)} * \vec{K}^{(22)}) \quad, \quad i = 1, 3 \quad, \quad j = 1, 3$$

(b)

Figure 4 Performing element matrix operations: (a) element by element; (b) on long vectors.

vectors and explicit loops over the elements are employed. The loop iterations can either be split equally among processors or vectorized.

4.3   Handling Data References and Sparsity

The m element matrices computed in Figure 4 are assembled into the global coefficient matrix according to pointer sets representing the element-coefficient matrix connections. A pointer vector can be defined for each of the vectors $\overrightarrow{D}^{(ij)}$ in Figure 4b. The values of the pointer vectors depend on the selected storage scheme and are functions of the index vectors $\overrightarrow{K}^1$, $\overrightarrow{K}^2$, and $\overrightarrow{K}^3$.

Figure 5 illustrates the sparsity of the coefficient matrix associated with the first aquifer of the system illustrated in Figure 1b. Calculation of the pointer vectors associated with some popular storage schemes, like the standard column scheme (Duff et al., 1989) or the compressed scheme (Radicati di Brozolo and Vitaletti, 1986), usually involves a large number of operations that are difficult to parallelize. To minimize the impact of this computation on the execution time, the definition of all index connections should be preprocessed (Peters, 1988b).

The efficiency of the assemblage step can be further improved by partitioning the coefficient matrix such that the assemblage process is executed in parallel on temporarily disconnected arrays. For the multilayered basin in Figure 1 a natural choice is to assemble concurrently the subgrids corresponding to the aquifers and then those corresponding to the aquitards.

A general procedure to reduce the number of node-element connections is to use either a smaller number of high-order elements or to create superelements in the discretization (Rice, 1986). Theoretically the number of superelements varies from 1 (the entire domain) to the number of finite elements. Rice (1986) argues that for every application and given architectural constraint there is a natural optimal granularity of partition.

An additional indexing problem arises during the solution of the linear system of equations. There are two basic steps of the PCG method that involve matrix operations (for a complete description of the PCG method see, for example, Van der Vorst, 1990):

Matrix-vector multiplication
Inverting the preconditioning matrix

An efficient multiplication algorithm for compressed matrices is presented by Peters et al. (1988). With $\mathbf{A}$ a matrix with n rows and m columns and $\overrightarrow{X}$ a vector with n elements, the multiplication $\mathbf{A} \times \overrightarrow{X}$ involves the following steps:

The matrix is stored according to the compressed scheme (Radicati di Brozolo and Vitaletti, 1986) into a vector $\overrightarrow{C}$ with $n \times c$ elements, matched by the index map $\overrightarrow{I}$. Figure 6 illustrates the compressed scheme for a small matrix, where $n = 4$, $m = 6$, and $c = 2$.

The elements of $\vec{X}$ are rearranged into a temporary vector $\vec{t}$ using a Gather type of operation and the index map $\vec{I}$:

$$T(e) = X(I(e)) \qquad e = 1, \ldots, c \times n \tag{6a}$$

The result vector $\vec{B}$ is obtained by a series of multiply-add operations:

$$B(e) = C(e) \times T(e) + C(e + n) \times T(e + n) + \ldots +$$
$$C(e + (c - 1) \times n) \times T(e + (c - 1) \times n) \qquad e = 1, n \tag{6b}$$

Each step of expression (6) can be either performed in parallel or vectorized.

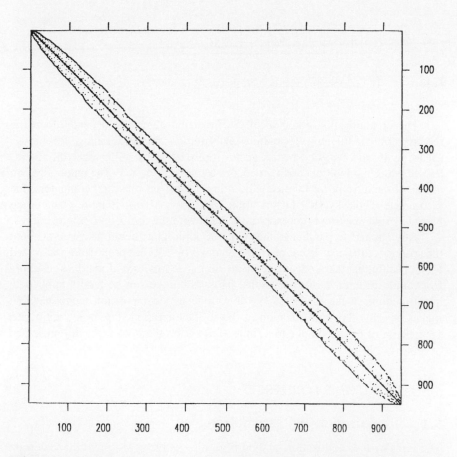

Figure 5   Sparsity representation of the coefficient matrix associated with the discretization in Figure 2. The graphic tool for matrix representations was provided by Paolini and Santangelo (1989).

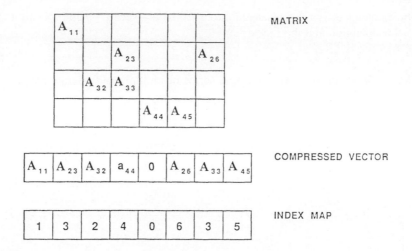

Figure 6   The compressed matrix representation (example).

The parallelization potential of the inverting algorithm depends upon the selected preconditioning technique. Polynomial preconditioning techniques (Saad, 1985) are based on matrix-vector multiplications and can be fully parallelized using the multiplication procedure already described. The parallelization of one of the most powerful preconditioning techniques on scalar computers, the incomplete Cholesky (IC) factorization (Kershaw, 1978), is more challenging because of the recurrent processes associated with the resubstitution procedures.

Any attempt to parallelize the IC preconditioning should take into account the sparsity of the coefficient matrix. Van der Vorst (1982) proposed a vectorized IC procedure that takes advantage of the diagonal structure of matrices resulting from finite difference discretizations. This technique can be hardly applied to quasi-random sparse matrices. We introduce a decomposition technique that makes possible the parallelization of the IC factorization step in Section 5. An application of this technique to a finite element problem was recently presented by Peters (1990).

5   ALGORITHMS FOR LESCO

5.1   Discretization and Assemblage

The least-squares collocation method is an alternative approach to the Galerkin-FEM that allows one to employ orthogonal discretizations and the PCG method for symmetric systems of linear equations to the solution of the advection-diffusion equation [Eq. (1)]. The philosophy behind the use of least squares as a method of solution for partial differential equations is to find an approximate solution that results in a minimum of residual errors. Several formulations are possible depending upon the method of minimization of the errors over the domain and the form of the approximate solution (Eason, 1976).

In the LESCO approach (Joos, 1986; Bentley et al., 1989; Laible and Pinder, 1990) the problem domain is divided into subregions (elements). Similarly to the Galerkin-FEM, a trial function $\hat{c}(x_i,t)$ is defined by basis functions ϕ_m (X_i), piecewise continuous over the elements, and by unknown coefficients $A_m(t)$ associated with the grid nodes:

$$\hat{c}(x_i, t) = \sum_{m=1}^{M} A_m(t)\phi_m(x_i) \tag{7}$$

The trial function $\hat{c}$ is substituted into Eq. (1), and the errors, that is, the difference between the exact and the approximate solutions, are evaluated at selected collocation points in the interior of the domain and on the boundaries. The least-squares residual function of the errors is constructed and minimized with respect to the unknown coefficients A_m. This yields a system of M linear equations:

$$[C_{lm}]\{A_m(t)\} = \{E_l\} \qquad \text{for} \qquad l, m = 1, M \tag{8}$$

where $[C_{lm}]$ is the coefficient matrix, $\{A_m\}$ is the vector of the unknown coefficients, and $\{E_l\}$ is the right-hand-side vector (Peters and Pinder 1990).

LESCO has several attractive properties with respect to parallel processing. The collocation points can be selected such that shape irregularities and the locations of the singularities are accurately described. The representation of randomly distributed singularities and irregularly shaped boundaries through collocation points does not affect the sparsity of the coefficient matrix of the global system [Eq. (8)]. This is because the least-squares residual function is evaluated at collocation points and the unknown coefficients A_m are associated with grid nodes. The data are carried between collocation points and grid nodes by the basis functions (Laible and Pinder, 1989).

A consequence is that the locations of the grid nodes can be selected based on purely geometric criteria. Laible and Pinder (1989) have shown that LESCO coupled with orthogonal discretizations displays highly accurate solutions for the advection-diffusion equation defined over irregularly shaped domains. They have covered several irregular domains with orthogonal meshes and have introduced the boundary conditions at their precise locations in space using collocation points. To obtain regular data structures some of the nodes that fall outside the domain are retained in the system during the computation. This approach significantly simplifies the discretization phase since it eliminates parametric transformations. Moreover, the sparsity of the coefficient matrix resulting from orthogonal discretizations has a simple diagonal structure that virtually eliminates indirect address computations during the assemblage and solution steps.

Another advantage of the LESCO approach is that the matrix of the linear system of equations [Eq. (8)] is symmetric and positive definite. Consequently Eq. (8) can be solved by the powerful PCG method.

5.2 PCG and Grid Refinement

It is well known that the accuracy of the solution of Eq. (1) critically depends upon the selected temporal and spacial discretization. Most of the existing local refinement techniques use tree data structures based on a large number of indirect index operations. An attractive alternative to these techniques is the approach termed "patch refinement" (Bramble et al., 1988; Ewing, 1990). In this technique the subregions to be refined are covered with uniform grids. When the coarse grid is also orthogonal the overhead due to indirect indexing is virtually eliminated.

We present a PCG-based patch refinement technique that makes possible vectorization of the solution on subdomains and parallelization of the entire solution (Peters and Pinder, 1990). The technique presented here can be applied to problems that inlude dynamic or multiple refinements or can be used to parallelize the solution only.

For the sake of simplicity we consider a composite discretization that consists of a coarse "parent" grid over the entire domain V and of a patch refinement over V^*. An example of a grid consisting of a coarse subgrid and a refinement is illustrated in Figure 7. The nodes of V and V^* are numbered independently. Since the intersection boundary is part of each grid, the nodes of the internal boundary are numbered twice. The steps of the algorithm are as follows:

LESCO is applied to Eq. (1) defined over V and V^*. The resulting system of equations is put together in the following matrix form:

$$
\begin{bmatrix} [C_{ij}] & [0_{lj}] \\ [0_{im}] & [C_{lm}^*] \end{bmatrix} \begin{Bmatrix} \{A_j\} \\ \{A_m^*\} \end{Bmatrix} = \begin{Bmatrix} \{E_i\} \\ \{E_l^*\} \end{Bmatrix} \quad \text{for} \quad i, j = 1, M \quad l, m = 1, M^* \quad (9)
$$

where M and M^* are the number of basis functions defined over V and V^*. $[0_{ij}]$ and $[0_{im}^*]$ contain only zero elements. If two processors are available the discretization phase and the assemblage of the system of linear equations can be performed for V and V^* concurrently.

The coefficients of $[C_{ij}]$ and $\{E_i\}$ corresponding to the nodes covered by the patch refinement are removed. The resulting gaps of the main diagonal are filled with unity (Ewing, 1990).

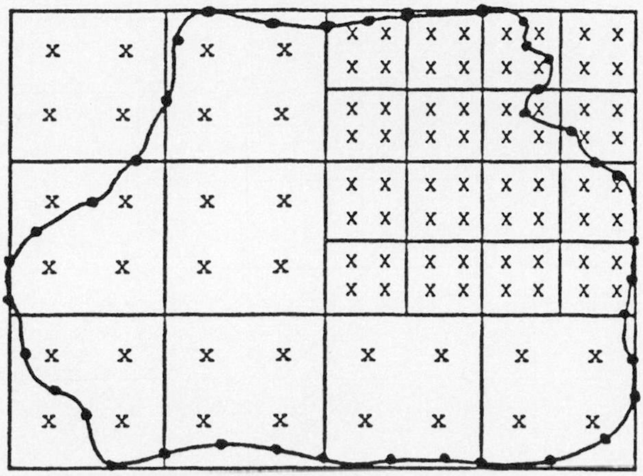

Figure 7 LESCO discretization of an irregularly shaped domain (example): (×) internal colocation points, and (●) boundary colocation points.

The nodes n and n^* of the intersection boundaries between V and V^* are connected by the set of equations

$$A_n - A_{n^*}^* = 0 \tag{10}$$

Continuity is enforced at the "slave" nodes as well. Slave nodes are grid nodes of the intersection boundary between V and V^* that belong only to V^*. The LESCO formulation of this continuity constraint is presented by Laible and Pinder (1990).

With the constraint imposed on the internal boundary, the global system of linear equations becomes:

$$\begin{bmatrix} [\tilde{C}_{ij}] & [\tilde{0}_{lj}] \\ [0_{im}] & [\tilde{C}_{lm}^*] \end{bmatrix} \begin{Bmatrix} \{A_j\} \\ \{A_m^*\} \end{Bmatrix} = \begin{Bmatrix} \{\tilde{E}_i\} \\ \{\tilde{E}_l^*\} \end{Bmatrix} \quad \text{for} \quad i, j = 1, M \quad l, m = 1, M^* \tag{11}$$

where $[0_{ij}]$ and $[0_{im}^*]$ contain nonzero elements only at the intersection of the rows and columns n and n^* and $[\tilde{C}_{ij}]$ and $[\tilde{C}_{im}^*]$ have the same sparsity as $[C_{ij}]$ $C_{lm}^*]$, respectively. To maintain the symmetry of the coefficient matrix Eq. (10) is added twice to the coefficient matrix: once to the row n of $[\tilde{C}_{ij}]$ and $[\tilde{0}_{lj}]$ and once to the row n^* of $[\tilde{C}_{lm}^*]$ and $[\tilde{0}_{im}^*]$. Figure 8 illustrates the sparsity of the resulting coefficient matrix corresponding to the discretization illustrated in Figure 7. In this example serendipitous Hermitean basis functions, which introduce three unknown per grid node, were employed (Lapidus and Pinder, 1982).

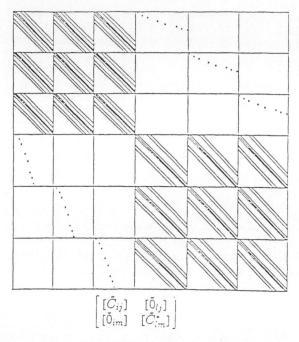

$$\begin{bmatrix} [\bar{C}_{ij}] & [\bar{0}_{lj}] \\ [\bar{0}_{im}] & [\hat{\bar{C}}_{lm}^*] \end{bmatrix}$$

Figure 8 Sparsity representation for the coefficient matrix associated with the discretization in Figure 7.

The PCG method is applied to the solution of Eq. (11). This is equivalent to solving the system

$$\begin{bmatrix} [H_{ij}] & [0_{lj}] \\ [0_{im}] & [H_{lm}^*] \end{bmatrix}^{-1} \begin{bmatrix} [\tilde{C}_{ij}] & [\tilde{0}_{lj}] \\ [\tilde{0}_{im}] & [\tilde{C}_{lm}^*] \end{bmatrix} \begin{Bmatrix} \{A_j\} \\ \{A_m^*\} \end{Bmatrix} = \begin{bmatrix} [H_{ij}] & [0_{lj}] \\ [0_{im}] & [H_{lm}^*] \end{bmatrix}^{-1} \begin{Bmatrix} \{\tilde{E}_i\} \\ \{\tilde{E}_l^*\} \end{Bmatrix} \quad (12)$$

by the standard conjugate gradients approach, where

$$\begin{bmatrix} [H_{ij}] & [0_{lj}] \\ [0_{im}] & [H_{lm}^*] \end{bmatrix}^{-1} = \Psi\left(\begin{bmatrix} [\tilde{C}_{ij}] & [0_{lj}] \\ [0_{im}] & [\tilde{C}_{lm}^*] \end{bmatrix} \right) \quad \text{for} \quad i, j = 1, M \quad l, m = 1, M^* \quad (12a)$$

is the preconditioning matrix. The sparsity of the preconditioning matrix is defined such that the nonzero coefficients of the submatrices $[\tilde{0}_{lj}]$ and $[\tilde{0}_{im}^*]$ are ignored (see Fig. 9).

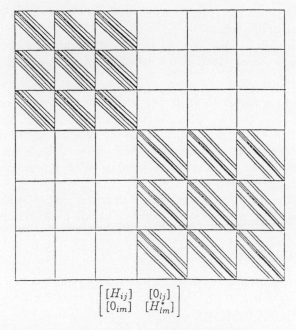

$$\begin{bmatrix} [H_{ij}] & [0_{lj}] \\ [0_{im}] & [H^*_{lm}] \end{bmatrix}$$

Figure 9 Sparsity representation for the preconditioning matrix.

The steps of the PCG involving matrix operations are parallelized, taking into account the special sparsity patterns of the coefficient and preconditioning matrices:

The matrix vector multiplication is performed concurrently for each block of the matrix, and the final result is obtained by adding the partial results. Each block of $[\tilde{C}^*_{ij}]$ and $[\tilde{C}^*_{lm}]$ is multiplied using the vectorized approach by diagonals of Madsen et al. (1976). The blocks of $[\tilde{0}^*_{im}]$ and $[\tilde{0}^*_{ij}]$ are multiplied using the approach for random sparse matrices illustrated in Figure 6.

The inversion of the preconditioning matrix can be performed in parallel even for the recursive algorithms, like the IC factorization, since there is no connection between the blocks $[H_{ij}]$ and $[H^*_{im}]$.

6 SUMMARY

This chapter summarizes our experience with the implementation of finite element and least-squares collocation methods for the solution of groundwater contamination problems on parallel processors. The discussion focuses on vector processing and multiprocessing because most of the commercially available computers are of this variety.

To obtain efficient solutions for the advection-diffusion equation defined over irregularly shaped domains, the numerical algorithms have been selected such that

The discretization algorithms yield symmetric, positive definite coefficient matrices.

The most efficient conjugate gradient-based methods are applied to the solution of the global system of linear equations.

The assemblage of the global system of linear equations is based on a reduced set of indirect index operations.

The grid-related sparsity of the coefficient matrix is easy to handle.

The potential for parallelization of the discretization phase is fully exploited.

ACKNOWLEDGMENT

This work was performed while A. Peters was visiting Princeton University. It was supported in part by NATO through a grant from the DAAD (Deutscher Akademischer Austauschdienst).

REFERENCES

Bentley, L. R., Pinder, G. F., and Herrera, I. (1989). Solution of the advective-dispersive transport equation using a least squares collocation Eulerian Lagrangean method. to appear in Int. Journal Numerical Methods for Partial Differential Equations. No. 5, pp. 227—240.

Bramble, J. H., Ewing, R. E., Pasciak, J. E., and Schatz, A. H. (1988). A preconditioning technique for the efficient solution of problems with local grid refinement. Computer Methods in Applied Mechanics and Engineering, 67:149–159.

Duff, I. S., Grimes, R. G., and Lewis, J. G. (1989). Sparse matrix test problems. ACM Transactions on Mathematical Software, Vol. 15, No. 1, pp. 1–14.

Eason, E. D. (1976): A review of least squares methods for solving partial differential equations. Int. Journal Numerical Methods in Engineering, 10:1021–1046.

Ewing, R. E. (1990). Domain decomposition techniques and their efficient application on supercomputers. Int. Journal Advances in Water Resources, Special Issue on Supercomputing, No. 3, pp. 117–125, Vol. 13, 1990.

Hackbusch, W. (1985). *Multi-Grid Methods and Applications*. Springer-Verlag, Heidelberg.

Hockney, R. W., and Jesshope, C. R. (1981). *Parallel Computers*. Adam Hilger, Bristol.

Jamieson, L. H. (1986): Characterizing parallel algorithms. In: *The Characteristics of Parallel Algorithms*. Jamieson, L. H., et al., Eds. MIT Press, Cambridge, Massachusetts, pp. 65–100.

Joos, B. (1986). The least squares collocation method for solving partial differential equations. Doctoral Dissertation, Princeton University.

Kershaw, D. S. (1978). The incomplete Cholesky conjugate gradient method for the iterative solution of systems of linear equations. Journal of Computational Physics No. 26, pp. 43–65.

Laible, J. P., and Pinder, G. F. (1990). Least square collocation solution of differential equations in irregularly shaped domains using orthogonal meshes. International Journal of Numerical Methods for Partial Differential Equations, Ab. 5, pp. 347—361 (1989).

Lapidus, L., and Pinder, F. G. (1982). *Numerical Solution of Partial Differential Equations in Science and Engineering*. John Wiley, New York.

Leismann, H. M., and Frind, E. O. (1989). A symmetric matrix technique for the efficient solution of advection-dispersion problems. Water Resources Research, Vol. 25, No. 6, pp. 1133–1139.

Madsen, N. K., Rodrigue, G. H., and Karush, I. (1976). Matrix multiplication by diagonals on a vector/parallel processor. Information Processing Letters, No. 5, pp. 41–45.

Nelson, P. A., and Snyder, L. (1986). Programming paradigms for nonshared memory parallel computers. In: *The Characteristics of Parallel Algorithms*. Jamieson, L. H.., et al., Eds. MIT Press, Cambridge, Massachusetts, pp. 3–20.

Obeyskare, U. R. B., Allen, M. B., Ewing, R. E., and George, J. H. (1987). Application of conjugate gradient-like methods to a hyperbolic problem in porous media. International Journal of Numerical Methods in Fluids, 7:551–566.

Ortega, J., and Voight, R. (1985). Solution of partial differential equations on vector and parallel computers. SIAM Review 27:149–240.

Paolini, G. V., and Santangelo, P. (1989). A graphic tool to plot the structure of large sparse matrices. IBM Technical Report ICE-0034, Rome.

Pelka, W., and Peters, A. (1986). FE groundwater models implemented on vector computers. Int. Journal of Numerical Methods in Fluids, 6:913–925.

Peters, A. (1988a). Vektorisierte Behandlung von Finiten Elementen fuer die Grundwassermodellierung. Mitteilungen Institut fuer Wasserbau RWTH Aachen, Heft 69.

Peters, A. (1988b). Vectorized programming issues for FE models. In: *Proceedings of the VII International Conference Computational Methods in Water Resources*, MIT Cambridge, Vol. 1. Celia, M., et al. Eds. CMI-Elsevier, Amsterdam-Southampton, pp. 23–32.

Peters, A. (1990). A domain decomposition technique for the parallel solution of partial differential equations. Gambolati, G., et al., Eds. *Proceedings of the VIII International Conference Computational Methods in Water Resources, Venice*. CMI Publications, Southampton, Vol. 2, pp. 479–486.

Peters, A., and Pinder, G. F. (1990). A parallel least squares collocation conjugate gradient approach for the advection diffusion equation. Int. Journal Advances in Water Resources, Special Issue on Supercomputing, No. 3, Vol. 13, pp. 126–136, 1990.

Peters, A., Romunde, B., and Sartoretto, F. (1988). Vectorized implementation of some MCG codes for the FE solution of large groundwater flow problems. In: Niki, H., and Kawahara, M., Eds. *Proceedings of the International Conference on Computational Methods in Flow Analysis*. Okayama, pp. 123–130, Okayama University of Science.

Pini, G., and Gambolati, G. (1990). Is a simple diagonal scaling the best preconditioner for conjugate gradients on supercomputers? Int. Journal Advances in Water Resources, Special Issue on Supercomputing, No. 3, Vol. 13, pp. 147–155, 1990.

Pini, G., Gambolati, G., and Galeati, G. (1988). 3-D finite element transport models by upwind preconditioned conjugate gradients. In: Celia, M., et al., Eds. *Proceedings of the VII International Conference Computational Methods in Water Resources*, Cambridge, Vol. 2. CMI-Elsevier, Amsterdam-Southampton, pp. 35–43.

Press, W. H., Flannery, B. P., Teukolsky, S. A., and Vetterling, W. T. (1986). *Numerical Recipes*. Cambridge University Press, London.

Radicati Di Brozolo, G., and Vitaletti, M. (1986). Sparse matrix-vector product and storage representations on the IBM 3090 with vector facility. IBM Technical Report G513-4098, Rome.

Rice, J. R. (1986). Parallel methods for partial differential equations. In: Jamieson, L. H., et al. Eds. *The Characteristics of Parallel Algorithms*, MIT Press, Cambridge, Massachusetts, pp. 209–230.

Saad, Y. (1985). Pratical use of polynomial preconditionings for the conjugate gradient method. SIAM Journal Scientific Computation, No. 6, pp. 856–881.

Sameh, A. H. (1983). Numerical parallel algorithms—A survey. In: *High Speed Computer and Algorithm Organizations*. E. D. Kuck et al., eds. Academic Press, New York, pp. 207–228.

Van Der Vorst, H. (1982). A vectorizable variant of some ICCG methods. SIAM Journal Scientific Computation, No. 3, pp. 350–356.

Van Der Vorst, H. (1990). Iterative methods for the solution of large systems of Equations on supercomputers. to appear in Int. Journal Advances in Water Resources, Special Issue on Supercomputing, No. 3, Vol. 13, pp. 137–146, 1990.

Varga, R. S. (1962). *Matrix Interative Analysis*. Prentice-Hall, Englewood Cliffs, New Jersey.

Index